The New Naturalist Library
A SURVEY OF BRITISH NATURAL HISTORY

LICHENS

Editors
Sarah A. Corbet ScD
S.M.Walters, ScD, VMH
Prof. Richard West, ScD, FRS, FGS
David Streeter, FIBiol
Derek A. Ratcliffe

The aim of this series is to interest the general reader in the wildlife of Britain by recapturing the enquiring spirit of the old naturalists. The editors believe that the natural pride of the British public in the native flora and fauna, to which must be added concern for their conservation, is best fostered by maintaining a high standard of accuracy combined with clarity of exposition in presenting the results of modern scientific research.

The New Naturalist

LICHENS

Oliver Gilbert

'the still explosions on the rocks'
Elizabeth Bishop

With 16 colour plates and over 120 black
and white photographs and line drawings

HarperCollins*Publishers*

HarperCollins*Publishers*
77–85 Fulham Palace Road
Hammersmith
London W6 8JB

The HarperCollins website address is:
www.**fire**and**water**.com

To Natasha, Kate and Emma

First published 2000

01 03 05 04 02 00
1 3 5 7 9 10 8 6 4 2

© Oliver Gilbert 2000

The author asserts his moral rights to be identified as the author of this work

The copyright in the photographs and illustrations in this book belongs to Oliver Gilbert except where otherwise stated

ISBN 0 00 220081 3 (Hardback)
ISBN 0 00 220082 1 (Paperback)

Colour reproduction by Digital Imaging, Glasgow

Printed and bound in Great Britain by The Bath Press

Contents

List of Plates

Editors' Preface

Author's Foreword

1 Our Lichen Heritage
  Medicine
  Dyeing
  Food for man
  Well-dressing
  Aesthetics
  Perfumes
  Minor and curious uses

2 What is a Lichen?
  The fungal partner
  Algal and cyanobacterial partners
  Growth forms, reproduction and establishment
  Some peculiarities of lichens
  Lichen communities

3 Creatures That Need Lichens
  Reindeer
  Birds
  The tree trunk ecosystem
  Moths
  Slugs and snails
  Mites and nematodes

4 Lichens and Air Pollution
  Lichen deserts
  Effects of environmental and other factors
  Lichen reinvasion with declining air pollution
  An urban super-race of lichens?
  Pollution other than sulphur dioxide

5 Trees, Woods and People
  Factors affecting lichen diversity
  The New Forest
  Loch Sunart, Ardgour
  Scottish pinewoods
  Coniferous plantations
  Western hazel woods
  The cropping ash of the Lake District
  Parkland

6 Acid Rock
Acid rock in lowland Britain
Acid rock in upland Britain

7 Heaths and Moors
The lichen flora
The succession following heather burning
Persistent *Cladonia* patches in closed heathland
The difference between heaths and moors
The effect of grazing
Epiphytes on dwarf shrubs
Pebbles on the heath
Conservation

8 Chalk and Limestone
The chalk
Carboniferous limestone
The Magnesian or hidden limestone
The Scarplands
Low level limestone in Scotland

9 Village, Church and Farmland
The village
The church
A comparison of the village and the churchyard
Farmland
The two landscapes of lowland England

10 Work, Wealth and Wheels
Work
Wealth
Wheels

11 Discovering the Montane Lichen Flora
The Crombie rarities
The Ben Alder treasure trove
The Cairngorm plateau
The Far North
Glory days on Ben Lawers
Caenlochan
Western Scotland
West of Ireland
The Lake District and Pennines
Snowdonia
Comparing different areas

12 Rivers and Lakes
The zonation across streams and rivers
Variation from source to estuary
Water chemistry or substratum?

The effect of shade on rivers
  Water quality and rivers
  Lakes and tarns
  Reservoirs

13 Coastal Habitats
  Lichen zones on rocky shores
  Limestone shores
  Sand dunes
  Shingle
  Saltmarshes and intertidal mud flats
  Saline lagoons
  Unprotected soft cliffs
  Coastal heath
  Maritime cliff grassland

Appendices

Glossary

Bibliography

Index

List of Plates

Plate 1
 (a) The lichenologist Ray Woods wearing a jersey knitted from lichen-dyed wool (J. Woods).
 (b) Lichen dyes. Cudbear (top) produces red dyes and crottle shades of brown (D.J. Hill).
 (c) A modern lichen dyer with dye-pot, lichens and treated wool, Strontian.
 (d) Central panel of a well-dressing tableau made from petals and several types of lichen, Derbyshire.
 (e) Canons Ashby Church, Northamptonshire, showing how lichens enhance architectural detail (D.J. Hill).

Plate 2
 (a) The lichen alga *Trebouxia* growing in culture (The Natural History Museum, London).
 (b) Jelly lichens contain the cyanobacterial partner *Nostoc* (J.M. Gray).
 (c) *Sticta canariensis* 'green algal morph' and *S. canariensis* 'cyanobacterial morph' contain the same fungus (J.M. Gray).
 (d) The fruticose growth form exhibited by a beard lichen (*Usnea articulata*) pendant to 60 cm (J.M. Gray).

Plate 3
 (a) Nest of a long-tailed tit (*Aegithales caudatus*) decorated with lichen to aid concealment by light reflection (B.J. Hatchwell).
 (b) Caterpillar of the light crimson underwing moth (*Catocala promissa*) mimicking a lichen-covered surface (J. Porter).
 (c) Caterpillar of the dotted carpet moth (*Alcis jubata*) feeding on *Usnea* (P.A. Ardron).
 (d) Autumn green carpet moth (*Chloroclysta miata*) at rest on a lichen-covered tree trunk (R.W. Barnes).

Plate 4
 (a) Thick sward of the pollution-tolerant lichen *Lecanora conizaeoides* on larch.
 (b) *Usnea florida*, a beard lichen that is highly sensitive to several forms of pollution (J.M. Gray).
 (c) Lichens on beech that have been killed by airborne fluorides, Invergordon, Scotland.

Plate 5

(a) Old beech woodland, New Forest, Hampshire (T. Heathcote).
(b) Lobaria pulmonaria festooning mossy Atlantic rainforest, Western Scotland (J.M. Gray).
(c) Atlantic oak woodland above Loch Sunart, Ardgour.
(d) A rich Lobarion community covering a bough at Loch Sunart; the lichen with golden soralia is *Pseudocyphellaria crocata* (F. Rose).

Plate 6

(a) Parmentaria chilensis, a strongly oceanic species known in Britain from one hazel wood at Loch Sunart (A.M. Coppins).
(b) Hypogymnia physodes, a common species of birch woods in the Highlands (J.M. Gray).
(c) A pin-head lichen, *Chaenotheca furfuracea* (J.M. Gray).
(d) A writing lichen, *Graphis scripta*, characteristic of smooth bark (J.M. Gray).
(e) Well-lit, smooth bark on many deciduous trees supports a mosaic of small crustose lichens (J.M. Gray).

Plate 7

(a) Ophioparma ventosum, a lichen of acid rocks (I.C. Munro).
(b) Lasallia pustulata, a gregarious lichen (I.C. Munro).
(c) Purple-stained lichen on quartzite, Foinavon (D. Miller).
(d) Ramalina polymorpha, a species typical of basalt tors.
(e) Damp, north-facing slabs of basalt are home to many rare lichens, Trapain Law, Lothian.

Plate 8

(a) Cladonia coccifera, abundant on acid soils (J.M. Gray).
(b) Cladonia floerkeana, the 'Bengal match lichen', abundant in acid habitats (J.M. Gray).
(c) Cladonia portentosa, the commonest of the 'Reindeer lichens' (P.A. Ardron).
(d) Close-up of the lichen carpet at Wangford Warren, Breckland.

Plate 9

(a) Lichen-rich chalk grassland has developed where the surface was scraped off in 1940 to form a shooting butt, Martin Down.
(b) Lichenologists inspecting a path on the chalk downs, Butser Hill.
(c) Lichenologists at work on a limestone pavement, Ingleborough.
(d) Synalissa symphorea and *Psora lurida* on the surface of a limestone pavement, Gait Barrows (J.M. Gray).
(e) Caloplaca aurantia, a species characteristic of Jurassic limestones (T.W. Chester).

Plate 10
   (a) Lecanora polytropa growing on iron railings.
   (b) Lecanora rupicola thickly encrusting a sandstone tombstone (T.W. Chester).
   (c) Rhizocarpon geographicum on a slate tombstone.
   (d) Timber-boarded Sussex barn carrying what is believed to be a unique assemblage of rare lichens, Parham Park.

Plate 11
   (a) The international community dominated by *Lecanora dispersa* (white) that is present on concrete (I.C. Munro).
   (b) Old open cast workings rich in heavy metals, Parys Mountain, Anglesey (O.W. Purvis).
   (c) Baeomyces roseus only fruits regularly at acid mine sites (P.W. James).
   (d) Baeomyces rufus is widespread at most acid mine sites (I.C. Munro).
   (e) The normally brown *Acarospora smaragdula* becomes green when growing on copper-rich rocks (O.W. Purvis).

Plate 12
   (a) Lower part of the Ben Alder buttress coloured yellow with *Fulgensia bracteata*.
   (b) The rare alpine *Lecanora epibryon* growing with *Salix reticulata*, Ben Alder (A.M. Fryday).
   (c) Margaret's Coffin almost devoid of snow, September 1996.

Plate 13
   (a) Solorina crocea has a thallus with an orange underside (I.C. Munro).
   (b) The rare alpine *Pertusaria glomerata*, Ben Lawers range (I.C. Munro).
   (c) Catolechia wahlenbergii (Goblin lights), a rare lichen centred on the Ben Nevis range (A.M. Fryday).

Plate 14
   (a) The River Jelly Lichen Steering Group at work by the River Eden, Cumbria (A.M. Coppins).
   (b) Sites where shelving beds of rock flank a river are usually rich in aquatic lichens, South Tyne above Hexham.
   (c) A mountain tarn with *Lecanora achariana* on the marginal boulders, Snowdonia.

Plate 15
   (a) Endocarpon adscendens on mossy lakeside boulders, Windermere, Cumbria.
   (b) Dermatocarpon intestiniforme dominates a high zone around Bassenthwaite Lake, Cumbria.
   (c) Lichen zonation on a sea stack showing the black, orange and grey bands, North Cornish coast.

(d) The tiny dot-like fruits of *Pyrenocollema halodytes* growing on barnacles (J.M. Gray).

Plate 16
(a) Colourful lichen assemblage of the grey zone (R.W. Barnes).
(b) The rare arctic-maritime species *Lecanora straminea* growing on a bird cliff, Flannan Isles.
(c) *Teloschistes flavicans* (Golden-hair lichen) grows on a few exposed cliff tops in southwest England and west Wales (P.A. Gainey).

Editors' Preface

Few groups of organisms have aroused such passionate controversy among naturalists as the lichens. That they represent a symbiotic relationship between a fungus and a photosynthetic partner is now so universally accepted that it is easy to forget that it is less than 150 years since the true nature of lichens was finally demonstrated and that the proposal was treated with derision by many of the leading botanists of the time. Recent years have seen an enormous increase in interest in these fascinating organisms. They are among the most ubiquitous on the planet, being one of the few forms of life able to tolerate both the icy wastes of the Antarctic and arid deserts and being found from rocky inter-tidal shores to the tops of mountains. In the British Isles they form an important and conspicuous component of the flora of many different habitats and it has long been the intention of the Editors to add a volume on the lichens to the New Naturalist library.

Oliver Gilbert is among the country's most distinguished lichenologists. But as well as a specialist, that he is also a naturalist in the true New Naturalist tradition will be evident from the pages of this book. His approach is habitat based and his evocative descriptions are those that can only be produced by one who has long and intimate experience in the field. In common with so many others, his interest in his subject was stimulated as a young man by the Field Studies Council, in his case at Malham Tarn among the stunning scenery of the Yorkshire Dales. That lichens were peculiarly sensitive to atmospheric pollution had been suspected for a long time, but it was Oliver Gilbert's studies of the lichens in the vicinity of Newcastle that first helped to put the subject on a firm footing and which led to the appreciation of the important role that lichens could play as biological indicators of atmospheric pollution.

It is probably fair to say that Oliver Gilbert has something of a reputation for searching for lichens in those places that other lichenologists do not reach. Hence, the reader will find descriptions of the lichens of such improbable situations as disused airfields, urban pavements, abandoned mineral workings and the shady banks of farm ditches as well as the classic habitats of ancient woodland, coastal cliffs, heath, mountain and moor.

To add to their fascination, lichens are extremely beautiful. The multicoloured mosaics that festoon both rocky coast and churchyard monument are one of the most striking features of the landscape. We hope that the wonderful colour photography that enriches the book will help to convey some of that beauty and stimulate a still greater interest in these extraordinary 'fungi'.

Author's Foreword

In 1856 Lauder Lindsay produced his taxonomically based *A Popular History of British Lichens*, aimed at the general botanist and still a good read. More recently, David Richardson's *The Vanishing Lichens* (1975) has filled a gap by reviewing, on a world scale, the interaction of lichens with man and beast. This present book is different again. Like other volumes in the New Naturalist series, it has in mind the British amateur, so I have sought to avoid giving long lists of Latin names, concentrating instead on more general matters. The present volume has an ecological bias, although a historical approach would have been equally appropriate. After a few introductory chapters, all major habitats in these islands are covered in turn. It has been a pleasure reviewing developments over the last 30 years, as lichenology takes its disciples to attractive places. If I have managed to convey the feel of exploring a habitat, particularly when discoveries are made, the rewards, the disappointments, the grovelling, the exhaustion, the elation and the companionship, I will be satisfied. The chapter covering montane lichens has been written in a more personal style to reveal these aspects.

I am often asked how my interest in lichens arose, and usually reply that it is in my genes, a perhaps unsatisfactory response. Since the age of three I have been passionately interested in plants, a thirst that as a schoolboy was satisfied by becoming acquainted with the flora of Britain and the Alps. At Exeter University I flirted with bryophytes and learnt to add *Cladonia* sp. to the end of my heathland quadrats. Following a postgraduate degree in plant pathology at Imperial College, where we were introduced to the precise discipline of identifying fungi from fruit body and spore characters, my first employment was as botanist on the staff at Malham Tarn Field Centre in Yorkshire. At that time Arthur Wade was running his celebrated annual, week-long lichen courses and, as he did not drive, I was detailed to transport its members around in the centre's Landrover. That is how I met my first real lichenologist and, as with so many of my generation, it was he who switched the light on.

After three happy years at Malham, during which I attended parts of three Wade courses, I joined the staff of the Botany Department at Newcastle University and completed a PhD on lichens and air pollution. Work on this brought me into contact with two more mentors of the day, Peter James and Ursula Duncan; I joined the British Lichen Society (BLS), and in 1966 hitchhiked to a field meeting in Connemara. In this way my interest in lichens evolved, there was no sudden conversion. They remain a very special part of a wider interest in all things botanical.

It is a privilege to have been involved in lichenology during the period when our lichen flora was being rediscovered after half a century of neglect. The years following the formation of the BLS were stirring times, almost a heroic age when reputations were made more easily than today. There was a pioneering spirit abroad as we helped each other with determinations, welcomed each new key, each new discovery, avidly discussing the twice-yearly issues of the *Lichenologist*, filling in mapping cards and above all meeting for field work, either on Society events or, becoming more independent, as groups of friends studying a particular habitat or region. There was a sense of brotherhood, any lichen enthusiast being accepted as a friend. For me, fieldwork in the company of kindred spirits was the very life blood, and I attempted to develop a new discipline, that of Adventure Lichenology. This is the use of expedition tactics to explore inaccessible and little-known habitats, chartering trawlers to land parties on remote islands, engaging porters to set up high camps, and on one occasion enlisting a helicopter to drop a team on top of Ben Nevis. It also involves studying neglected habitats which may take the practitioner to dramatic locations such as disused airfields, derelict industrial premises, reservoir draw-down zones or find them following pylon lines across the landscape.

Being by inclination more of an ecologist than a taxonomist, I have found it easiest to work a habitat for several years until familiarity with the lichens frees one from excessive collecting and enables their interrelationships with each other and the environment to be studied. Throughout this time my respect has mounted for friends and referees who have helped name difficult material.

This book is a tribute to those fellow lichenologists who, by sharing their interests and enthusiasms, their time and hospitality, have contributed towards its creation. It may seem invidious to mention individual names, but I count myself fortunate in having spent substantial time in the field with Brian Coppins, Peter James, Francis Rose, Alan Fryday, Vince Giavarini and Prof. Brian Fox. Parts of the manuscript have benefited from being read and commented on by Brian and Sandy Coppins, Tony Fletcher, Francis Rose, William Purvis and Vanessa Winchester, while Albert Henderson has been a constant and kindly critic commenting on each chapter as it was produced.

Progress has been facilitated by a swift response to requests for information. A list of names is not an adequate acknowledgement of help received but, regrettably, will have to suffice: in addition to those mentioned above it must include Barbara Benfield, Tom Chester, Ivan Day, Peter Earland-Bennett, Prof. David Hawksworth, David Hill, Prof. Richard Holmes, Peter Lambley, Jack Laundon, David Long, Prof. David Richardson, Neil Sanderson, Prof. Mark Seaward and Ray Woods. I have relied on the generosity of others for many of the photographs. Here I owe a special dept of gratitude to Jeremy Gray who generously allowed me access to his unrivalled slide collection. Others who have contributed slides or black and white photographs are acknowledged next to their contributions. To provide variety, line drawings have been commissioned from Michael Lindley and Paul Ardron, while Ken Alvin, Claire Dalby, Pat McCarthy and Alan Orange have allowed me to reproduce from their work. Nick Gibbons and Glyn Woods are thanked for preparing the computer-drawn text figures. Space, not ingratitude, precludes the naming of every person who has made a contribution of value.

Oliver L. Gilbert
Sheffield
September 1998

1

Our Lichen Heritage

The Anglo-Saxon Chronicle for 1066 describes events leading up to the Battle of Hastings. Prior to the conflict, King Harold instructed his noblemen to assemble with their armies at the *har* ('hoar') apple tree on Caldbec Hill. At that time *har* was the adjective used to describe a tree or stone that was grey and shaggy with lichen, and has given us our modern English words 'hoar' and 'hoary' occurring in such expressions as 'hoarfrost'. This lichen-covered apple tree must have been a well-known landmark on the open downs and the instructions as clear as it would be today to arrange to meet under the clock at Waterloo Station.

Though possibly the most celebrated early reference to lichens in Britain, the above is not the first. Rackham (1976; 1986) has drawn attention to the many Anglo-Saxon charters describing village bounds that refer to hoar apple trees, hoar maple trees, hoar thorn trees, hoar hazels, etc. Among the earliest lichen records of this kind is one referring to the bounds of Thorpe-by-Chertsy, Surrey, that dates from 675 AD. It is clear that lichen-clad trees and stones were widely used as landmarks and boundary indicators in Saxon times. It is still possible to identify particular lichen-covered stones first referred to over a thousand years ago. A Cornish charter of *c.* 967 AD describes the limits of Traboe-in-St Keverne on the Lizard Peninsula as: 'then along the way to *cru draenoc* (Thorny Barrow); then to *carrecwynn* (White Outcrop); and back again to *pollicerr* (now a farm called Polkerth)' (Davidson, 1883). The outcrop *carrecwynn* still carries a conspicuous white lichen, *Ochrolechia parella*, from which its name was doubtless derived. Later perambulations of the area refer to *main mellyn* (yellow rocks) which is interpreted as an allusion to the common yellow lichen *Xanthoria parietina*.

By the time of William Shakespeare (1564–1616), lichens were still unrecognised as such, being referred to as moss or stains on the rock. The closest link I have discovered between the Bard and lichens comes in *Sonnet 55*:

> 'Not marble, nor the gilded monuments
> Of princes, shall outlive this powerful rhyme;
> But you shall shine more bright in these contents
> Than unswept stone, besmear'd with sluttish time.'

During the Elizabethan period there was a great awakening of scientific interest all over Europe. The new attention paid to plants was at first entirely medicine-based and therefore the province of herbalists whose work was greatly aided by the invention of printing and woodcut illustrations. As far as herbals were concerned, the English were slow off the mark, relying on foreign productions until William Turner brought out his *A New Herball*, the third part of which (1568) includes several lichens; he appears to be the first author to mention a British lichen in print. Gerard's *The Herball or Generall Historie of*

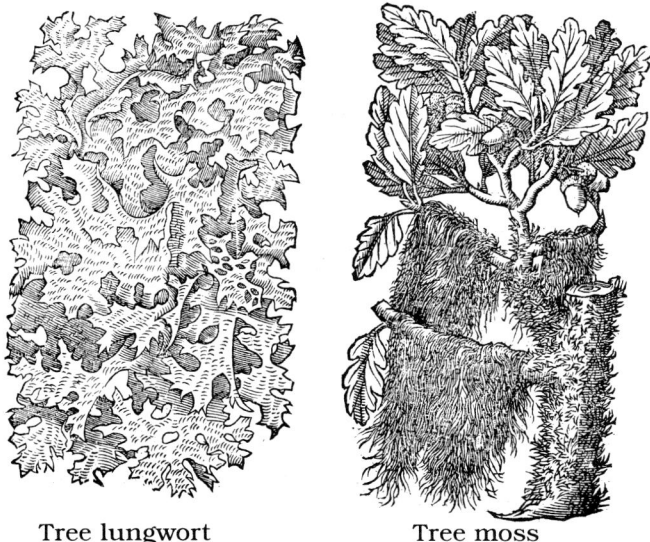

Tree lungwort   Tree moss

Fig. 1.1 Sixteenth-century woodcuts of lichens from Gerard's *Herball*: tree lungwort – *Lobaria pulmonaria*; tree moss – *Bryoria* sp.

Plantes (1597), a widely popular book in its day, recommends the lichens *Alectoria* (hair moss), *Cladonia pyxidata* (chalice moss), *Lobaria pulmonaria* (tree lungwort), *Sphaerophorus* (coral moss) and *Usnea* (moss of the trees) as medicinally valuable and provides charming woodcuts of them (Fig. 1.1). Johnson's enlarged edition of Gerard (1633) includes additional lichens, and more were added in publications by Parkinson (1640), How (1650), Merrett (1666) and Ray (1670; 1686; 1690), though the latter had rather little time for the group.

Until the end of the seventeenth century, lichens were classified and named as types of moss. It was the Frenchman, Tournefort (1694), who first distinguished them as a distinct group under the generic designation 'lichen'. Then a few years later the Oxford botanist Robert Morison (1699) classified them as 'Musco-fungus', thus emphasising their fungal nature. Following on from this the paths of herbalists and botanists started to divide, the latter studying lichens for their own sake.

Throughout most of the last century the botanical establishment treated with contemptuous disbelief de Bary's notion (1866) that lichens were dual organisms. Indeed, many refused to accept it right into the present century. Leading contemporary lichenologists including William Nylander in Finland and the Rev. James Crombie in London, bitterly opposed the hypothesis. Crombie characterised it as 'this sensational "Romance of Lichenology", or the unnatural union between captive algal damsel and tyrant fungal master' (Crombie 1874), while M. C. Cooke in 1879 asserted of the dual hypothesis that 'even if endorsed by the nineteenth century it will certainly be forgotten in the twentieth'. It was not forgotten and gradually became universally accepted.

The history of European lichenology is ably summarised by Smith (1921b), briefly by Ainsworth (1976), while Hawksworth & Seaward (1977) have published an enjoyable account of *Lichenology in the British Isles 1568–1975*.

Our lichen heritage, as far as the non-specialist is concerned, involves their use down the ages in medicine, dyeing, for food, decoration, perfume manufacture, and the way they add beauty and maturity to a landscape. These aspects will be examined in turn with respect to the British Isles.

Medicine

The Elizabethan herbalists initially relied heavily on information already available in continental herbals. These in turn trusted in the 'Doctrine of Signatures', which held that the Creator had marked those plants suitable for treating diseases by a resemblance to a specific part of the human body. Consequently, the tree lungwort, which superficially resembles the inside of a lung, was considered suitable for treating respiratory complaints, while the hair moss was thought to be effective against disorders of the scalp. One of the more bizarre beliefs was that lichen growing on human skulls was worth its weight in gold as a cure for epilepsy. Many of the remedies were taken after steeping the relevant lichen in wine or milk for several days or by drinking a concoction of the powdered lichen in water. There is not much evidence that prescribing these medicines was a success; even the most famous remedy, involving the common dog lichen (*Peltigera canina*) as 'a certain cure for the bite of a mad dog', had fallen into disfavour by the year 1800.

It is perhaps surprising that lichen remedies have endured and are still listed in standard pharmacopoeias such as *Martindale* (Reynolds, 1996), a reference book present in every chemist shop. They were rapidly falling out of favour earlier this century until interest in the group was revived by the discovery of antibiotics. Lichens owe their therapeutic properties to the presence of bitter and astringent substances, known as lichen acids, some of which have antibiotic properties. Usnic acid, for example, found in the lichen genera *Evernia* and *Usnea*, is a broad-based but weak and rather insoluble antibiotic. It is marketed as a cream effective against infectious skin disorders under the proprietary names Evosin, Usnagram and Usnaderm. It is also occasionally encountered on sale as the active ingredient in anti-dandruff hair shampoo, deodorants and foot powders (Fig. 1.2).

While usnic acid products usually need to be ordered, Iceland moss (*Cetraria islandica*) can be bought immediately over the counter at most herbal and natural remedy shops. It is sold dried, at around £2 per 50 g, for brewing as a tea effective against upper respiratory tract congestion. The therapeutic ingredients are protolichesterinic acid and its high mucilage content. Iceland moss is also marketed as a throat lozenge. Lungwort lichen (*Lobaria pulmonaria*), so-called in the price lists to separate it from the vascular plant *Pulmonaria* with the same English name, can be purchased dried at £2.50 per 50 g in most large towns where it is prescribed for asthma attacks, bladder complaints and as an aperitif to counter lack of appetite. A firm in Derbyshire markets a cough mixture named Lichenes Syrup that contains *Cetraria islandica, Cladonia rangiformis, Usnea barbata* and *Lobaria pulmonaria*.

Modern medical research, based on an understanding of molecular structure, is verifying many of the old lichen remedies. Investigation has shown that 50% of all species have antibiotic properties, and that *Cetraria islandica* is a rich

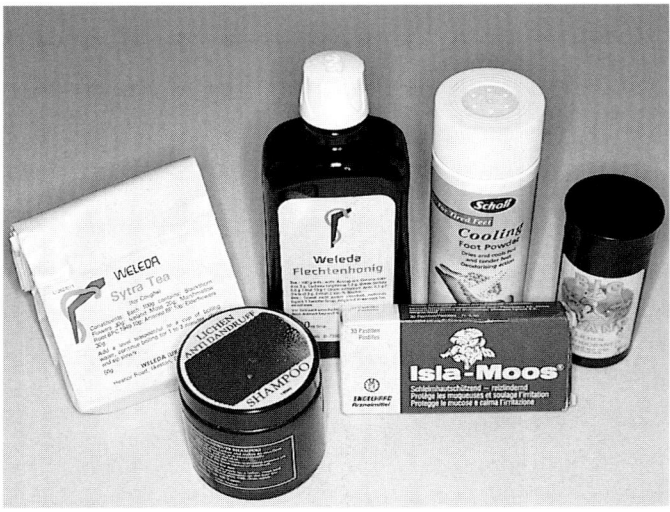

Fig. 1.2 Pharmaceutical products containing lichens currently on sale in Britain.

source of protolichesterinic acid which is active against cancer tumours. Currently a company in Slough is screening hundreds of British lichen-forming fungi for novel pharmaceutical products. The fungal partner is isolated, then grown in batch liquid culture and evaluated as a source of commercially exploitable metabolic products. One problem is the slow growth rate of the fungi. At the moment the jury is out with regard to the future role of lichen products in conventional medicine, but they have a firm following as folk medicines, natural remedies and homeopathic aids. This is based on several thousand years of trial and error.

In addition to curing diseases, lichens can cause them. Foresters and sawmill workers exposed to lichens and lichen dust are prone to develop a contact dermatitis known as 'woodcutters' eczema'. It affects the backs of the hands, the forearms and the waist; in fact wherever lichen dust collects on the body. It starts as an itching which leads to reddening, excoriation and lesions. The cause is not as simple as once thought; different patients react differently and exposure to sunlight is a factor. Cases in Britain have mostly related to *Lecanora conizaeoides*, which often forms a thick crust on tree bark. Patch testing has identified lichen acids as the allergens responsible.

Dyeing

By far the most important use of lichens in Britain has been for dyeing (Plate 1), first as a cottage industry and later on a commercial scale. Vivid descriptions of the process can be found in the writings of travellers in Wales and the Scottish Islands. An account from Shetland records:

> 'My aunt was always the one for making dyed yarn. I mind seeing her work with yon scrottyie, yon grey lichen you scrape from the stanes. She made up a brawly thick gruel, ye ken, and had it boiling abun the fire in a muckle three-taed kettle, with the layers of yarn packed

Fig. 1.3 Women gathering lichen using metal spoons at Roineval, Leverburgh, 1939 (A.M. MacDonald).

between. A few hanks came out soon and the rest she'd leave a while longer to get it a darker shade. She'd knit her stockings striped in different shades of brown, five gengs of each – that was the custom.' (Venables, 1956).

As late as the 1950s, black three-legged iron pots used for boiling lichens could still be seen outside many crofts in the Outer Hebrides, but I do not know whether they were still in use.

An eighteenth-century account from North Wales (Evans, 1800, in Vickery, 1995), describes another process thus:

'The poor people employ themselves in gathering them [lichens] at the low price of one penny per pound. They will, however, collect from 20 to 30 pounds a day. From these a beautiful dye called arcell is prepared. The lichens, when dry, are placed under a large indented stone, bruised and thrown into capacious vats with lime and urine. After six months the substance appears like the mire, afterwards like the husk of grapes. It is then dried and packed in barrels for use.'

In Pembrokeshire, dye lichens were indirectly responsible for a famous military victory. In February 1797, at the height of the Napoleonic Wars, a French army landed at Fishguard intent on wreaking havoc and destruction. This surprise attack went almost unopposed, but the French army mistook the red, lichen-dyed cloaks of a number of distant Welsh women mounted on hill ponies for the uniforms of advancing battalions of regular soldiers. This error led to their decision to surrender to Lord Cawdor without a shot being fired. Historians have largely overlooked the role of lichens in this bloodless victory.

Gaelic communities in the west of Britain and Ireland were chiefly involved in the production of yellow-brown or red-brown lichen dyes known as crottle or crotal (Plate 1b). The lichens used in its production are *Parmelia omphalodes* and *P. saxatilis*, both of which contain salazinic acid. These foliose lichens, which are still common, were scraped off the rock using metal hoops, spoons (Fig. 1.3) and, in the poorer districts, seashells. Damp weather was considered best for collecting and the most experienced gatherers were highly skilled at harvesting pure material which was dried and stored in sacks till needed. Crottle dyeing was never carried out on a commercial scale, the lichen being collected in small amounts by individual crofters. Alternate layers of wool and lichen were packed into the large iron three-legged pots (Fig. 1.4), using approximately equal quantities of each, and the pot was filled with peaty water and boiled for several hours until the correct shade was obtained. The colours, which ranged from golden to chocolate-brown, were glowing and durable, and the process gave the wool a delightful aroma. Crottle dyes have the advantage of being fast, they do not fade, and are cheap as they do not require a fixing agent to help them stick to the wool. After dyeing, the wool is carded, spun, woven on hand looms and the cloth waulked. This tedious process consists of dipping the cloth in soap suds and working it on a rough board so the wool fibres are partly felted; it was formerly engaged in communally by women singing specific Gaelic songs.

Fig. 1.4 Traditional iron pot used in the Outer Hebrides for dyeing wool with lichen. It stands on three legs so a fire can be lit underneath (D.H.S. Richardson).

The use of crottle in dyeing continued in Scotland long after its replacement elsewhere by imported dyestuffs. It enjoyed a revival in the mid-nineteenth century when it became fashionable to wear tartans and tweeds. The most famous Scottish tweed is Harris tweed, produced only in the Outer Hebrides, where over a million yards of this cloth are still woven annually. Originally it was dyed entirely with vegetable dyes, browns and fawns from crottle, green from heather, yellow from bracken roots, purple from elderberries and blue-grey from privet. In the 1920s mechanised looms replaced the old hand looms, and commercial dyes replaced those collected from the hills. A demand for traditionally produced tweed with its softness and fragrance was satisfied by a small band of workers who continued to use the old methods. In 1975 they numbered half a dozen, but the last two gave up in 1997 bringing to a close a cottage industry that had provided employment since at least the sixteenth century. A video of them at work reveals that they had their own pronunciation for the word lichen, calling it 'lickin', to rhyme with 'pickin(g)'. A few people have taken the old methods up again to cater for tourists (Plate 1c).

The other main lichen dye produced in Britain was orchil or cudbear, a purple and red pigment derived from erythrin, lecanoric and gyrophoric acids obtained from species of *Ochrolechia*, *Roccella*, *Lasallia* and *Umbilicaria* (Plate 1b). Orchil dyes have a longer recorded history than crottle, having been known to the ancient Egyptians, and receive a mention in the Old Testament when the prophet Ezekiel denounces the people of Tyre with the words 'blue and purple from the Isles of Elishah was that which covered thee' (Ezekiel 27, v. 7.). From the earliest times, as well as upholding a cottage industry it was also prized as a commercial dye and source of trade. When supplies of *Roccella* in the Mediterranean were becoming exhausted, merchant adventurers chartered ships to bring back lucrative cargoes of 'Canary weed' and 'Cape Verde weed' to satisfy the market.

The attraction of orchil dye was the brilliance, softness and lustre it imparted to wool and silk compared with other dyes; it was also one of the very few products that could produce the highly esteemed colour purple. The pigment was produced by steeping the lichen in a solution of ammonia or urine in an airtight container for several weeks. After drying out, the resulting paste or powder, now smelling of violets, could be exported as 'cakes'. The dye, which requires a mordant, is inclined to fade unless appropriate chemicals are added.

The use of orchil in Britain increased with the growth of the textile industry. A variety of the dye produced in Scotland, known as corkir or cudbear, was made from the thick, white, crustose lichens *Ochrolechia tartarea* and *O. androgyna*, and occasionally from *Lasallia pustulata* (Plate 7b). Initially it was used on a domestic scale in the Highlands and Islands where almost every farm and cottage had its barrel of 'graith' (putrid urine) and its 'lit-pig' (dye pot), but eventually commercial interests took over. The story of cudbear dye is that George Gordon, a coppersmith from Banffshire, was carrying out repairs in a London dyehouse when he noticed similarities between the preparation of orchil and the preparation of corkir back home. Seeing commercial possibilities, he set up a factory in Edinburgh to manufacture the dye which he called 'cuthbert', his mother's maiden name, (later contracted to cudbear). Production next switched to a site in Glasgow which expanded to cover 17 acres, which gives some idea of the scale of operations. The works bought in 250 tons of lichen each year. Other smaller cudbear and orchil manufacturers were to be found in textile centres such as Bristol, Leith, London, Norwich, Manchester and Leeds. Supplying them with the 400 tons of dry lichen they required annually provided considerable employment, and gave rise to the Scottish proverb 'Better the rough stone that yields something, than the smooth stone that yields nothing' and the lines of poetry (in Petch, 1984):

> 'Cattle on the hills,
> Gold on the stones.'

When Scottish sources and the small English reserves in Cumbria, Derbyshire, Devonshire and Lancashire became exhausted, the lichen was imported from Scandinavia. The Glasgow works closed in 1856, but small amounts of orchil continued to be manufactured by Yorkshire Chemicals Ltd. in Leeds until 1940. The history of the Leeds works, which were centred on Cudbear Street (Fig. 1.5) and Orchella Place in Hunslet, has been researched by Henderson (1984; 1985a; 1985b). The dyers of silk, cotton, linen and wool were extremely grateful to Cuthbert Gordon for his invention, as Canaries orchil cost between £150 and £300 a ton, while a ton of home-grown English orchil, which would do the same job, cost only £30 to £36 (Anon, 1786). Further information on the use of lichens for dyeing, including recipes, can be found in Bolton (1960), Kok (1966), Richardson (1975) and Smith (1921b).

Food for man

Lichens form a regular part of the human diet only in the case of semi-nomadic tribes living in arctic areas such as Lapland and North America. Britons eat them only as emergency rations, when species of *Umbilicaria*, known as 'rock tripe', have saved lives. The most famous occasion involved a party from Sir John Franklin's 1820 expedition to discover a Northwest Passage who marched for 11 days across northern Canada eating only boiled lichen. It is recorded that while allaying the appetite the lichen was inefficient at recruiting strength and contained a bitter principle that produced severe bowel complaints. Survival manuals issued to the armed forces still contain a section on the use of lichen as food. They advise that the bitterness, caused by lichen

Fig. 1.5 Street sign in the industrial area of Leeds where dyeing manufacture took place (D.J. Hacket).

acids, can be removed by parboiling in an alkaline solution, created by adding wood-ash to the water. Modern studies have shown that, weight for weight, lichens are slightly more nutritious than cornflakes, being rich in a carbohydrate called lichenin.

In the past, lichens were imported into Britain for use in brewing and as a food additive to prolong the life of products. Johnston (1831) mentions that powdered *Cetraria islandica* was added to ship-biscuits to prevent them being attacked by worms. Accounts of the use of lichens as food on a world scale are provided by Smith (1921b) and Richardson (1975).

Well-dressing

Compared with Scandinavian countries, there is little folklore associated with lichens in Britain. Few have English names and superstitions regarding them are rare, though I have heard of them being used to frighten off elves. In one area of plant folklore, however, they do play a part. In the limestone district of Derbyshire it is traditional to dress the village wells (springs) by creating floral pictures on large screens which are erected nearby to honour them and prevail upon them to keep flowing. This custom, probably pagan in origin, is practised in 50 villages. The vivid pictorial scenes, usually depicting religious themes, are created by pressing plant material onto large wooden frames (*c.* 3.7 x 2.5 m) that have been filled with clay and soaked in the river for a week. Materials commonly used include hydrangea and pelargonium flowers, alder fruits, various seeds and quantities of lichen (Plate 1d).

Lichen is popular because it keeps its colour and does not deteriorate. One of the most frequently used species, as it comes in many shades, is *Xanthoria parietina*, referred to as golden lichen in some villages, bronze moss in others. Most popular of all is *Parmelia saxatilis*; if the dressers ask for grey moss they get handed a piece of the lichen the right way up; if they request black moss the same lichen is passed over, but upside down. Up to a third of each tableau may be composed of lichen, which is used to depict buildings, walls, the sky, clothes, beards and the border. The programme of well-dressing starts on Ascension Day and progresses around the villages until the end of August. Considerable tourist interest is aroused which should ensure that this long-standing lichen use survives. Further details can be found in Vickery (1975; 1995), but better still, visit Derbyshire during the well-dressing season.

Aesthetics

Lichens impart colour and character to the objects on which they grow, helping them to blend with their surroundings, and providing a touch of immortality. In these days of rapid change there is a need for 'time statements' in our environment and there are few better candidates to provide these than lichens. Poets and writers down the ages have been fully aware of this function. Henderson (1981) sought out the following examples. An unknown Anglo-Saxon elegist meditating among Roman ruins in the city of Bath presents the durability of lichens and stone as a model of stoicism in the face of changing fortune, and the reddish hues of the lichen assemblage described are compared with the bloodstains left by passing battles.

> 'Time and again this wall endured,
> Lichen grey and red stained, as kingdom followed kingdom.'

Fig. 1.6 A lichen-covered god, one of the lost statues of Stowe, Buckinghamshire (J. Gibson).

In his diary of 1904, A. C. Benson asks:

'Beauty, beauty? What is it? Is it only a trick of old stone and lichen in sunlight?'

More recently Henry Moore quoted approvingly Leonardo's observation that 'in the lichen lines on a wall an artist should be able to discover a whole landscape'. Lichens induce similar feelings in many people even if they are unable to articulate them so expressively. Who could fail to respond to the lichen-covered gods from Stowe Park (Fig. 1.6) or the lichen-clad graves of Scottish Kings at Iona Abbey?

Nearer at hand the interplay of lichens and stonework can be observed in country churches, on stately homes and along many village streets. The distribution of lichens on a building is anything but random; it is controlled by a range of interacting factors that fit the lichen cover to the structure in a most natural, satisfying and aesthetically pleasing manner. Figure 1.7 illustrates some of the variation in lichen cover on a typical church. Lichens respond to architectural detail by picking out and emphasising certain features. Plate 1e shows how the arcaded front of Canons Ashby Church in Nottinghamshire is rendered more emphatic by the striking contrast between the dark ironstone columns and the ashlar arches which have been preferentially colonised by the white, shade-loving lichen *Dirina massiliensis*. The figure also illustrates how enhanced lichen colonisation of sloping surfaces, such as window sills and the slabs at the foot of the arches, helps highlight these features, providing a new dimension to the architecture. As conditions on the building determine the distribution of the lichen cover they are automatically well related.

In addition to giving buildings and walls a pleasing appearance, lichens can help to interpret their history. Extensions, due to their later date of construction, or use of a different stone, support a dissimilar lichen cover to the original structure. Areas where repair or renovation have taken place also stand out.

Lichenologists are frequently asked how lichen colonisation can be accelerated to help buildings, especially their roofs, fit more harmoniously into the landscape. There have been almost no scientific studies of this topic, but conventional recipes are to 'paint' the surfaces with substances such as yoghurt, milk, beer and particularly a dilute solution of farmyard manure in urine. A few counties have the power to compel new structures to be washed down with manure in the expectation of speeding up lichen colonisation. When enforced, this often leads to humorous comment in the press. Removing lichens is easier than encouraging them; it involves the use of various chemicals such as tar wash and bleach, but before you remove them remember they may have been there several hundred years and are unlikely to return within your lifetime.

For some people the ultimate beauty of lichens is only revealed with a hand lens or stereoscopic microscope. Under magnification all species, but particularly those in fruit, exhibit a ravishing display of colour, texture, symmetry, neatness and sculptural form – the kind of perfection normally associated with a butterfly's wing, a recently expanded toadstool, or a newly opened flower. Henderson has explored this aspect of lichen aesthetics using the Thomistic approach, which involves an analysis of thalli in terms of wholeness, harmony and clarity (Henderson 1987a,b; 1989; 1990; 1992).

Perfumes

The major commercial exploiter of lichens is the French perfume industry. In Britain we experience the end product when we buy high class toilet soaps, perfumery or other cosmetics. The raw materials, oak moss (*Evernia prunastri*) and the slightly inferior tree moss (*Pseudevernia furfuracea*), are gathered in

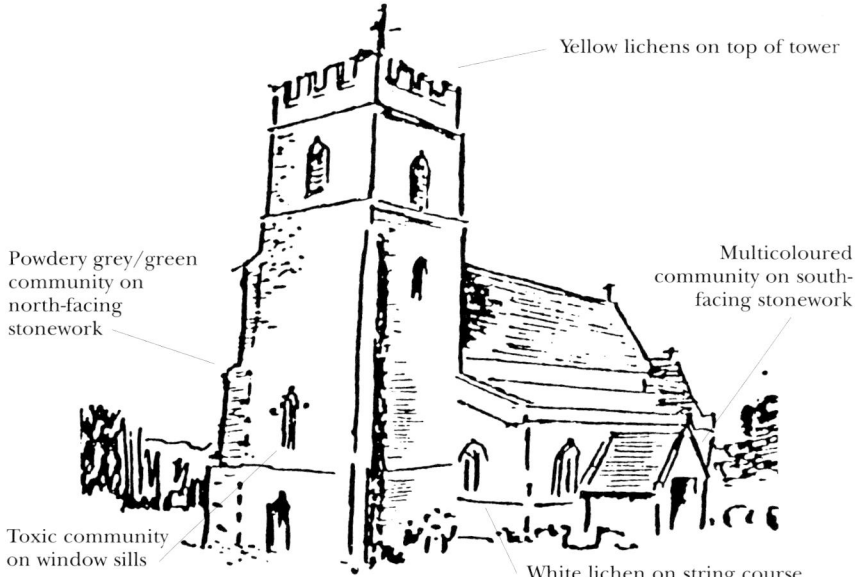

Fig. 1.7 Sketch illustrating some of the variation in lichen cover on the small limestone church at Sulgrave, Northamptonshire.

southern Europe and North Africa. It was estimated that in 1980 between 8,000 and 9,000 metric tonnes were collected by peasants whose poverty and low pay contrasts dramatically with the wealth of the consumers. The main concern of the manufacturers is not that the supply of lichen will become exhausted, but rather that the availability of gatherers will diminish as more profitable opportunities for employment arise.

Both lichens possess a subtle musky fragrance, but their main attraction is that on processing they yield extracts, known in the trade as 'concrete', which act as a fixative, allowing added scents to be released slowly. Most perfume companies have inventories listing 15 to 20 qualities of oak moss extract. Oak moss absolute is sold only in dilution or already compounded into finished retail products. It is used as a base in many perfumes, but if you wish to experience it almost unadulterated it is the main ingredient of Fougère and Chypre to which it imparts a haunting mossy odour. Further details of manufacture together with photographs can be found in Richardson (1975) and Moxham (1980).

Lichens can sometimes be encountered in gift shops, even supermarkets, as a component of potpourri. These may contain up to 20% of *Pseudevernia furfuracea* and *Cladonia* (reindeer moss) mixed with aromatic flower heads, leaves, seeds and coloured wood shavings. The fragrance, often supplemented by oil of lavender, is absorbed by the lichen then slowly released over a long period. This fixative property has been known since at least the eighteenth century when it was fashionable to sprinkle one's hair and wig with 'Cyprus powder' which was made to a secret formula. Eventually it was divulged that the main ingredient was finely ground *Evernia* and *Usnea*, impregnated with fragrance through lying in contact with whole flowers of rose, jasmine and orange.

Minor and curious uses

There are a number of minor and quirky uses that lichens have been put to down the ages. Michael Rogan (1833–1905), the great Irish dresser of salmon and trout flies, relied on natural materials, including lichens, to achieve a delicacy of shade and a subtlety of colour that has rarely been equalled. For example, he employed crottle dyes to create his Fiery Brown fly which was of unprecedented brilliance. Still on the subject of dyes, Johnston (1831) describes how children in the Scottish Borders used *Xanthoria parietina* to colour their eggs yellow at Easter. Industry has also shown an interest in lichen extracts which have been used by food manufacturers as a colouring agent for soft drinks and jam, by physicists to colour the alcohol in thermometers and by biologists for staining microscope preparations. For a while during and after the Second World War, Johnson's of Hendon manufactured litmus from imported *Roccella*, but production has now moved to the Netherlands.

The imported reindeer lichen *Cladonia stellaris* has found a number of uses in modern Britain. Small amounts are dyed green, soaked in glycerol to render them pliable, then employed to deputise for trees in architects models and along toy railways. In the 1970s yellow-dyed reindeer lichens were reported from seaside amusement arcades where they were being used to decorate 'penny falls' gambling machines. In their natural condition they have been used in window dressing to help advertise outdoor clothing and as background in taxidermic displays. The German custom of using *C. stellaris* as a basis for constructing funeral wreaths and crosses utilises thousands of tonnes annually, and the fashion is starting to spread to the UK. Such wreaths remain in good condition for many months and are traditionally replaced on All Saints Day. A walk round any large metropolitan cemetery in Britain should afford examples. The lichen is imported from Finland.

2

What is a Lichen?

In the old days, when life was simple, every school child knew that a lichen was a symbiotic association between a fungus and an alga. Then it was realised that the so-called blue-green algae, which can be one of the partners, are not algae at all, but belong to a more primitive group of organisms, the cyanobacteria. So the definition was changed to 'A lichen is an association of a fungus and a photosynthetic symbiont'. Later, this description too was found unsatisfactory as certain familiar brown seaweeds such as knotted wrack (*Ascophyllum nodosum*) and channel wrack (*Pelvetia canaliculata*) invariably support fungi that do them no harm, but these associations are not considered lichens. Such partnerships can be excluded by redefinition. The winner in a poll of lichenologists held in 1981, for instance, was: 'A lichen is an association of a fungus and a photosynthetic symbiont resulting in a stable thallus with a specific structure'.

Science moves on. It was not long before the expression 'a thallus of specific structure', was being questioned as excluding some structurally very simple lichens. In desperation, David Hawksworth, who was at the forefront of these attempts at a definition, at one point suggested that a lichen be defined as an organism studied by lichenologists! Two definitions are given in the current *Dictionary of the Fungi* (Hawksworth et al., 1995). One of them is simple: 'A lichen is a stable self-supporting association of a fungus (mycobiont) and an alga or cyanobacterium (photobiont)'. For the pedant there is a more complex and precise definition: 'A lichen is an ecologically obligate, stable mutualism between an exhabitant fungal partner and an inhabitant population of extracellularly located unicellular or filamentous algal or cyanobacterial cells'. Take your choice.

Problems over delimitation have been laboured so that the following crucial point can be made. Lichens are not a taxonomic group like the ferns or the liverworts; they are a fungal lifestyle equivalent to parasitism or saprophytism. Since lichenisation is a nutritional option, different groups of fungi have, over time, independently developed the ability to form lichen associations and the difficulty of arriving at a clear-cut definition arises because lichenisation has reached a higher level of complexity in some groups than others. A by-product of the lifestyle is that the thallus usually contains unique secondary compounds known as lichen acids and some lichenologists prefer to see this key feature included in the definition. For further discussion of this challenging topic, including descriptions of lichen symbioses involving three, four or even five partners, see Hawksworth (1988) or Hawksworth & Hill (1984).

The fungal partner

A fifth of all fungi are lichenised; this gives a current world total of around 13,500 lichens of which *c.* 1,700 are found in Great Britain and Ireland. Though the fungal and photosynthetic partners each have separate names, the name given to the fungus is also the name of the lichen as a whole organism; as a consequence the classification of the lichens can be integrated into that of the fungi. Molecular studies have revealed that lichens are polyphyletic in origin; that means they have evolved from or are descended from more than one ancestral group. The lichen lifestyle has arisen most frequently among the Ascomycetes (98% of all lichens), but a few Basidiomycetes (*Dictyonema, Multiclavula, Omphalina*) and Fungi Imperfecti (*Cystocoleus, Lepraria, Leproplaca, Thamnolia*) have adopted the lichen mode of nutrition. As if 1,700 taxa were not enough, British lichenologists also study some fungi that are not lichenised; 43 of these are included in the current lichen checklist (Purvis *et al.*, 1994a). The reason for this is historical; they are mostly small epiphytes once thought to have a lichenised thallus, but now known to be free-living (*Arthopyrenia, Leptorhaphis, Mycoporum, Stenocybe*, etc.). Some belong to primarily fungal genera which include a few species that are lichen-forming, others have a saprophytic or parasitic mode of nutrition or may even be optionally (facultatively) lichenised. Some genera that cross biological boundaries in this way are listed in Table 2.1.

Table 2.1 Examples of fungal genera, studied by lichenologists, that include a mixture of lichens and other nutritional lifestyles.

Fungal genus	*Arthonia*	*Chaenothecopsis*	*Lecidea*	*Omphalina*
Lichen-forming	*cinnabarina*	*pusiola*	*lithophila*	*luteovitellina*
Lichenicolous*			*insularis*	
Parasitic	*fuscopurpurea*	*parasitaster*	*vitellinaria*	*cupulatoides*
Saprophytic		*debilis*		*chrysophylla*

*Growing on lichens

Though not found free-living under natural conditions, many lichen fungi have been grown in culture in the laboratory where they are characterised by a slow growth rate, have little organised structure and do not produce fruit bodies; in other words they fail to resemble a lichen. The algal partner clearly has a major influence on thallus development and architecture.

Algal and cyanobacterial partners

While each lichen contains a different fungus only a small number of photosynthetic partners are involved in lichen formation. On a world scale around 40 genera are represented, 25 algal and 15 cyanobacterial. Most of the photobionts belong to groups which are also found free-living, for example the cyanobacteria *Calothrix, Gloeocapsa, Nostoc, Scytonema* and the green algae *Coccomyxa, Stichococcus* and *Trentepohlia* (Fig. 2.1). Only the green algae *Trebouxia* and *Pseudotrebouxia* appear to be primarily lichen-formers; there is currently much discussion as to whether they have ever been found free-living. *Trebouxia* is far and away the most widespread algal genus in British lichens, where it is believed to occur as a number of different species. The taxonomy of lichen algae and cyanobacteria is very underdeveloped, one problem being their totally different appearance when growing inside a lichen and in pure culture. For example, filamentous types become almost unicellular within a thallus. The same algal species often occurs in a wide range of taxonomically unrelated lichens, for instance, *Trentepohlia umbrina* is the algal partner in

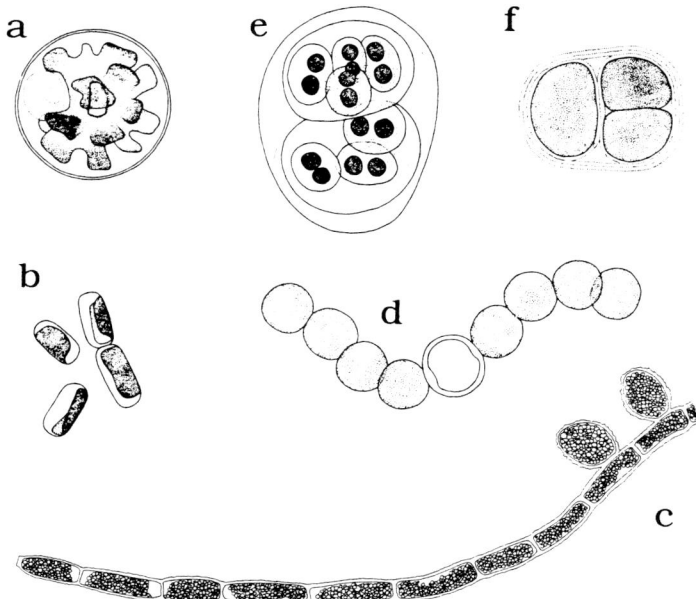

Fig. 2.1 Examples of green algae and cyanobacteria that take part in lichen formation. Green algae: a) *Trebouxia*, b) *Stichococcus*, c) *Trentepohlia*. Cyanobacteria: d) *Nostoc*, e) *Gloeocapsa*, f) *Chroococcus*. From Purvis *et al.*, 1992.

Arthonia, *Chaenotheca* and *Graphis*, while *Trebouxia glomerata* is found in *Cladonia*, *Porpidia* and *Stereocaulon* (Plate 2a).

The photosynthetic partner may be distributed evenly throughout the thallus but, more often, is limited to a layer just below the upper surface, where it can be exposed by scratching with a finger nail, when *Trebouxia* appears green and *Trentepohlia* orange. A number of species containing the cyanobacterial genus *Nostoc* as the algal partner are distinctive as they swell up when wet; for this reason those belonging to the family Collemataceae are called jelly lichens (Plate 2b). A few lichens regularly contain two photosynthetic partners, a primary layer of green algae and a subsidiary cyanobacterium distributed in discrete clumps called cephalodia which form pink, grey or brown warts on the upper or lower surface of the thallus or 'packets' inside it. They are a useful taxonomic character. Their biological importance is that they are thought to help with the nitrogen metabolism of the lichens in which they occur such as *Lobaria*, *Placopsis*, *Solorina* and *Stereocaulon*.

For a long time it was believed that algae were the subordinate partners in the symbiosis, as the fungus contributes the bulk of the tissue. However, Peter James made a discovery in New Zealand, followed by one in Scotland, which changed all that (Anon, 1975; James & Henssen, 1976). In sheltered, high humidity habitats such as gorges he found strange thalli in which different lobes belonged to different lichen species. For example, in Scotland *Sticta dufourii*, a species with a dark brown leafy thallus containing *Nostoc*, was found bearing lobes of the bright green *Sticta canariensis* which has a green algal symbiont (Plate 2c). Such composite thalli are the response of a single fungus to two different photosynthetic partners. In the New Zealand discovery the same fungus produced a leafy grey-green growth form (*Sticta filix*) or a brown shrubby one (*Dendriscocaulon* sp.) depending on whether the photobiont was green or blue-green (Fig. 2.2). Initially called 'chimeras' after the mythical Greek monster that had the head of a lion and the body of a goat, these composite thalli now tend to be called photomorphs or phycotypes. As all lichen names are based on the fungus alone a single name has to be applied to what were previously considered two distinct lichens. In the Scottish example they are now known as *Sticta canariensis* 'green algal morph' and *S. canariensis* 'cyanobacterial morph'. The discovery of chimeras was a triumph of careful field work and has been described as one of the most profound discoveries since 1867 when it was realised that algae were present in lichens. Chimeras demonstrate that the photosynthetic partner strongly influences thallus morphology, ecology, the range of chemical substances produced, and whether or

Fig. 2.2 The distribution of green and cyanobacterial phototypes of *Sticta filix* growing on the side of a humid gorge in New Zealand. Similar phenomena should be looked for in the west of Britain. From James & Henssen, 1976.

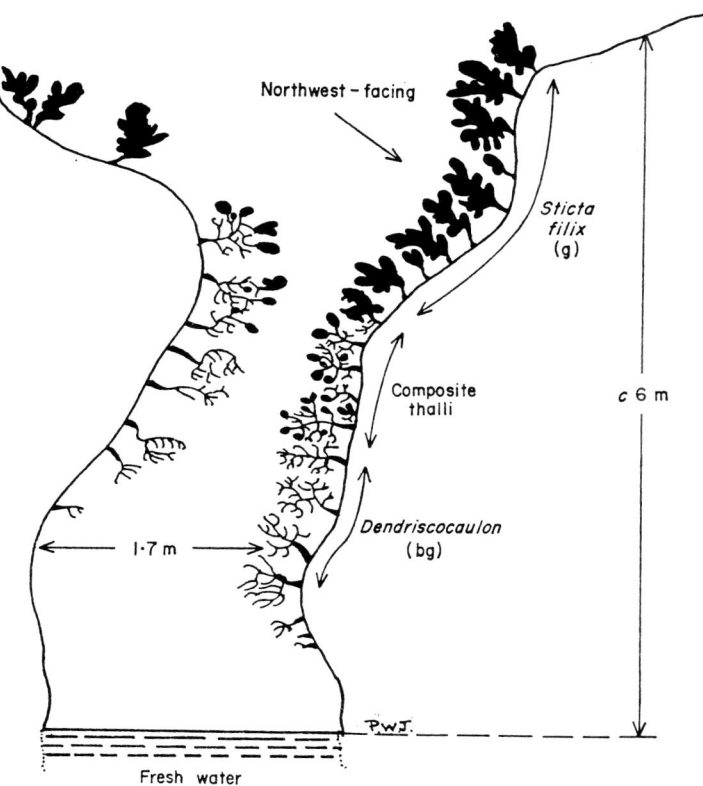

not reproductive bodies are formed. Anyone scrutinising the families Pannariaceae, Peltigeraceae and Stictaceae in the west of Britain could turn up additional examples.

Lichen names

In Britain we have never been sufficiently involved with lichens for them to acquire English names, apart from the very few that are of economic importance, and since several of these names refer to the lichen as moss (reindeer moss, Iceland moss), lichenologists are reluctant to use them. However, as any lichen listed on Schedule 8 of *The Wildlife and Countryside Act, 1981* must by law have an English name, epithets have been invented for the 28 rare species involved. Mostly they are a poor lot lacking in poetry. The forked hair-lichen (*Bryoria furcellata*), southern grey physcia (*Physcia tribacioides*), oil-stain parmentaria (*Parmentaria chilensis*) and upright mountain cladonia (*Cladonia stricta*) are instantly forgettable. Snow caloplaca (*Caloplaca nivalis*) and river jelly lichen (*Collema dichotomum*) are better. Two that might achieve common usage are the evocative golden hair-lichen (*Teloschistes flavicans*), tresses of which blow around on windy cliff tops, and goblin lights (*Catolechia wahlenbergii*), which gleams in dark crevices.

Albert Henderson (1986a, b; 1987a, b; 1988a, b) has researched the meaning behind the Latin names of a number of our lichens and discovered a cornucopia of information on the appearance and habits of individual species. Many names are teeming with scientific, historical and aesthetic significance, sometimes decidedly quirky. A selection follows:

Absconditella celata	A tiny hidden secret.
Bryophagus gloeocapsa	Moss devourer, with sticky cups.
Caloplaca teicholyta	A beautiful flat plate destructive of walls.
Candelariella aurella	A small candle holder, with minute yellow flame.
Cetraria cucullata	Like a Spanish shield adorned with hoods.
Cladonia coccifera	Branching bearing carmine grains or berries.
Gyalecta jenensis	With fruits like empty bowls and coming from Jena (Germany).
Haematomma ventosum	With blood-red eyes and puffed up.
Icmadophila ericetorum	Lover of moisture and dweller in heaths.
Parmelia sulcata	A small, round, furrowed shield.
Scoliciosporum umbrinum	With spores like worms, and of an umber hue.
Stereocaulon pileatum	Solid stemmed, tiny as a dwarf.
Umbilicaria hyperborea	Springing from a navel and found at the back of the north wind.

The following examples demonstrate how Henderson, who is well versed in classical tongues, works on a derivation:

Orphniospora atrata	Dusky spored, apparelled as for a funeral. *orphnos* (Greek) = dark, dusky. *spora* (Latin) = spore, seed. *atratus* (Latin)= dressed in black.

Acarospora sinopica   With extremely short, mite-like spores, associated with sinopite.
akares (Greek) = very short like a mite or a tick.
spora (Latin) = spore, seed. *sinope* (Greek) = Sinope, the Greek colony on the Black Sea, where inhabitants use red clay earth (sinopite) for a paint.

Growth form, reproduction and establishment

When the two partners come together they form a lichen thallus, the name given to a vegetative plant body not differentiated into leaf, stem and root. In primitive forms they just establish a tangled mass of cells, but more usually a thallus with a distinctly layered structure, exhibiting division of labour, is produced (Fig. 2.3A). It involves an upper cortex of densely packed cells that has

Fig. 2.3 a) Thallus section of *Parmelia*, b) section of apothecium with a lecideine margin, c) section of apothecium with a thalline margin, d) section of a perithecium, e) asci containing spores (various sources).

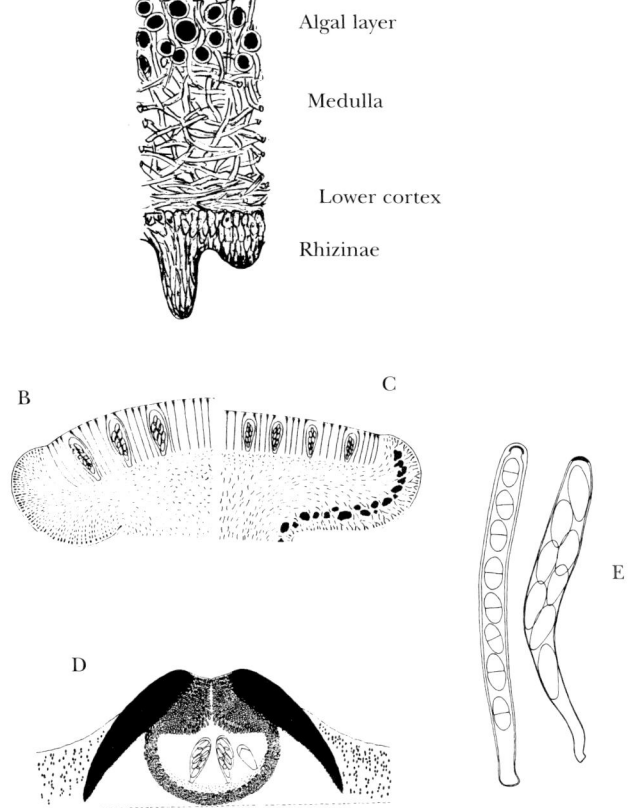

a protective function and may incorporate pores to facilitate gas exchange. Below this an algal layer is found, conveniently situated to receive plenty of light. This important layer occupies less than 10% of the thallus volume. Next there is a wide medulla formed of loosely packed fungal hyphae. This acts as a reservoir for water and a storage area for carbohydrates produced by the algae. These are conducted, or diffuse, as soluble glucose (cyanobacteria) or polyols (green algae) and are then stored as sugar alcohols which may benefit lichens in cold weather by acting as antifreeze. In crustose lichens the medulla also serves to attach the lichen to the substratum while foliose species have an additional lower cortex the underside of which bears swards of root-like structures (rhizinae) that act as attachment organs. There are, as to be expected, variations on this model.

A glance through the colour illustrations will show that lichens exhibit a great variety of life forms. Six categories are generally recognised. The fruticose or shrubby lichens and the foliose or flat leafy ones together constitute the macrolichens (Fig. 2.4). They are, in general, conspicuous (Plate 2d), easy to identify, and comprise around 20% of the British flora. The remaining 80% are variations on the crustose growth form where the underside of the thallus is closely attached to the substratum. The range of crustose morphology encompasses squamulose mats, consisting of small scales; elegant rosettes that are lobed towards the margin (placoid, Fig. 2.4); true crusts (crustose) in which the upper surface may be broken up to resemble crazy paving or is continuous; or, lastly, the surface may be a diffuse powdery mass of small granules (leprose). No classification is absolutely comprehensive so intermediate forms like sub-fruticose have had to be recognised, and in genera such as *Cladonia* and *Stereocaulon*, each thallus is composed of a basal squamulose system from which an erect fruticose one arises.

Reproduction

Lichens have evolved several ingenious methods of vegetative reproduction using propagules containing both partners. The simplest form is thallus fragmentation; dry lichens are quite brittle and parts that break off are capable of

Fig. 2.4 Lichen growth forms: fruticose (left), foliose (centre), placoid (right) (Claire Dalby).

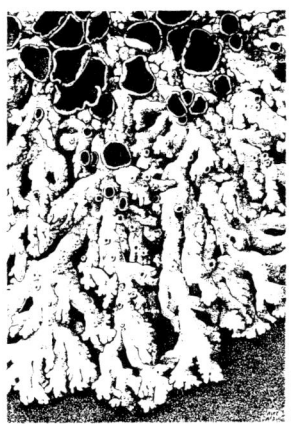
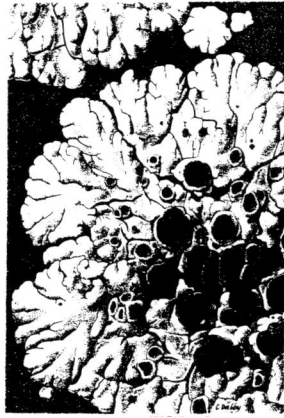

WHAT IS A LICHEN?

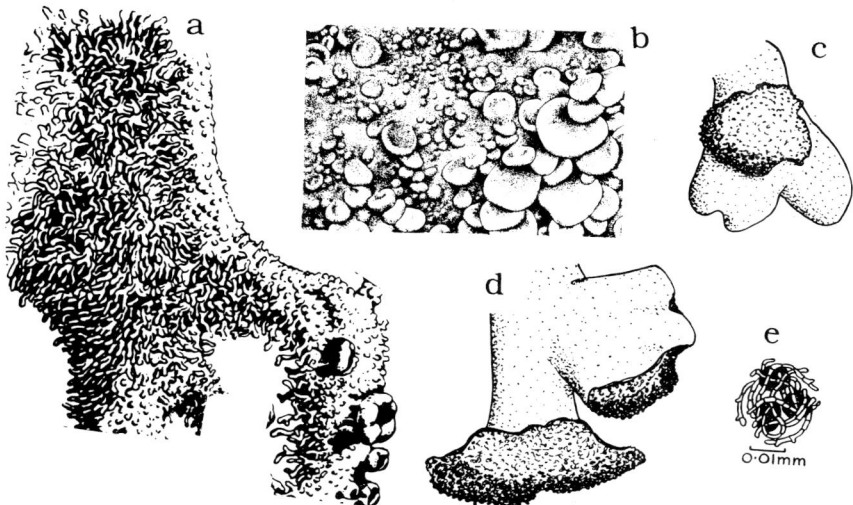

Fig. 2.5 Types of vegetative propagule: a) sward of coralloid isidia, b) foliose isidia, c) surface soralium, d) terminal soralium, f) greatly magnified detached soredium (various sources).

growing into new plants. Rather than rely on chance fragmentation many species produce special structures known as isidia. These small outgrowths on the upper surface of the thallus, being somewhat constricted at the base, are easily broken off when touched by animals, other vegetation or by gale force winds. Isidia may be rod-shaped, club-shaped, cigar-shaped, branched, coralloid, scale-like or warty (Fig. 2.5). Once dispersed, they attach themselves to the substratum by producing establishment hyphae, then growth proceeds.

Even commoner than the formation of isidia is the production of powdery granules called soredia. These structures, which are unique to lichens, are borne either in extensive patches on the upper surface or edges of a thallus, or in a more organised fashion in which case the discrete area manufacturing them is called a soralium (Fig. 2.5). Soredia originate from the algal layer of the thallus which is exposed by a rupturing of the upper cortex; under the microscope each soredium is seen to consist of a few algal cells bound together by a weft of hyphae. They are mainly dispersed locally by rain water trickling over the thallus or by the action of invertebrates, or, more widely, by wind; after stormy weather sticky slides exposed to monitor the pollen count are covered in them.

An alternative method of reproduction is by sexual means, through spores produced by perennial fruit bodies found on the surface of the thallus. Only the fungal partner reproduces sexually, the structure of the fruit body depending upon the taxonomic order to which the fungus belongs. The two commonest types are apothecia, which are cups or discs, and perithecia, which are flask-shaped and closed except for a tiny pore at the top. Several kinds of each

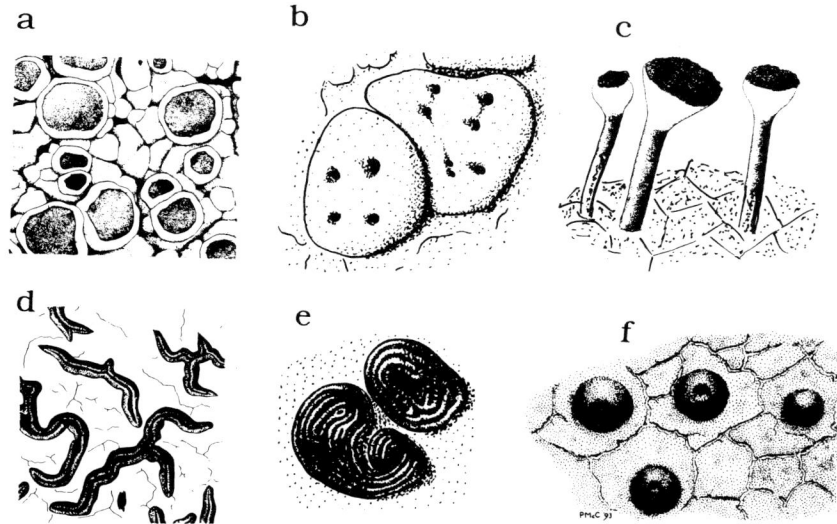

Fig. 2.6 Types of fruit body produced by lichens: a) disc-shaped apothecia, b) small immersed apothecia, c) pin-head apothecia, d) lirellate apothecia of the 'writing lichens', e) gyrose apothecia, f) superficial perithecia (various sources).

occur (Fig. 2.6), and characters associated with them are very helpful in assigning lichens to a genus. Some of the most spectacular apothecia are linear, forked or stellate and look like ancient script on the bark. These are known as lirellae and the groups producing them as 'writing lichens'. Others, the 'pin-head lichens', form swards of tiny 'indian clubs' on deeply shaded bark; they are often dusted with yellow, white, or brown powder, and are so appealing they are many people's favourite group. As Figure 2.6 shows, the fruiting bodies of lichens can be very beautiful.

The structure of lichen fruit bodies is of great taxonomic significance, which is why lichenologists spend much time at their microscopes. A section of an apothecium (Fig. 2.3B and C) shows that it is composed of a margin (exciple) that may or may not contain algal cells and can be coloured in various ways. A fundamental field character is whether the margin is composed entirely of fungal tissue (lecideine margin) or if it contains algae and is of the same colour and consistency as the thallus (thalline margin). The colour of the lower part (hypothecium) is also significant. The really important component is the hymenium, a layer of vertically orientated tissue in which sac-like cells (asci), containing usually eight spores, are packed around by slender filaments (paraphyses). At the generic level it is the structure of the ascus apex viewed after staining in iodine which is critical. At the species level the size, septation and number of spores in each ascus is often crucial to an identification (Fig. 2.3E).

To make a slide preparation two or three thin vertical slices of a moistened apothecium are cut using a razor blade or scalpel, mounted in water or dilute potassium hydroxide, pressure applied to the coverslip to spread the tissues, and the specimen examined under first low then high power. A graticule in the eye piece is useful for measuring dimensions. For the first few months, or even a year, a beginner may get by with a x10 hand lens as the majority of British lichens can be named on sight, but eventually, as the smaller species begin to intrigue, a microscope becomes essential.

The second type of lichen fruit body is the flask-shaped perithecium (Fig. 2.3). Lichens bearing them are known as pyrenocarps. A perithecium contains asci with or without paraphyses. Though usually small and requiring examination under a microscope, pyrenocarps are, on the whole, easy to identify after examining either a section or a squash of a whole fruit body. The secret of a good squash is to use very little material and apply even pressure.

As lichen genera that have arisen from the Fungi Imperfecti would, by definition, not be expected to produce fruit bodies, the news that an apothecium was present on the type specimen of *Leproloma membranacea* (Laundon, 1989a) was greeted with some astonishment. Since Laundon had rocked the lichen world in the past, particularly with respect to nomenclatural matters, this was accepted as a further example of his diligence and there was a tentative suggestion that *Leproloma* should be removed to the Pannariaceae. However, normality was restored eight years later when further examination revealed that the *Leproloma* was growing in an intimate mixture with fertile *Parmelia discordans* to which the apothecium belonged (Tønsberg & Jørgensen, 1997).

Fruit bodies release their spores throughout the year, the rate of discharge being greatest in winter and higher in wet weather than dry. Discharge distance is only a few millimetres, but this is sufficient to project the spores above the boundary layer of still air into turbulent air. Little is known about spore longevity. For a germinating ascospore to form a lichen, it needs to meet a suitable photosynthetic partner. The photobiont may be 1) acquired from chance encounter with a suitable free-living algae, though the commonest lichen symbionts, *Trebouxia* and *Pseudotrebouxia*, are very rare in this state; 2) stolen from an existing lichen, for example *Diploschistes muscorum* frequently establishes itself as a parasite on *Cladonia pocillum* before becoming independent; or 3) acquired from stray vegetative propagules, for example the soredia, of other lichens. Despite these three routes to success, it is believed that most ascospores fail to germinate because they find themselves in unfavourable environments and of those that do germinate only a few meet up with a suitable algal host.

Fungal recognition and selection of photobionts is a grey area. The fungi, most of which have a high degree of specificity, do not appear to be attracted to the right algae; they merely envelop any candidate photobiont cells encountered, parasitising and killing inappropriate species, while the few that can tolerate being attacked enter into mutualistic symbiosis and lichen tissues begin to differentiate.

Some peculiarities of lichens

Water relations

Compared with vascular plants and many other fungi, lichens have a remarkable ability to tolerate complete drying out for long periods, resuming normal activity once re-wetted. In the dry state they can survive extremes of heat and cold, but are incapable of any growth, being in a state of suspended animation that can last for over a year. Lichens take up liquid water i.e. rain or dew, extremely rapidly by absorbing it all over their surface like blotting paper, the thallus becoming fully saturated in just a few minutes. Gelatinous species like *Collema tenax* absorb particularly large quantities. Water loss by drying is much slower, taking hours rather than minutes, and is controlled by external conditions such as wind speed, air humidity and temperature.

Lichens that grow in underhangs never get directly wetted, but have the ability to obtain all the water they need from vapour. This is quite a slow process; it takes from three to six days for a thallus to reach equilibrium in a saturated atmosphere. One of the lichens most efficient at absorbing water vapour is the common *Dermatocarpon miniatum*, which can achieve a level of thallus hydration which is 68% of that reached by immersion in water (Kershaw, 1985). Some of these species have a surface that is covered with non-wettable substances that may help to prevent water loss.

The water relations of a lichen used to be compared to that of a sheet of filter paper; purely physical, with no special organs such as roots for absorption, and no features like a cuticle or stomata to retard water loss. Recent studies have suggested that things are not quite so simple; fruticose lichens at least, can slow down passive water loss through morphological adaptations that increase evaporative resistance. These modifications include adopting a curled-up, clumped or mat-forming growth habit (*Alectoria, Cladonia*), the production of a convoluted surface (*Lobaria pulmonaria*), and a high degree of branching (*Bryoria*).

Lichens are opportunistic; metabolic activity commences swiftly upon hydration and ceases rapidly with drying, in which state they maintain a very low level of activity. Under British conditions many lichens thrive best under strong wetting/drying cycles as prolonged saturation is fatal to all but aquatic species.

Legendary slow growth and great age

Lichens are reckoned to be amongst the slowest-growing organisms in the world; an audacious statement that needs qualifying. In such a varied group a considerable range of growth rates might be expected and this is exactly what has been reported. Growth rate studies have concentrated on measuring the radial increase of species growing on stone. Table 2.2 shows a range of annual growth rates from Britain which can be seen to vary from a rapid 17 mm in young *Peltigera canina* to a hardly measurable 0.09 mm in old *Rhizocarpon geographicum*. What is frequently overlooked is that many of the lichens with lower growth rates show an early burst of fast radial growth, lasting for c. 10 to c. 40 years depending on the species (the non-linear phase), which is succeeded by a long period in which radial growth is constant (the linear or low growth rate phase). Consequently, a simple statement of growth rate for a species is meaningless unless it is linked to the age of the thallus. This should be clear from Table 2.2 and an examination of the growth curves in Figure 2.7.

Growth rate is also controlled by environmental conditions. Topham (1977), for example, measured the annual growth rate of *Rhizocarpon geographicum* on dated granite tombstones across Scotland and found a considerable increase in growth from the dry east to the rainier west. Most investigators agree that on both long and short timescales growth rate is linearly related to total rainfall. Seaward (1976) correlated a highly significant reduction in the radial growth of *Lecanora muralis* on asbestos roofs (0.78 mm/km) as the centre of Leeds was approached with increasing levels of pollution.

Table 2.2 A range of growth rates of lichens in the British Isles. Figures are the annual radial increase in mm (various sources).

Foliose species	Y.T.	O.T.	Placoid and crustose species	Y.T.	O.T.
Peltigera canina	17		Lecanora muralis	2.14	
Lobaria pulmonaria	16		Diploicia canescens	0.2	
Parmelia caperata	6.7		Rhizocarpon geographicum		0.09
Parmelia sulcata	4.8		Lecanora campestris	2.3	0.12
Xanthoria parietina	3.1	0.12	Aspicilia calcarea	2.3	0.24
Parmelia saxatilis	3.8		Verrucaria nigrescens	1.0	0.27
Parmelia glabratula	2.4		Porpidia tuberculosa	0.63	0.12
Lobaria virens		1.0	Placynthium nigrum	1.66	0.08

Y.T. = Young thalli in the steady high growth phase
O.T. = Older thalli in the steady slow growth phase

In temperate climates the growth of crustose species can be maintained for a few hundred years unless it is interrupted by competition, rock weathering or thallus disintegration. The oldest lichen measured in Britain is a specimen of *Aspicilia calcarea* on the Rollright stone circle in Oxfordshire. It has a radius of 240 mm, and is estimated to have begun life in 1195 AD. It is accompanied by a companion dating from 1366 (Winchester, 1988). On a world scale it is just a youngster. Specimens of *Rhizocarpon geographicum* in the Alps have been estimated at 1,300 years of age and in West Greenland at 4,500 years, while extrapolations have suggested ages of up to 9,000 years for *Rhizocarpon alpicola* (480 mm diameter) in Swedish Lapland (Denton & Karlén, 1973). This, representing a mean radial expansion of 2.65 mm a century, is slow growth indeed.

Lichenometry, the use of lichens as dating tools, is most frequently employed to estimate the age of glacial moraines, stone monuments and the return period of earthquakes. Using the largest thalli available on surfaces of known age, calibration curves are constructed (Fig. 2.7) from which surfaces of unknown age can be dated. This technique has been used with remarkable success by Winchester (1984; 1988). She assembled a growth curve for *R. geographicum* by measuring colonies on gravestones in Crosthwaite churchyard near Keswick. Using the extrapolated growth curve, colonies on the roof of the church were dated to around 1525; it was subsequently discovered that the church received its first slate roof in 1523. In another test, five of the growth curves in Figure 2.7 were used to determine when boulders dug out of a field had been deposited beside the River Kennet near Avebury; they picked out the exact year of the event (see Appendix 1). Lichenometry has recently been used in an attempt to determine whether high corries in the Cairngorms were reoccupied by glaciers during the 'Little Ice Age' of the seventeenth century, but the results were inconclusive.

In the 1980s the Historic Buildings and Monuments Commission used lichenometry to shed light on recent stone movements at the popular stone circles of Castlerigg in Cumbria and Rollright in the Cotswolds (Fig. 2.8). The results provided evidence of heightened lichen initiation at dates coinciding with major surveys of the stones and publications on them which would have stimulated such conservation activities as the re-erection of fallen stones, clearance of vegetation and possibly stone cleaning (Winchester, 1988). The methods employed are described in Appendix 1.

Where are Britain's largest lichens? Few observations have been made on this simple topic. Some of the biggest I have encountered include *Caloplaca flavescens* forming a hollow circle 27 cm in diameter on a seventeenth-century limestone tomb in the Cotswolds; an intact thallus of *Ochrolechia parella* 54 cm in diameter in Caenlochan Glen; and on a British Lichen Society field meeting to Pistyll Rhaeadr waterfall in the Berwyn Mountains, thalli of *Lasallia pustulata*, over 15 cm across, were the largest anyone in the party had ever seen. These should not be difficult to beat as colonies of *Anaptychia runcinata* up to 75 cm diameter and *Alantoparmelia alpicola* c. 70 cm across have been reported from Norway.

Why the pretty colours?

One of the attractions of lichens is the range of colours they exhibit, which may be why poets and the general public notice them more than they do bryophytes. All shades from white to grey, palest green to apple-green, cream through yellow to deep orange and rust, black, brown and navy blue are represented. Some of these colours are due to pigments, others are structural, caused by the differential scattering of visible light. White or grey thalli have a surface covered with hyphal secretions that selectively refract the light, while the protruding tips of cortical hyphae are responsible for the milky appearance of some species. A fine powder of very small particles, known as pruina, is found on many species, it reflects short wave components, giving a bluish hue or bloom to thalline lobes and apothecial discs.

Lichens are famous for the wide range of secondary metabolites (lichen acids) they produce. Many of these are coloured and are responsible for the spectacular range of hues to be seen on a lichen-covered boulder or tree trunk.

Fig. 2.7 Growth curves for a number of lichens established from measurements made in Avebury churchyard. From Winchester, 1984.

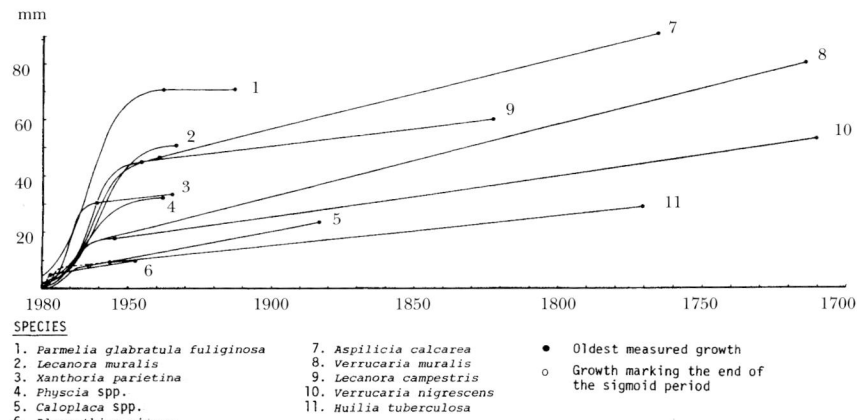

SPECIES
1. *Parmelia glabratula fuliginosa*
2. *Lecanora muralis*
3. *Xanthoria parietina*
4. *Physcia* spp.
5. *Caloplaca* spp.
6. *Placynthium nigrum*
7. *Aspilicia calcarea*
8. *Verrucaria muralis*
9. *Lecanora campestris*
10. *Verrucaria nigrescens*
11. *Huilia tuberculosa*

● Oldest measured growth
○ Growth marking the end of the sigmoid period

Fig. 2.8 The Rollright Stones as illustrated by Plot (1677). They support the oldest lichen colony so far measured in Britain.

These substances occur as water-insoluble crystals on the surface of the hyphae and regularly form from 2 to 5% (-10%) of thalline dry weight. Chemists are attracted to them because of their unique chemistry. Around 400 lichen acids are known, each lichen species normally containing one to three different kinds, though up to 40 have been recorded from certain foreign species. While many lichen acids are colourless, others are gaudy such as rhodocladonic acid seen in scarlet-fruited *Cladonia*, parietin in the orange-yellow hues of *Xanthoria*, rhizocarpic acid in the emerald-green of *Rhizocarpon geographicum*, calycin in the egg-yellow of *Candelariella*, haemoventosin in the purple apothecia of *Ophioparma ventosum* and usnic acid in the sulphur-yellow of many *Usnea* species.

The uniqueness of these substances and the restriction of most of them to lichens, has stimulated much speculation as to their role. Only dusty answers have been forthcoming. They increase the opacity of the upper cortex, reducing the amount of light reaching the algal layer, which might benefit *Trebouxia* under certain circumstances such as high altitude. Lichen substances have proven antibiotic properties so they could help to counter attacks by bacteria and fungi, and their bitter taste reduces palatability to slugs and other invertebrates. Some lichen acids appear to exert an allelopathic effect reducing competition from higher plants and bryophytes. This can be observed on sand dunes, heather moors (p. 123) and tree trunks.

While the importance of lichen acids to the functioning of the thallus is unclear, their value in lichen taxonomy is beyond doubt. In addition to giving many species a distinctive colour, they confer a chemical fingerprint that is often unique. Lichen acids can be identified by a number of techniques that are becoming ever more sophisticated and, unfortunately, ever more beyond the scope of the amateur who is stuck with the crude but still helpful 'spot tests'. These involve applying a dab of chemical reagent to the thallus and observing what colour it turns. In the 1930s the Japanese lichenologist Asahina perfected a more accurate technique in which the lichen acids are crystallised out on a microscope slide and identified by comparison with drawings or photographs. As there is no easily obtainable summary of these tests in English they have not been widely used, but their effectiveness is beyond doubt. It is said that Asahina never misidentified a lichen acid during a lifetime of involvement; he died aged 94. Since the 1960s professional lichenologists have used thin layer chromatography to identify lichen acids; this is a considerable advance on previous methods, but requires laboratory facilities. For the time being amateurs have to rely on 'spot tests'.

Spot tests have been used by lichenologists since the 1860s. Their beauty lies in their simplicity – they can be employed under field conditions – and in their unambiguous nature – there are few grey areas. The main reagents used are an aqueous 10–35% solution of potassium hydroxide known as K, fresh undiluted domestic bleach (e.g. domestos, parazone) called C, both of which can be stored in dropper bottles, and paraphenylenediamine crystals (Pd) freshly dissolved in alcohol (Pd must be used with care being a possible carcinogen). The colours produced by a drop of reagent vary from red, through rose, pink, orange and yellow to green. A positive test is indicated by +, negative by -, and variable by ±. So the reagent tests for the common epiphyte *Parmelia sulcata* would be written: medulla Pd+ orange, K+ orange, C-, while those of its relative *P. revoluta* run: medulla Pd-, K-, C+ red. Lichen acids have a characteristic fluorescence under ultraviolet light (UV) which is increasingly being added to species descriptions e.g. UV+ white or UV-. A valuable introduction to spot and other tests is provided by White and James (1985). Directions can also be found in most floras. The best flora for beginners is *Lichens: An illustrated guide* (Dobson 2000); this, the fourth edition, contains keys, distribution maps, line drawings and colour photographs of around 40% of British lichens. After this has been mastered devotees will wish to move on to the standard work *The Lichen Flora of Great Britain and Ireland* (Purvis *et al.* 1992) which contains descriptions, but few illustrations, of 1,650 British species.

So important are lichen acids in taxonomy that a species description lacking chemical data is considered incomplete. There has been endless discussion on how much importance to attach to chemical characters. For a while, particularly in North America, there was a tendency to recognise each chemical variant as a separate species. Today entities which are morphologically uniform but chemically different are called chemotypes and not given formal rank.

Finally it is worth noting that almost all the uses to which lichens have been put – medicine, perfume, dyeing – depend on lichen acids.

Lichen Communities

As with other plants, lichens form communities which can be remarkably uniform over wide areas. Field workers have developed names for the most distinctive of them. The discipline of classifying and naming plant communities is known as phytosociology. It is practised mainly on the Continent, particularly by the Germans. In Britain we have lagged behind or not seen the need for a detailed hierarchical nomenclature. However, a few of the higher units regularly in use are mentioned in this book, where they form a convenient shorthand for referring to widespread assemblages. Only two categories are relevant; alliances ending in the suffix '*-ion*', which are composed of one or more associations which end in '*-etum*'. The name of a community is latinised and obtained by adding the appropriate ending to the name of the most dominant or characteristic lichen.

The principal communities (alliances) mentioned in this book are the *Lobarion pulmonariae, Graphidion scriptae, Usnion barbatae* and *Xanthorion parietinae*, which are almost always shortened to Lobarion, Graphidion, Usnion and Xanthorion. A conspectus of lichen communities in the British Isles was published some years ago (James *et al.* 1977) but has not been developed. It would appear pedantic to use many of the names outside academic papers but examples of their usefulness can be seen in Figure 8.9 and Table 10.2.

Collecting lichens

Lichenology can be pursued with a minimum of paraphernalia. Often, during a walk, species are collected in an impromptu fashion, wrapped in a handkerchief, or if small, in a banknote, and conveyed home. However, when undertaking serious field work, it is normal to carry a lichenologist's bag. For many years ex-Army gas mask cases were ideal, being cheap, just the right size, and with the interior conveniently divided into compartments. Unfortunately supplies ran out in the 1980s, since when fishing bags have proved a more expensive alternative, but have the advantage of being waterproof. A basic collecting kit includes the following items.

A stout knife is essential for collecting off trees and cutting small squares out of soil or moss. Corticolous specialists prefer a sheath knife, but as this can be heavy; if you do not anticipate using it much, a penknife with a blade-locking device is less cumbersome. The other indispensable equipment is a hammer and cold chisel for use on rock and hard lignum. As I like to keep my bag light I carry a 400 g carpenter's hammer, but some people are never seen with anything less than a 1.5 kg club hammer. The chief advice over chisels is that they should be short enough to keep the weight down and regularly sharpened, so that potential specimens are not reduced to powder. It is useful to carry a spare as they are easily mislaid.

Once collected, specimens are placed in stout paper envelopes, often several from the same site per envelope. Shattered specimens, fragile specimens, those on soil or delicate ones that might get rubbed are best wrapped in tissue and placed in rectangular tobacco tins. For the spot tests described above, it is also normal to carry two or three reagents in the field. Small (*c.* 10 ml) brown eye-dropper bottles with a hexagonal cross section obtainable from any chemist are ideal. As already discussed, potassium hydroxide solution (K) and fresh bleach (C) are essential; paraphenylenediamine (P) is handy, but in practice difficult to keep in condition.

The above are the essential tools of the trade. In addition I carry a notebook and half a dozen biros/pencils, a spare hand lens, a 3 m flexible steel tape, a compass, a Mars Bar and, in summer, a tube of anti-midge cream. It is becoming fashionable to carry a mat to avoid any hesitancy or discomfort in kneeling on wet ground if the habitat demands. Some people add a razor blade for sectioning apothecia. This can, for example, save the trouble of bringing back a range of semi-aquatic *Porpidia* species to see if any have the blue-green interior of *P. hydrophila*.

It is interesting to see how fashions change. A 1958 note on collecting advised beginners in lichenology to carry a 2 kg pick-head geological hammer,

30 cm long chisels, a pair of secateurs and a small trowel. The perfect bag is light enough not to restrict movement, but has sufficient capacity to accommodate a day's finds.

3
Creatures That Need Lichens

Studies relating the present distribution of lichens to plate tectonics suggest that they are at least twice as old as the flowering plants. As a consequence, the animal kingdom has had plenty of time to learn how to exploit them for food, for concealment and for shelter, while lichens have fought back, developing a range of defences. In this struggle, lichens have come off best. It is rare to see badly damaged specimens; however, during an average year, a field lichenologist can expect regular encounters with cryptically coloured moths, to meet feeding caterpillars, to come across numerous thalli damaged by slugs, snails and other invertebrates, and, in particular, to find the underside of many specimens 'alive' with congregations of springtails (Collembola) and tiny brown mites (Acari). This is the superficial side; over the years, specialists have uncovered a fascinating range of interactions between lichens and mammals, birds, molluscs, insects, arachnids and microscopic animalcules such as tardigrades.

Reindeer

According to the Orkneyinga Saga (Harting, 1880), reindeer (*Rangifer tarandus*) were hunted in Caithness 800 years ago by the Jarls of Orkney, but apart from this reference the evidence is that they have been extinct in Britain for nearly 10,000 years. In 1952 a domestic herd was introduced to the Cairngorms under the auspices of the Reindeer Company Ltd., the 120 beasts being equally divided between Tomintoul and the Glenmore area. The latter spend the summer in a 400 ha enclosure, accessible to the public, but during the winter range freely in the mountains supporting themselves on a diet of lichen (Fig. 3.1). Walkers are frequently surprised when confronted by them.

A survey of the reindeer enclosure has shown that the standing crop of ground-dwelling lichens (*Cetraria islandica, Cladonia arbuscula, C. portentosa, C. rangiferina* and *C. uncialis*) is greatly reduced compared with matched plots outside the fence, the reduction in biomass being about five-fold. The percentage cover of ground lichens, however, was reduced only by 10–20% and diversity was unaffected. In summer the reindeer browse on young heather, sedges and grasses. In this part of Scotland boulders normally support big peeling mats and cushions of lichen composed of, for example, *Cornicularia normoerica, Parmelia omphalodes* and *Sphaerophorus globosus*, but in the enclosure these have been largely nibbled and licked off leaving only closely adpressed specimens. A small area of relic pine forest in the enclosure shows a grazing line on the trunks at 2 m; above this height *Bryoria, Pseudevernia* and *Usnea* hang in big shaggy bunches, while below, the boles carry only a close cover of *Hypogymnia physodes*.

Overgrazing and trampling by introduced reindeer has been detrimental to lichen communities in Alaska and South Georgia, but in Scotland they appear to be in balance with the communities they are feeding on. Reindeer require about 3 kg of moist lichen per animal daily; as it has only a low nutritive value

Fig. 3.1 Reindeer grazing lichen heath on the slopes of Cairn Gorm, Scotland (E. Smith).

with respect to protein, calcium and phosphorus, the diet of the feral animals is supplemented with high protein 'reindeer nuts'. Additional small herds exist in zoos and wildlife parks (Bedfordshire, Derbyshire, Norfolk, Worcestershire, Cotswolds) where they are kept in paddocks and fed on cattle concentrate supplemented with imported Iceland moss (*Cetraria islandica*) which is fed to the beasts at the rate of a 25-litre bucketful per animal per day. The Norfolk herd was fed on locally collected lichen until the supply ran out.

Birds

A number of British birds use pieces of lichen to decorate the outside of their nests. Among the most skilled at this are long-tailed tits (*Aegithalos caudatus*) which, on average, employ just short of 3,000 flakes of lichen, placed pale surface outermost, to give their orbicular nest a grey-green colour (Plate 3a). This largely conceals the main structure which is composed of moss and cobwebs with an inner lining of feathers. The birds seem particular in their choice of lichen; in a sample of 17 nests from all over Britain, three-quarters of the lichen covering was composed of the three species *Parmelia perlata*, *P. sulcata* and *Physcia tenella* (Hansell, 1994). The birds are ingenious in their method of attaching the lichen, using the 'Velcro' principle in which the rhizinae on the underside of the lichen flakes form a hooked surface binding onto the structural layers. Chaffinch, hawfinch and goldcrest are also skilled at incorporating lichen material into their nests.

What is the lichen covering for? Most lichen-covered nests are found in low scrub where there are virtually no lichens. Guest (1995) has suggested that the original habitat of the long-tailed tit was woodland and that nests in the fork of trees were the rule. In the past these would blend beautifully into the lichen-covered trunk, appearing like a thickening at the fork. Nests on Merseyside, where there is a dearth of lichens both on the trees and for decoration, are poorly camouflaged and mostly ripped out by predators within a few days of construction. Guest has correlated the breeding success of chaffinch and long-tailed tit in that area with the availability of suitable nest-building material. Hansell (1996) found little evidence to support the branch matching hypothesis, believing instead that a lichen cover enhances concealment by light reflection which makes the nest dissolve into the background. This is supported by the fact that white materials like polystyrene are sometimes substituted for lichen flakes and by the general absence of lichens on branches to which the nests are attached.

Jonathan Guest also has ideas about a possible interrelationship between tree creepers (*Certhia familiaris*) and lichens. These birds climb trees starting at the base and working their way up the trunk, before flying on to the base of the next tree. In doing this they must dislodge soredia and other lichen fragments transferring them to positions higher up the trunk. In Cheshire most of the early lichen colonisers are sorediate and species such as *Evernia prunastri* have been observed spreading upwards. The nuthatch (*Sitta europaea*) which works along branches may carry out the same function. To test this theory I borrowed five frozen corpses of each species from Weston Park Museum, Sheffield and brushed the breast feathers and feet over a dish of water. The arisings were then concentrated using a centrifuge. The deposit contained pollen, fungal spores, general debris and scores of lichen soredia. Whether this method of dispersal is biologically significant compared to say, strong winds or trickling water, remains to be determined, but it certainly occurs.

The tree trunk ecosystem

A tree trunk is a convenient habitat in which to start unravelling the interrelationships between lichens and the organisms that have come to depend on them. As a habitat it appears fairly straightforward; lichens have no regular periodic pattern of growth, and a line of lichen clad trees provides standard replicated plots. With this in mind, over a summer, I paid a series of visits to five mature ash (*Fraxinus excelsior*) trees in the Northumberland countryside to see what could be found. At each visit about 30 minutes was spent collecting off each bole at a height of between 0.8 and 1.8 m.

The basic components of the ecosystem were:

1) Autotrophs; green plants such as algae and lichens (bryophytes were absent) forming the base of the food chain.
2) Herbivores browsing on the above.
3) Omnivores and decomposers.
4) Carnivores preying on the herbivores, omnivores and decomposers.
5) Travellers going up and down the trunk to feed in the canopy.
6) Resting animals basking in sun or resting in shade.
7) Accidental species there by chance, for example aphids and capsid bugs that had fallen out of the canopy.

The lichen cover that provided food, refuge and camouflage for the large number of invertebrates present was a varied and uneven growth of tufted and leafy forms, any gaps being filled by crustose species, particularly *Lecanora conizaeoides*. The lichens were concentrated on the 'weather' side of the trunk

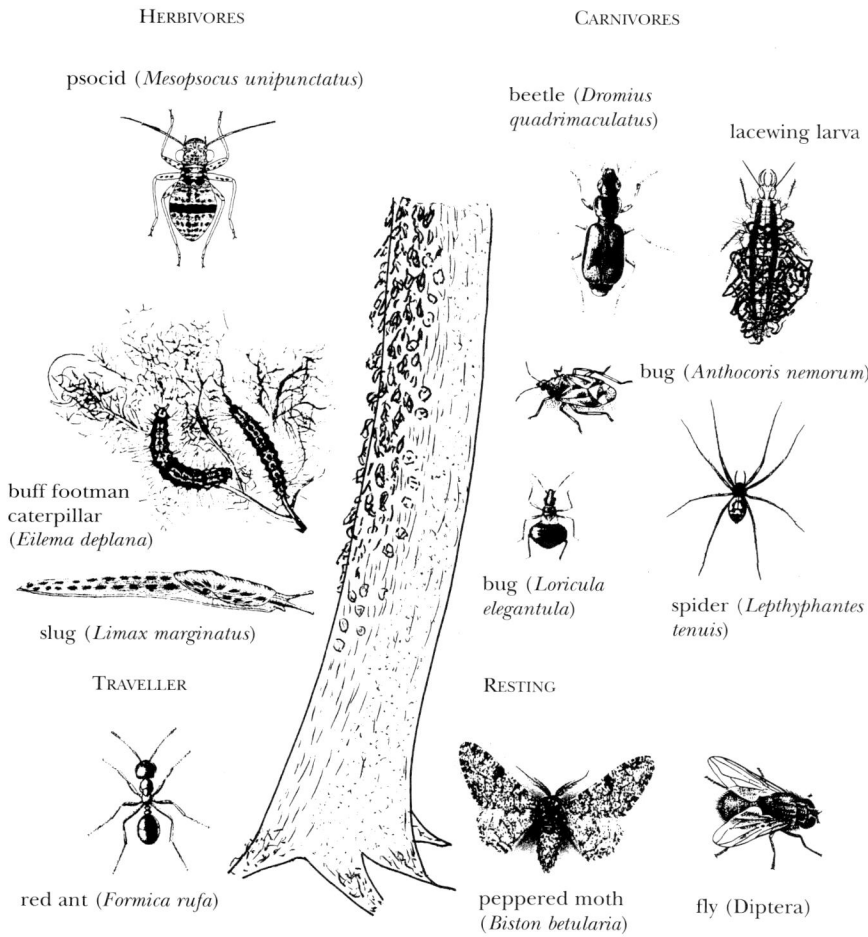

Fig. 3.2 The fauna associated with the bole of a lichen-covered ash tree.

with the opposite side either bare or covered with a green powdery growth of the alga, *Desmococcus* (Fig. 3.2).

The boles of the ash trees supported considerable populations of bark lice (Psocids). Most are wingless, about half the size of ants, and scurry about; they are by far the most important grazers. Five species were present, together with egg batches and nymphal instars, which suggests they complete their life cycle on the trunks. Broadhead (1958) has studied the feeding preferences of the five species involved and showed that *Elipsocus mclachlani* and *Reuterella helvimacula* prefer lichens, while *Amphigerontia bifasciata*, *Loensia fasciata* and *Mesopsocus unipunctatus* favour the alga *Desmococcus*, taking lichens only when other food is scarce. By August up to a hundred psocids were present in the sample area on each tree bole. Occasionally plague numbers are reported causing widespread damage, but this is exceptional. Few other resident herbivores were encountered during the survey, none of them regularly. Five caterpillars of a footman moth (Lithosiinae) were found feeding on *Ramalina farinacea*. No gastropods were observed, though their slime trails were noted. Mites were present but not studied; it is arguable whether they are herbivores or detritus feeders.

Carnivores were well represented. Among the leading species were plant bugs (Heteroptera) of which *Anthocoris nemorum*, *Loricula elegantula* and *Phytocoris populi* were the commonest. Five species of harvestmen (Opiliones) were present, the lichen cover providing them with refuge and protection from desiccation during the day; they then hunt over the tree at night. Spiders were present in about the same numbers as harvestmen. Other resident carnivores included ferocious lacewing larvae and the handsome yellow-spotted beetle *Dromius quadrimaculatus*, while dolichopodid and empid flies hunt round the trunks on hot days. Some lacewing larvae (Chrysopidae) had developed the odd habit of entangling debris, including fragments of lichen, among the spines that cover their bodies, so acquiring camouflage.

In terrestrial ecosystems the detrital food chain is usually more important in terms of energy flow than the grazing food chain, but on a vertical tree trunk dead material has little opportunity to collect. Consequently, large decomposers are scarce. Earwigs (*Forficula auricularia*), woodlice, springtails and psychodid flies belonged to this group as did various small beetles encountered in pockets of rotten bark.

Travellers were not common; during the period of observation only geometrid larvae, empty chrysalids and red ants were noted.

Resting animals comprised chiefly flies sunning themselves, especially if there were cattle in the vicinity. Small numbers of nocturnal moths and caddis flies were regularly encountered, which spent the day resting on shaded areas of trunk. These were so beautifully camouflaged against the lichen cover that many will have been missed.

Without an epiphyte cover the tree trunks would have been largely devoid of their rich invertebrate fauna. This was put to the test by carrying out a parallel survey of bare ash trunks in the nearby conurbation of Newcastle-upon-Tyne, where, not unsurprisingly, herbivores were all but absent, the few present surviving on small nodules of algae and fungal spores (Gilbert, 1970).

The degree of dependency revealed, and the variety of ways the fauna interacted with the lichen cover, was unexpected, even the carnivores using lichens for cover, camouflage and provision of a suitable microclimate. A recent study in Sweden found that natural forests had nearly five times more invertebrates per branch than managed woodland and correlated this with the richer lichen cover (Pettersson *et al.*, 1995).

Moths

A number of Lepidoptera have evolved beautiful and close relationships with lichens. The scalloped hazel (*Odontopera bidentata*), for example, has a 'stick' caterpillar that feeds on a range of trees. When nearing maturity the larva has the ability to turn either a mottled bluish-green, mimicking a lichen-covered twig, or dark purple, resembling a bare one, depending on its surroundings. In addition, the adult moth, which spends the day resting on trees or fences, is well camouflaged against a lichen-covered background. The Brussels lace (*Cleorodes lichenaria*) is even more closely adapted to a lichen-dominated environment, with a greenish-grey caterpillar that feeds on lichen and resembles a lichen in colour, form and attitude so precisely that it is almost impossible to detect, while the adult moth displays an elegant mimicry of a lichen-covered surface. The caterpillar of the light crimson underwing (*Catocala promissa*) is similarly camouflaged (Plate 3b).

The 17 British members of the subfamily Lithosiinae, all types of footman moth, have larvae that feed exclusively on lichens ranging, according to moth species, from epiphytes such as *Evernia, Hypogymnia, Peltigera* and *Usnea* to saxicolous lichens. The caterpillars are not often observed feeding, as they hide during the day. If lichens on a wall are suspected of being grazed by the common footman (*Eilema lurideola*), its dark grey larvae with tufts of black and yellow hairs can sometimes be found by looking under the capstones.

Elsewhere among the larger moths, species with caterpillars that feed on lichens occur sporadically, for example the beautiful hook-tip (*Laspeyria flexula*), dotted carpet (*Alcis jubata*) (Plate 3c), marbled beauty (*Cryphia domestica*), marbled green (*C. muralis*), marbled grey (*C. raptricula*) and speckled beauty (*Fagivorina arenaria*). These extend the range of lichen food plants to *Diploica canescens, Lecidea confluens, Lobaria pulmonaria, Physcia* spp. and *Xanthoria parietina*, though most frequently the lichen is described in the literature as 'various' or 'unspecified'. Since the majority of these moths have imagoes that mimic lichen-covered tree trunks, their connection with lichens is twofold.

The most widespread relationship of moths with lichens is simple camouflage of the resting adult against a lichen-covered background, whether on trees (Plate 3d) or rock. Such camouflage is a reminder of how universal a dense lichen cover must once have been. The species that have evolved intricate, mottled, cryptic patterns on their forewings are too numerous to mention, but include some of our most attractive moths, such as the merveille du jour (*Dichonia aprilina*), frosted green (*Polyploia ridens*), oak beauty (*Biston strataria*), mottled beauty (*Alcis repandata*), grey chi (*Antitype chi*) and brindled beauty (*Lycia hirtaria*). The names speak for themselves.

The most famous example of lichen mimicry and its consequences is that of the peppered moth (*Biston betularia*), which has brought lichens to the attention of people who might not otherwise have concerned themselves with them. The moth, which spends the day at rest on tree boles, occurs in two common forms; f. *typica* simulates lichen-covered tree bark while f. *carbonaria* is completely black. The distribution of these two forms is related to the colour of their background habitat. In towns where tree bark is devoid of lichens and blackened, up to 93% of the peppered moth population is f. *carbonaria*. By contrast, in rural areas the pale form predominates. This difference is the result of strong natural selection pressure exerted by birds which prey on the resting moths, but fail to recognise easily those with appropriate camouflage. The phenomenon, known as industrial melanism, is often quoted as an example of evolution in action. The response of peppered moth populations to changing conditions is quite fast. As pollution levels have fallen and lichens have begun to return to our cities the proportion of f. *typica* has increased rapidly.

Industrial melanism is widespread among moths and occasionally affects those spiders and beetles that use a lichen-covered background as camouflage. The lichen-feeding psocid *Mesopsocus unipunctatus* exhibits abdominal

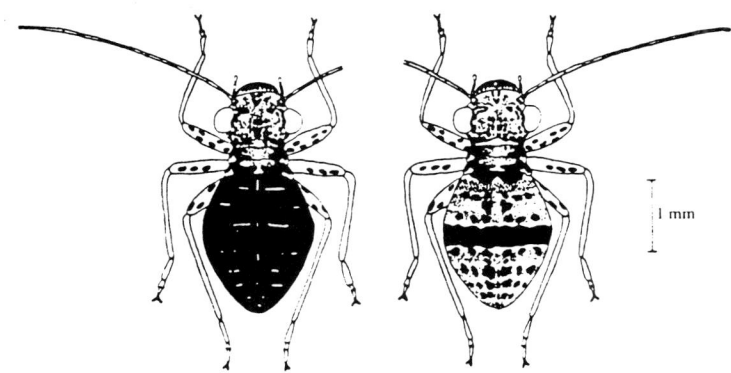

Fig. 3.3 Melanic and non-melanic forms of the lichen-feeding psocid *Mesopsocus unipunctatus*. From Popescu *et al.*, 1978.

melanism (Fig. 3.3), which prevents the formation of a 'searching image' by birds that feed on it. Melanism has an enormous literature that has been reviewed by Kettlewell (1973) and Majerus (1998). A variety of twists and turns complicate the classic theory outlined above.

Slugs and snails

Slugs and land snails are active mainly at night so are not often observed eating lichens. Old mortared garden walls support huge populations of snails which graze after dark, influencing lichen community composition by reducing the lichen cover to a monotonous low 'planed-off' sward of just a few species. They obtain their food by scraping the thallus with their tongue, which bears hooked cutting teeth that continually regrow to replace any that get damaged. This operation leaves parallel scratch marks on the surface of the lichen where white medulla shows through. Typically the gouges have a zigzag form, which reflects the side to side movement of the mouth parts. Ingested material is readily absorbed as the alimentary tract secretes enzymes capable of breaking down lichenin, the main carbohydrate present in lichens.

A large number of molluscs have been observed in close association with lichens, including the common snails *Candidula intersecta, Cepaea nemoralis, Clausilia bidentata, Discus rotundatus* and *Helix aspersa*, and slugs belonging to the genera *Arion* and *Limax*.

A habitat that experiences widespread damage from gastropods is limestone pavement. Here the entire community on the clint surface is crisscrossed by the grazing trails of molluscs that create shallow trenches in the bottom of which pale underlying limestone shows through. This is responsible for the community being dominated by species with immersed thalli, the high proportion of empty fruit bodies and the absence of many common lime-loving species. Preferred species have superficial thalli and include *Aspicilia calcarea, Caloplaca* spp., *Lecanora albescens, Verrucaria nigrescens* and free-living cyanobacteria. The common snail *Clausilia bidentata* is a particularly catholic feeder that will even devour carbonised perithecia, which are avoided by other species. It has been claimed that lichens regenerating following heavy grazing can exhibit new characters; for example, *Caloplaca variabilis* may transform from a superficial to an immersed appearance. Could *C. chalybaea* with its innate apothecia be merely snail-grazed *C. variabilis*?

Another habitat where grazing molluscs sometimes give cause for concern is woodland. The culprit here is our only arboreal slug, *Limax marginatus*. It occurs throughout Great Britain and Ireland, is up to 8 cm long, and can be recognised by its greyish colour, translucent, gelatinous appearance and ability to exude masses of watery mucus when disturbed. It has been observed feeding on a wide range of lichens, including *Hypogymnia physodes*, *Lobaria pulmonaria* and *Pertusaria pertusa*. Lichenologists engaged in monitoring are increasingly reporting that slug/snail grazing is responsible for the demise of thalli of species at the edge of their range (*Lobaria amplissima*, *L. pulmonaria* and *Teloschistes flavicans*), though in one case complete recovery from grazing was recorded. *Parmelia acetabulum* is particularly susceptible to grazing by molluscs; it has been speculated that its western limit in the UK is determined by the rainfall level that allows a critical level of slug damage to be exceeded.

For over a hundred years controversy has raged over the role of lichen acids in protecting thalli from invertebrate grazers. Zukal (1895) argued for their effectiveness in herbivore defence, while Zopf (1896) held that they provided very little protection. Both carried out experiments to prove their point. As recently as 1977 Gerson & Seaward wrote 'the whole subject is full of contradictions'. Since then Lawery (1984) has gone some way to resolving the Zukal-Zopf controversy. By offering slugs yeast-baited discs of filter paper soaked in extracts from various lichens he demonstrated that lichens with a high nutritive value contained lichen acids, which rendered these normally attractive baits unpalatable. Discs soaked in extracts from certain other lichens were eagerly consumed. This tied in with field observations and supports the avoidance hypothesis. The avoided species contained a much wider diversity of secondary compounds and had a higher total phenolic content than the palatable species. It seems that slugs, like children, pay more attention to the taste of their food than its nutritive value.

Mites and nematodes

Hard-bodied oribatid mites are sometimes found in large numbers feeding on, in and under lichens (Fig. 3.4). In Britain over 80 species are involved; though some are general feeders at least 25 are lichen specialists completing their life cycle in association with thalli (Seyd & Seaward, 1984). Though individual mites are usually less than 1 mm in size they demand attention because of their abundance. They eat either the whole lichen or just certain parts, such as asci and spores. They prefer lichen material that has been softened and partly decayed by fungi. Their food preferences pose a number of problems; *Parmelia*

Fig. 3.4 Oribatid mites living under a colony of foliose lichens.

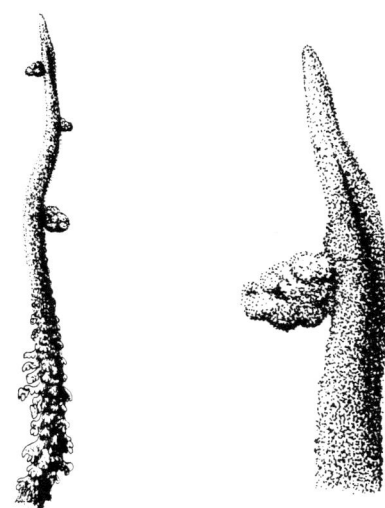

Fig. 3.5 Galls on the podetia of *Cladonia glauca* formed by a nematode worm. P.A. Ardron after Siddiqi & Hawksworth, 1982.

saxatilis, for example, supports numerous species, while *P. omphalodes*, with an identical set of lichen acids, has not yielded a single mite. Sometimes a lichen will be collected that appears intact, but inside is seething with brown mites. These attract carnivores, such as Hemiptera, which in turn may be eaten by birds. Thus, infested lichens support a considerable food chain. It has been suggested that on a world scale, mites are the main consumers of lichens.

Galls are occasionally found on lichens. Figure 3.5 shows swellings, the largest about half the size of a peppercorn, on the podetia of *Cladonia glauca* formed by nematode worms. Eriophiid mites can also induce the formation of swollen galls; an infected specimen of *Ramalina siliquosa* from Harlech Castle had such an unusual appearance that an unsuspecting lichenologist described it as a new variety.

Research into the often complex interrelationships between lichens and their fauna is a largely undeveloped field. It has been held back because both lichenologists and zoologists already deal with more than enough organisms without taking on additional unfamiliar groups. A few further examples of animal/lichen interrelationships are recounted in Chapter 13 on coastal habitats.

4

Lichens and Air Pollution

In the dark days of the 1960s, the far-sighted ecologist Kenneth Mellanby brought the link between lichens and air pollution to the public's attention by throwing down the environmental challenge 'Air fit for lichens, water fit for trout'. Prior to this only specialists had been aware of the connection.

The first casual references to this association can be found in nineteenth-century floras and the journals of local natural history societies; the following are typical. Grindon, writing in *The Manchester Flora* (1859), stated 'the quantity [of lichens] has been much lessened of late years through the cutting down of old woods and the influx of factory smoke, which seems to be singularly prejudicial to these lovers of pure atmosphere'. In 1879 the Rev. W. Johnson visited Gibside Woods, 8 km from Newcastle-upon-Tyne, and on his return wrote 'The lichens which flourished here in the fine condition spoken of by Winch (1831) have perished and this obviously from the pollution of the atmosphere by the smoke and fumes of Tyneside and the collieries of the surrounding district'. Erasmus Darwin can probably claim the earliest mention of the phenomenon in his poem the *Botanic Garden* (1790), written following a visit to the copper mine and smelter on Parys Mountain, Anglesey:

'No grassy mantle hides the sable hills,
No flowery chaplet crowns the trickling rills,
Nor tufted moss nor leathery lichen creeps,
In russet tapestry o'er the crumbling steeps.'

The first scientific paper on the topic was written by the renowned Finnish lichenologist, Nylander (1866), and a Scandinavian (Sernander, 1926) coined the evocative term 'lichen desert' to describe the centre of towns where trees are devoid of foliose and fruticose lichens. It gradually became a popular exercise to map the lichens around towns to demonstrate how far the influence of air pollution extended. Jones (1952), faced with the vast conurbation of Birmingham, made a careful study of lichens on trees along a 64 km transect cutting through the centre of the city. He found that the drop in diversity commenced a long way out, was continuous, and that the extinction point of individual lichen species depended partly on tree species. Fenton (1960) used the same technique in Belfast. Throughout this period it was gradually becoming clear that sulphur dioxide (SO_2) was the pollutant involved in the formation of lichen deserts, an important advance being made when workers in Britain, with the advantage of a national network of nearly 1,300 gauges measuring SO_2 levels, felt able to attach tentative SO_2 levels to some of the boundaries they were mapping. The production followed of the first biological scales for the estimation of air pollution (Gilbert, 1970; Hawksworth & Rose, 1970), which provided a great deal of publicity for lichenology. The Hawksworth-Rose scale is reproduced in Appendix 2. For a while, Britain became a hotbed of lichen/air pollution studies with at least seven PhD theses being undertaken on the subject. Though it was not fully appreciated at the time, stable SO_2 emission rates aided the calibration and usefulness of the scales.

Lichen deserts

A rapid method of investigating lichen deterioration around your home town or local industrial complex is to examine a series of standardised habitats, such as mature ash trees or asbestos roofs, along a transect stretching upwind from the centre of pollution. This will enable species to be placed in approximate order of sensitivity and at the same time the effect of substratum on survival can be studied. The results derived from such a transect (Fig. 4.1) show a gradual decline in the number of species of lichen in each habitat as pollution increases, until in the city centre only a few remain. Though diversity is strongly affected in all habitats, sensitive species are seen to disappear first from trees and then sandstone walls, while asbestos roofs retain their flora the longest. So the behaviour of lichens under pollution stress is to some extent governed by the nature of the substratum. A further example comes from Leeds where Seaward (1976) found that the widespread species *Lecanora muralis* had three fronts. Proceeding in from the suburbs it disappeared first from the capstones of sandstone walls, persisting a further 1,900 m into the city on concrete and mortar, while its ultimate limit was 550 m further inward on asbestos cement roofs (Fig. 4.2). Under pollution stress this generalist became a strict calcicole.

Despite the variety shown by lichens the process of extinction is strikingly similar. Diminishing luxuriance is followed sooner (*Parmelia sulcata*, *Evernia prunastri*) or later (*Xanthoria parietina*) by a depression of fruiting. At their

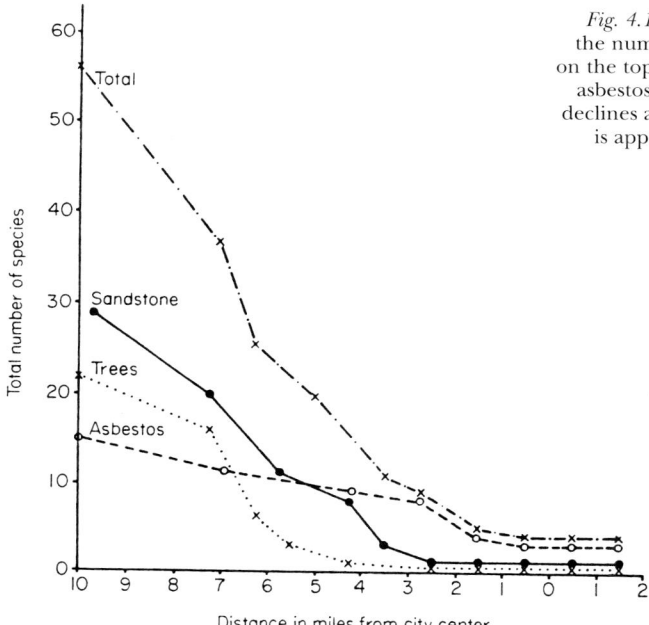

Fig. 4.1 Transect showing how the number of lichens growing on the top of sandstone walls, on asbestos roofs and on ash trees declines as Newcastle-upon-Tyne is approached from the west. From Gilbert, 1965.

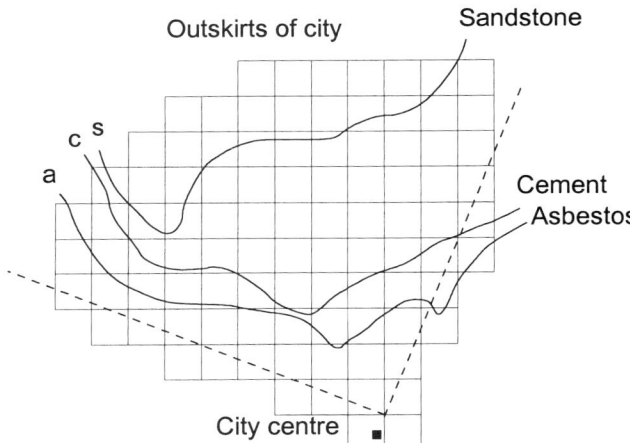

Fig. 4.2 Map of the northern sector of Leeds showing the inner limit of *Lecanora muralis* on asbestos roofs (a), on cement, concrete and mortar (c), and on sandstone wall tops (s), in 1970. The grid lines are 1 km apart. From Seaward, 1976.

inner limit species are sterile and compact, individuals tending to be small and with a low cover. The gradual suppression of the shrubby lichen *Evernia prunastri* around Newcastle-upon-Tyne is shown in Figure 4.3, which demonstrates how biomass starts to decrease long before the species is eliminated. Herbarium specimens reveal that the old lichenologists regularly collected this species in fruit, but no-one has seen it in this condition in northeast England for over 100 years. The centre of towns and industrial complexes never lose their lichen flora entirely; a few species, such as *Candelariella aurella* and *Lecanora dispersa*, persists on concrete, and *Lepraria incana* survives in sheltered acid niches. Trees, however, are usually completely devoid of lichens.

A few species do not follow the pattern of decline described above, one being the crustose epiphytic lichen *Lecanora conizaeoides* (Plate 4a). This species increases in abundance as towns are approached and where the last sensitive lichen disappears, usually in the outer suburbs, it reaches its maximum cover, colouring tree boles from top to bottom with a thick, grey-green crust. It has never been satisfactorily determined whether it has a requirement for SO_2 or is responding to the lack of competition, but field evidence in the form of its absence from areas free of SO_2 pollution suggests the former. Nonetheless, though toxitolerant, it is absent from the centre of towns where the annual average level of SO_2 exceeds 170 µg/m^3.

The species has an interesting history. The earliest British specimen is a collection made by the Rev. Bloxam from a fir tree at Twycross in Leicestershire around 1861; by 1870 it had been recorded from Buxton, Manchester, London and Hampshire, spreading rapidly with the rise in background air pollution as the Industrial Revolution gathered pace. It is probably an introduced species whose natural habitat, near sulphur springs in Iceland, preadapted it for a rapid invasion of coal and oil-burning areas. In parts of the world which it has not yet reached, the most resistant corticolous species are *Lecanora hagenii* (*L. dispersa* complex) and *Scoliciosporum chlorococcum*. Both are present in Britain, but remain subordinate to the newcomer.

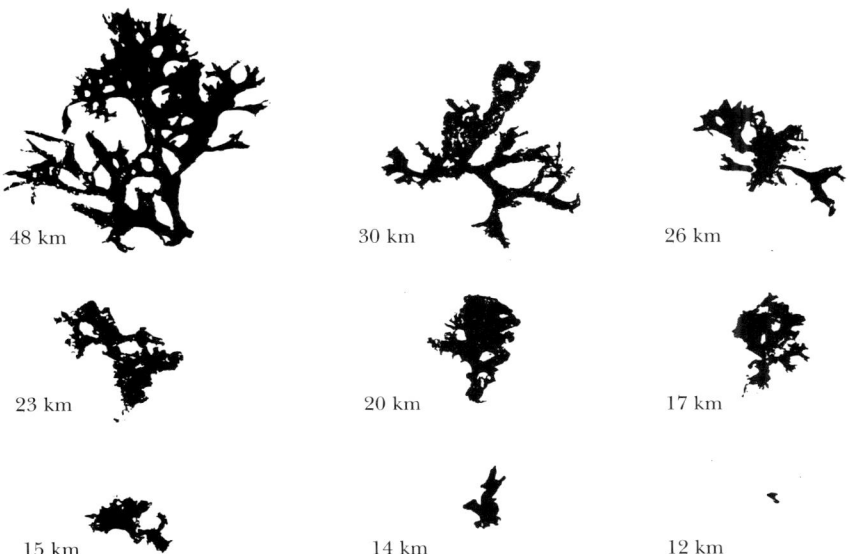

Fig. 4.3 Typical thalli of *Evernia prunastri* collected from ash trees on a 48 km-long transect out from the centre of Newcastle-upon-Tyne; luxuriance increases as pollution levels fall.

Through its acidification of tree bark, SO_2 has encouraged the spread of such acid-loving species as *Chaenotheca ferruginea* and *Parmeliopsis ambigua*, which have increased spectacularly throughout England this century. So the effects of air pollution are more complex than might be first thought, with most species declining, but a few expanding, and the response of individual species being tempered by substratum characteristics.

An area affected by pollution can be delineated rapidly and cheaply by mapping indicator species. A good indicator species should be widespread, easy to recognise in the field, and should have shown a sharp extinction point in any transect studies. To obtain meaningful maps it is important to standardise the habitats examined as field observation has shown that shelter, pH, nutrient flushing and age of habitat can have a pronounced effect on survival. By using different species and selecting a range of substrates it is possible to determine a boundary at almost any distance from a pollution source. The early workers mapped the inner limit of foliose lichens on trees, but many other parameters can be used. In 1972 a national survey by 15,000 schoolchildren using a simple scale and observing lichens on deciduous trees, acid stone and concrete, produced a 'mucky air map of Britain' that compared favourably with those produced by professional lichenologists (Fig. 4.4).

During the 1970s, when lichen/SO_2 studies in Britain were at their height, lichen/pollution maps were commissioned by a range of organisations to help identify areas of deprivation and to guide forestry planting. Maps were prepared covering the whole of South Yorkshire, Merseyside, the Peak District National Park, Tyneside and the Leeds-Bradford conurbation, together with

Fig. 4.4 Mucky air map of Britain produced by school children who mapped lichens around their homes using a simple scale. From Gilbert, 1974.

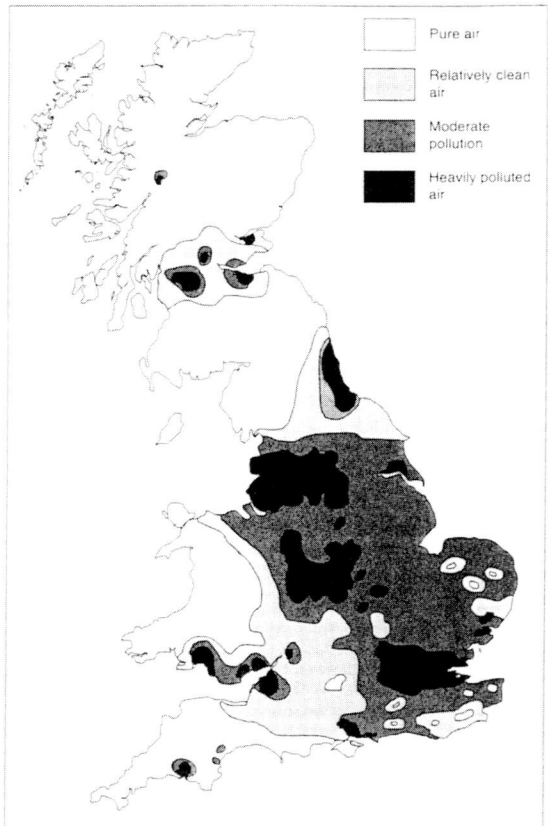

Greater London, Glasgow and the Southampton area. The maps were accurate within the range of SO_2 levels from pure air to 170 µg/m^3 at which level virtually all epiphytic lichens have disappeared.

Effect of environmental and other factors

The bald statement that 'lichen diversity declines as levels of SO_2 increase' hides a wealth of fascinating detail involving interactions between lichens and the rest of the environment. These can accentuate the effects of pollution so they are still detectable 50 or 100 miles away, or minimise them so that certain attenuated lichen communities survive close to a source.

Most lichenologists are aware that sheltered sites such as dense woodland, tall herbage, deep valleys and crevices in stonework regularly carry sensitive species to well inside their general limit. The more sheltered the site the further the species come in, the combination of woodland in a deep ravine, where much pollution is blown over the top, being especially favourable. The survival of *Cladonia* spp. and *Coelocaulon aculeatum* among heather on the commons of southwest London is an example of shelter involving tall herbage. The converse is also true: high ground catches pollution. In any landscape coming under pollution stress, exposed trees on ridges lose their lichen diversity first, starting with twig assemblages which rapidly succumb to invasion by *Lecanora conizaeoides*.

Examples have already been given of the way in which a high pH apparently reduces the influence of SO_2 pollution, normally wide-ranging species such as *L. muralis* behaving as calcicoles at their inner limit. Epiphytes show the same phenomenon, persisting longest on trees with a less acid bark such as ash, poplar, sycamore and willow (pH 4.5 to 6.0), and on those parts of the bole with the highest pH, for example around bark wounds and on bosses and buttresses. Acid-barked trees such as pine, birch, alder and oak are the first to lose their lichens. Not only is the bark of these trees innately acid (pH 2.5 to 3.8), but it has a lower buffering capacity against acidic substances than that of ash, willow and elm.

It has repeatedly been observed that the survival of lichens in polluted areas is enhanced by dry nutrient flushing. The spectrum of nutrients provided by roadside dust encourages the persistence of *Acarospora fuscata* and *Candelariella vitellina* on sandstone walls, while even more effective is the stronger flushing produced by bird droppings. Before lichens disappear from roofs they become restricted to places where birds perch such as the edge, the ridge, gable ends and below TV aerials; such niches carry lichens into the innermost suburbs. Further dramatic examples can be found associated with bird 'song posts'. Some old deciduous trees carry nutrient-rich streaks below decaying bracket fungi and rotting limbs, which retain a lichen flora when it has disappeared from other parts of the bole. These sites invariably support nitrophilous vegetation including species of *Physcia* and *Xanthoria*.

When preparing distribution maps around pollution sources, some workers observed that lichens could occasionally be found inside their normal limit on older trees and walls. This anomaly was eventually solved by Laundon (1967). Working on dated limestone memorials in London he realised that the lichen *Caloplaca flavescens* had persisted for over 70 years in an area where rising levels of air pollution made the colonisation of new surfaces impossible. Once recognised this phenomenon of relic lichen communities has been of immense help in interpreting field data. It has been observed on asbestos roofs, brick walls, sandstone walls and trees, but is at its most spectacular on calcareous substrata.

Lichen reinvasion with declining air pollution

Since the late 1960s there has been a steady decline in ground level concentrations of SO_2 in urban areas. The course of this decline is linked to the increasing use of sulphur-free fuels such as natural gas, lower industrial energy demands and energy conservation. Average urban concentrations, as measured by the network of gauges, have fallen from over 200 µg /m^3 to around 30 µg /m^3 and the difference between summer and winter levels has almost been eliminated. It would be valuable to know about trends in rural areas, but the national network has little information on, for example, SO_2 concentrations in East Anglia, over the Pennine moorlands or in the Lake District. It is believed that due to the tall stack policy of spreading dilute pollution over a wide area, levels outside towns are falling more slowly or not at all. Remote areas may even be experiencing slight increases as highly pollution-sensitive lichens such as *Usnea articulata* and *U. florida* (Plate 4b), are continuing to decline in southwest England. In most places, however, the story is one of recovery.

A study that illustrates the reinvasion scene involved a 125 km transect stretching westward from Liverpool into the clean air of North Wales (Cooke et al., 1990). The epiphytic cover of mature, freestanding oaks was assessed at sites along the transect in 1973 and again in 1986. Changes during the time interval were not uniform throughout its length. There was a marked increase in the cover and extent of *Lecanora conizaeoides* at the urban end, a small increase in lichen diversity along the middle stretch, and a possible decline of large foliose species at the western, least polluted end. Most of the new species in the middle part were juvenile thalli of *Evernia prunastri*, *Hypogymnia physodes*, *Parmelia saxatilis*, *P. sulcata* and *Physcia tenella*. When sites were classified according to the Hawksworth-Rose 10 point pollution scale (Appendix 2), it was apparent that over the 13 years the zones had moved inwards considerably at the eastern end while at the western end they remained unchanged (Fig. 4.5). These results are consistent with estimated changes in pollution levels along the transect. The most noteworthy points are the long time-lag between pollution levels falling sufficiently for foliose and fruticose lichens to colonise the trees and their actual appearance and build-up into communities, and the suggestion that long distance transport of atmospheric pollutants may be affecting lichens at the rural end.

Another long transect involving oaks, this time extending 70 km SSW from central London, was recorded annually between 1979 and 1990 (Bates et al., 1990). Their results differ markedly from those quoted above. Despite gauge measurements of SO_2 levels having declined greatly prior to and during the course of observation no firm evidence of recolonisation was obtained, even deep into Sussex. The conclusion was that lichen communities on oak in the London area continue to indicate the high SO_2 levels that prevailed 30 years ago.

The results of the London oak transect are all the more perplexing in the light of other reinvasion studies that have been undertaken in the capital. Rose & Hawksworth (1981) and Hawksworth & McManus (1989) have reported a general reappearance of foliose and fruticose lichens commencing around 1970 and gathering pace through the decade. By 1988, 49 epiphytes were known in the capital, 25 of which had not been

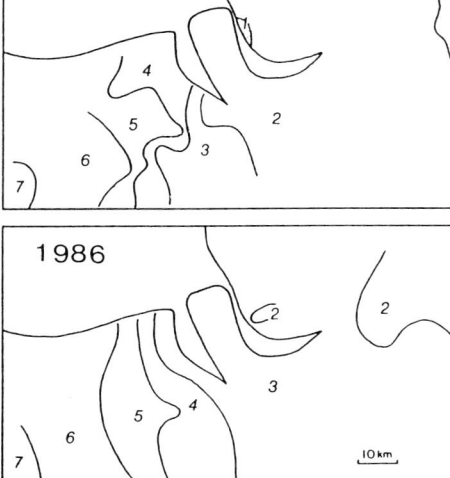

Fig. 4.5 Map of the Liverpool-North Wales area showing how the Hawksworth-Rose lichen zones altered between 1973 and 1986. Changes reflect amelioration on the east side, but no apparent improvement in the originally less polluted western areas. From Cooke et al., 1990.

seen within 16 km of the centre of London this century; *Evernia prunastri* and *Ramalina farinacea* were in many central parks, while the even more sensitive *Parmelia caperata* was expanding in the outer suburbs.

In London, and elsewhere, the recolonisation process, instead of following an orderly reverse sequence through the Hawksworth-Rose zones, was controlled by dispersal efficiency rather than SO_2 sensitivity, as central London was now available to all species in zones 1–7 (Appendix 2). This led to some unexpected discoveries, with highly sensitive species such as *Candelaria concolor* in Chelsea Physic Garden and *Usnea subfloridana* at Dollis Hill. This phenomenon was christened 'zone skipping', and defined as 'recolonisation without the orderly return of species progressively lost under conditions of gradually rising ambient air pollution' (Table 4.1). By contrast, species with a poor recolonising ability have been called 'zone dawdlers' (Gilbert, 1992), these tending to be crustose lichens characteristic of dry bark rarely wetted by rain. Such sites retain acidity acquired in the past. The two groups appear to be ecologically distinct, showing that habitat is as important in determining the detail of reinvasion as it was in affecting extinction.

Table 4.1 List of epiphytic lichens that are either 'zone skippers' or 'zone dawdlers'.

Zone skippers	Zone dawdlers
Candelaria concolor	*Calicium* spp.
Evernia prunastri	*Chaenotheca* spp.
Parmelia caperata	*Chrysothrix candelaris*
P. perlata	*Diploica canescens*
P. revoluta	*Graphis scripta*
P. subrudecta	*Hypocenomyce scalaris*
Physcia aipolia	*Lecanactis abietina*
Ramalina farinacea	*Opegrapha vulgata*
Usnea subfloridana	*Pertusaria hymenea*
Xanthoria polycarpa	*Ramalina fraxinea*

Studies in Cheshire have confirmed that, initially at least, lichen reinvasion is habitat-dependent. In the late 1970s Brian Fox discovered that epiphytes were returning to this polluted county, but only in one specialised habitat, willow carr. In such sites he observed the return first of *Hypogymnia physodes*, *H. tubulosa* and *Parmelia sulcata*, soon to be followed by *Evernia prunastri*, *P. glabratula*, *P. subaurifera*, *Platismatia glauca* and *Ramalina farinacea*. As pollution abated further *Parmelia revoluta*, *P. subrudecta*, *Physcia aipolia* and *Usnea subfloridana* joined what were now quite rich assemblages. The latest species to arrive have been *Parmelia caperata* and *P. perlata*. During the last few years a number of the species have started to spread from the carrs to the base of field trees, especially ash and sycamore. These studies raise the question of where the propagules originate from. Current opinion favours long distance wind dispersal from populations in Wales, the Lake District and Ireland.

Several features of reinvasion are not fully understood and are suitable for investigation by amateurs. One poser is the time-lag required between pollution abatement and the establishment of lichen thalli. Several variables appear to be involved and even the onset of colonisation is hard to define, as juvenile thalli often disappear after a year or two, presumably casualties of peaks in fluctuating pollution levels or possibly eliminated by mollusc grazing. Most workers agree that pH is of fundamental importance in controlling re-establishment, which is not unexpected as it is also a key factor in the expansion of lichen deserts. Experimental work has shown that sulphur present in low pH habitats is far more toxic than when the pH is high; currently this is the best explanation we have as to why lichens return first to trees with basic bark such as willow. Bark throughout the London oak transect was extremely acid (pH 2.9 to 4.0). This acidity, acquired during years of SO_2 exposure, is due to leaching by sulphurous acid and persists following pollution abatement, so that trunks remain poorly colonised. Often smooth-barked young oaks which have not experienced the conditions that acidified the parent trees are the first to be recolonised.

There is a suggestion that a new and different lichen flora from that destroyed by pollution earlier this century may be filling up at least parts of the lichen deserts. This is a result of urban dust enrichment, the eutrophication of farmland by fertiliser use and the emission of ammonia by intensive cattle and pig-rearing units. Evidence from Holland and East Anglia indicates that acid-loving assemblages (*Evernia prunastri*, *Hypogymnia physodes*, *Lepraria incana*) are being replaced by nitrophilous communities in which *Candelariella vitellina*, *Physcia* spp. and *Xanthoria* spp. are prominent. Over much of lowland Britain the formerly rather local species, *Xanthoria polycarpa*, is spreading nearly as rapidly as *Lecanora conizaeoides* did last century.

Reinvasion studies have been hampered by SO_2 levels falling more rapidly than the rate at which many lichens can reinvade. For this reason workers have been reluctant to attach SO_2 levels to the colonisation gradients which have developed, as at best they will be four to five years out of date, and at worst meaningless. While lichens can provide a general picture of pollution abatement, organisms with an annual life cycle, such as tar-spot fungus (*Rhytisma acerinum*), a leaf pathogen of sycamore, are more appropriate indicators of rapidly ameliorating conditions. Another difficulty is deciding which pollution criterion controls establishment and subsequent growth. Lichenologists have never been very certain about this, but in the UK have tended to use the mean level of the six winter months. There is some evidence from America, however, that a decrease in peak levels is sufficient to trigger reinvasion (Showman, 1981). The hard lesson for lichenologists is that under the present ameliorating pollution climate of Britain lichens are not particularly valuable as pollution monitors.

An aspect that requires further study is the performance, under declining SO_2, of species originally favoured by it. For decades *Lecanora conizaeoides*, a prime favourite, formed a monotonous cover on tree trunks and siliceous habitats in and downwind of urban areas. It was everywhere, even being reported growing densely on a mackintosh worn by a scarecrow. It was too common to bother with, an accepted background against which other changes occurred. Casual observation suggests that recently its swards have experienced a general disintegration. It is now rather uncommon on Millstone grit in the Pennines, where it is becoming restricted to vertical faces, often below bird perches. Trees that were formerly smothered in it, today support a discontinuous cover; only on conifers and lignum is it still abundant. The mechanism of decline involves fungal attack of epidemic proportions that produces large circular patches of dead lichen which are then overgrown by the alga *Desmococcus*.

An urban super-race of lichens?

Seaward has made a long-term study of *Lecanora muralis* in Leeds. Although the investigation was originally set up to follow decline under increasing pollution stress, by 1970 it became apparent that reinvasion was occurring. From 1970 to 1980 *L. muralis* spread towards the centre of Leeds at an average rate of 150 m per annum by colonising asbestos. The invading front was characterised by a dense splatter of very small colonies which suggests that the inoculum pressure of propagules was high. Even so, the time-lag between SO_2 falling below the threshold limit for establishment on asbestos roofs (*c.* 200 µg/m^3) and juvenile thalli appearing was about five years.

Quadrat studies revealed a phenomenon still not fully explained. The first three annual cohorts of *Lecanora muralis* to invade asbestos roofs as pollution abated differed in a number of respects from later cohorts (Fig. 4.6). In addition to colonising earlier, they displayed a faster radial growth rate, averaging 3.87 mm per year, and apparently became established during the winter months; their subsequent spread on the roofs showed no aggregation of colonies. Thalli establishing subsequently had a significantly slower mean annual growth rate of 2.50 mm; they first appeared in May, and their ensuing spread showed aggregation, suggesting the colonies were exhibiting local dispersal. The 'urban super-race' may be an example of a resistant lichen variety evolving under the impact of pollution stress (Seaward, 1976; 1982).

Pollution other than sulphur dioxide

For 30 years there has been universal agreement that SO_2 is the major cause of lichen deserts worldwide, but what of other types of air pollution? There is surprisingly little information about the sensitivity of lichens to these. So far,

Fig. 4.6 Mean growth of successive cohorts of *Lecanora muralis* thalli colonising asbestos roofs in Leeds during the period 1970–80. Cohorts A, B and C have been designated the 'urban super race'. From Seaward, 1982.

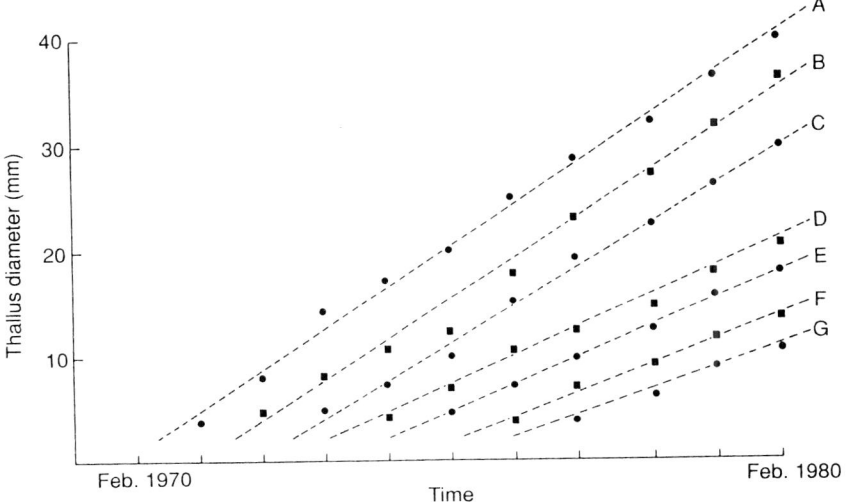

fluorine, alkaline dust, acid rain and airborne eutrophication (dealt with in Chapter 8) have been identified as having effects that are discernible in the field, while current levels of traffic fumes, ozone, carbon monoxide and organic toxins do not appear to have a direct affect on lichen distribution in the UK.

Fluorides

Only a few parts of Britain are exposed to airborne fluorides, the chief producers being aluminium smelting, brickworks, potteries and the manufacture of glass and steel. Many of these industries are situated in conurbations where SO_2 levels mask the flouride's effects, but fortunately for research workers there are isolated aluminium smelters at Fort William, on Anglesey, and between 1971 and 1981 at Invergordon in northeast Scotland.

A lichen survey around the Fort William factory (Gilbert, 1971b) found a pattern of deterioration that was concentric round the works and conspicuously elongated downwind. By observing lichens in a variety of standardised habitats a 'lichen desert', 'transition zone' and 'normal zone' were distinguished, the affected area being 4 km at its longest diameter. Within it, the tops of wooden fence posts formerly supported a luxuriant 'cap' of lichens, but this became progressively reduced until in the vicinity of the factory the posts were bare. Close to the works exposed acid rock is so clean as to appear scrubbed; the first lichen to recolonise the schist was *Stereocaulon pileatum*, followed in sequence

Fig. 4.7 The area over which severe damage to lichens had developed one (1972) and four (1975) years after the aluminium smelter at Invergordon commenced production. The locations of the smelter and 18 recording stations are shown together with a wind-rose to illustrate the direction in which the fumes would be blown. Numbers indicate mean herbage fluoride levels ($\mu g\ g^{-1}$) during summer 1975. From Gilbert, 1985.

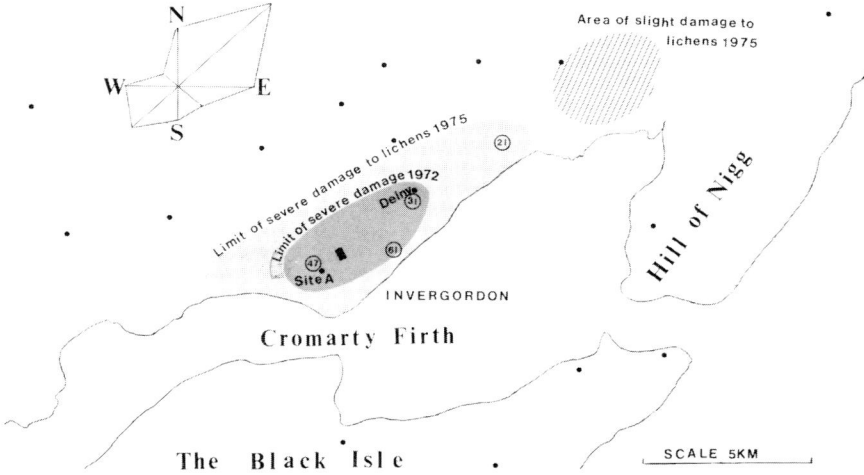

by *Candelariella vitellina*, *Parmelia saxatilis* and *Rhizocarpon geographicum*. A small unidentified *Buellia* was found to be frequent on conifers in the 'transition zone', where its abundance recalled that of *Lecanora conizaeoides* around towns; it was later described as a new species, *B. pulverea*. The pattern of deterioration caused by fluorides is greatly influenced by shelter whether from topography, aspect or dense vegetation; this results in an irregular margin to the lichen desert.

The environmental effects of the aluminium smelter at Invergordon were closely monitored by a number of concerned parties during its operational life of 11 years. Prior to start-up, a lichen monitoring scheme involving 20 recording sites was established to act as an early warning system, as it was estimated that in addition to livestock, 16,900 ha of coniferous forestry was at risk. Less than three months after the smelter had commenced limited production spectacular damage to lichens was reported (Plate 4c). Full resurveys after one and four years showed a progressive enlargement of the affected area which then stabilised (Fig. 4.7). The monitoring revealed that epiphytic lichens were among the most sensitive components of the landscape; at its maximum, severe damage (> 50% injury) extended over an area of 25 km^2 and up to 11 km downwind. This corresponded closely with the area over which fluorosis to cattle was reported. Adverse effects on commercial forestry and amenity trees were minor and mostly of a temporary nature. Since aluminium production ceased in 1981 the lichen flora has started to recover.

Alkaline dust

Alkaline dust emanating from quarry crushing machinery, lime kilns and cement works falls like a white pall on the surrounding countryside. In the Peak District of Derbyshire, where 31 quarries produce a fifth of the national limestone output, it is regarded as a particular nuisance as it gives rise to numerous complaints. Dust levels have proved difficult to measure using gauges, but its effect on epiphytic lichens provides a semi-quantitative yardstick. The mechanism by which the dust affects lichens involves the surface mineralisation and elevation of bark pH to over 6.5, which favours the mobilisation of various nutrients that cause hypertrophication.

A study around the Hope Valley Cement Works in Derbyshire (Gilbert, 1976) revealed that the twin effects of enhanced pH and hypertrophication resulted in a complete replacement of acid-loving assemblages by ones rich in *Xanthoria* species (Xanthorion alliance). With increasing distance from the source, pH and the strength of eutrophication gradually fell off and transitional communities occurred so a pattern of concentric zones could be distinguished. In this part of Derbyshire, at the time of the survey, levels of SO_2 were such that the ubiquitous community on ash trees was dominated by *Lecanora conizaeoides* (the Conizaeoidion) (average 4 spp./tree), around the works this was substituted by a Xanthorion containing up to 30 spp./tree (Fig. 4.8). The consequence of this is that in areas experiencing significant SO_2 pollution, trees growing near alkaline dust sources frequently carry the richest epiphytic assemblages in the region. Unexpected epiphytes occurring in the Hope Valley include *Caloplaca flavorubescens*, *C. cerina* and *Ramalina fraxinea*.

A survey of lichens on mature ash trees around the works revealed an inner zone of dense, smothering, dust deposition where lichens are rare. This is followed by a wide zone in which the mineralising and alkaline properties of the dust produces the curious sight of lichens that are normally saxicolous growing on trees, for example *Aspicilia calcarea, Caloplaca decipiens, Lecanora campestris* and *Lecidella scabra*; these occur alongside epiphytes typical of strongly hypertrophicated bark. In this zone the Xanthorion extends all round the trunk and throughout the canopy, while the saxicolous element is restricted to the trunk and major horizontal limbs. In the zone external to this, diversity is reduced as saxicolous species fall off and the Xanthorion becomes restricted to the side of the trunk facing the works. At this distance (2 km) *Candelariella vitellina, Caloplaca holocarpa, Scoliciosporum umbrinum, Xanthoria parietina* and three or four *Physcia* spp. are conspicuous.

Three to four kilometres downwind there is a return to trees with an acidified bark of pH *c.* 3.5 that support an enriched Conizaeoidion. The last traces of the dust syndrome are detectable at 5 km by the presence of enhanced amounts of *Buellia punctata, Lecanora chlarotera*, and on bosses, knees, crotches and leaning trunks, occasional small colonies of *Physcia* and *Xanthoria*.

It has been estimated that alkaline dust from the 700 limestone quarries in the UK affects epiphytic lichens over a total area slightly larger than County Durham. In districts experiencing significant pollution by SO_2 the replacement of species-poor, acidophilous communities by richer ones is a welcome diversification which, presumably, also encourages lichen-dependent invertebrates, but where mineral dusts are blowing into primeval woodland, as at Burnham Beeches, they raise problems for lichen conservation.

Acid rain

The two primary causes of acid rain are SO_2 and nitrogen dioxide (NO_2) which, when transported over long distances, oxidise to strong acids that dissolve in cloud water. Unpolluted rain has a pH of about 5.7, acid rain varies from 4.5 down to 2.0, the pH of vinegar. The most acid episodes are usually

Fig. 4.8 Relationship on mature, free-standing ash trees between lichen diversity, bark pH and distance downwind from the Hope Valley Cement Works. From Gilbert, 1976.

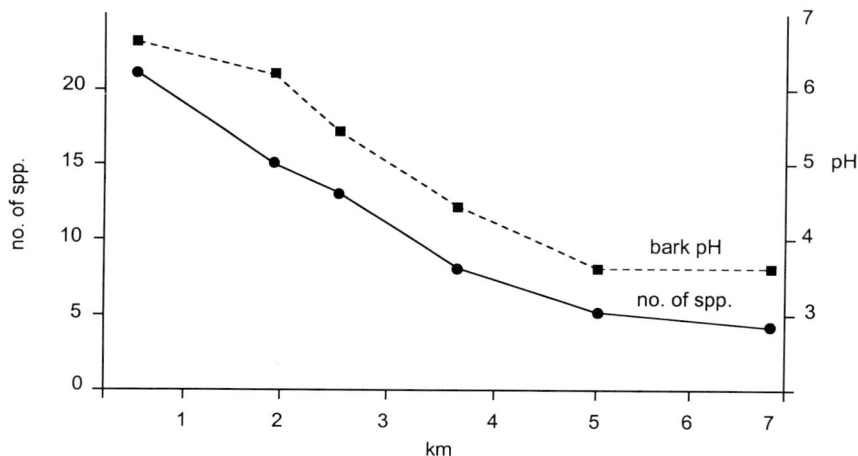

associated with upland mists which can deposit directly on surfaces. This is known as 'occult deposition'; as the sun begins to dry the 'dew', acid in the droplets become more concentrated and can burn tiny holes in plant material. Snow also contains high quantities of acidity which is released into soils and fresh water in a large pulse each spring.

Evidence for a harmful effect of acid precipitation on lichens comes from several sources, but the literature is nothing like as extensive as that on direct SO_2 effects. One difficulty in analysing its effects is that acidity is a natural phenomenon, so that increasing acidification will favour certain species, for example *Cladonia coniocraea*, *Parmelia saxatilis*, *Pertusaria amara*, while harming others, for example, those with a cyanobacterial partner. Lichens with a cyanobacterial partner appear to be particularly susceptible to acid rain as their photobionts require a high pH environment if they are to function effectively. Field evidence is quite compelling. For instance, in parts of Northumberland remote from SO_2 pollution, but subject to increasingly acid rain, well-established populations of *Lobaria pulmonaria* have declined on oak and *Sticta limbata* has become extinct on ash. These populations survived the SO_2 maxima that occurred earlier this century, but succumbed to acidification during the 1970s and 1980s (Gilbert, 1986). Ivan Day and Andrew Farmer have reported similar effects in the adjacent county of Cumbria, linking the extinctions to habitats poorly buffered against acidity. A government-funded, nationwide study to monitor the effects of acid rain on Lobaria-rich communities reported a loss of species, low relative growth rates and declining bark pH of the host trees from areas where acid deposition was particularly high (Looney, 1991; Wolseley & James, 1991).

All studies have shown that acid deposition causes a switch of substrate preference away from trees with bark that is readily acidified (e.g. oak) to those with better buffered bark (e.g. ash, elm). Arup *et al.* (1989) have reported that the saxicolous and terricolous lichen flora of southern Sweden is also showing signs of decline due to acidification. In Britain we have scarcely started to assess its effect in these habitats, though lowland heaths and Scottish snow-beds would be suitable places to start. The rapid decline of the Breckland rarities (Chapter 9) is the most convincing evidence we have of an effect on terricolous lichen communities in the UK. A further poorly studied aspect is the fertilising effect of acid rain caused by the nitrogen constituent. Could this be why *Hypogymnia tubulosa* and *Usnea florida* are apparently declining in mid-Devon? Acid rain is a complex subject; further information regarding its effects on lichens and bryophytes in western Europe is provided by Farmer *et al.* (1992).

Others

Other types of pollution that lichens are able to monitor include heavy metals, radionuclides and ultraviolet (UV) pollution associated with the ozone hole. For these, simple field survey is not sufficient, expensive apparatus capable of analysing quantities of metal or other substances in the thallus are required, which puts the techniques well beyond the resources of most amateur naturalists. Their value as indicators of carbon monoxide, ozone and photochemical oxidants is uncertain. So when lichens are promoted as a prime example of biological indicators it is as well to remember that they are somewhat selective at this task, but form a useful backup to gauge measurements.

Evidence of the continuing importance attached to lichen/air pollution studies is the size of the international literature; over the last decade this has averaged 120 papers a year.

5

Trees, Woods and People

Beginners usually start their career in lichenology by investigating woodland or wayside trees, as epiphytes are easy to collect and include many distinctive species. Their study may commence by trying to separate *Evernia prunastri* with its matt surface from glossy *Ramalina farinacea*, proceed with attempts to distinguish between the grey *Parmelia* species, and go on to discover *Hypogymnia physodes* with its lack of rhizinae, followed by the golden *Xanthoria* genus and the apparently straightforward beard-lichen *Usnea subfloridana*. A few crustose species may also be collected and taken back for closer examination including spot testing with chemicals. Collections can conveniently be stored dry in paper envelopes on which name, habitat, locality and date are recorded, then accumulated in empty shoe boxes, which hold up to 50 packets each. That is how many of us started.

I remember the exact tree that launched me into a lifetime of involvement with lichens; it was in the early 1960s when lichenology was starting to recover from 50 years of neglect. The new wave of recruits was thinly scattered, heavily reliant on the expertise resident in The Natural History Museum, London, and was having to learn everything from scratch. In those days most people were satisfied when they had examined the 'weatherside' of a trunk where macrolichens grew thickly, then moving on to the next tree. Brian Coppins was the first person I saw examining a tree thoroughly. He paid equal attention to the trunk, roots and buttresses, sticking his head into cavities to look for pinhead lichens (Caliciales), scrutinising dry shaded bark, scanning twigs with a lens and lying flat to investigate the base of hazel coppice. He set new standards, though through concentrating on inconspicuous species he occasionally missed large ones such as *Parmelia caperata*!

Only a small number of trees can be investigated in such detail, but experience has shown that it is preferable to spend 40 minutes on one good tree rather than five minutes on eight, so it is important to choose the most promising. It gradually became apparent that certain trees were conspicuously richer than their companions; these were, in general, old trees of great girth standing in sheltered but well-lit positions. Where there was low level air pollution, ash and elm were invariably found to be richer than other species.

Of even greater importance is site selection. Initially we went anywhere with trees. Some woods were interesting, others proved disappointing. In the business of picking good sites we all learnt from Francis Rose (Fig. 5.1) who gained a legendary reputation for recognising promising sites from the ground, from the car and from the map. This apparent 'clairvoyance' was a highly developed ecological awareness which was to elevate woodland/lichen studies into one of the most valuable conservation tools available to the forest ecologist. Rose's repertoire including seeking out relic populations of the spectacular *Lobaria pulmonaria*, at that time regarded as virtually extinct in England. He was the first person to recognise its links with ancient woodland and medieval deer parks. Throughout the decade 1968–78, searching for this species became a minor obsession, parties setting out armed with books such as *The Botanist's Guide through England and Wales* (Turner & Dillwyn, 1805) and *The Deer Paddocks and Parks of England* (Whitaker, 1862) with their suggestions of possible localities. The thrill of these times is in danger of being forgotten as, through letters, postcards and phone calls, news of each discovery spread. Something of the spirit of this age is captured in the account of a nine-day excursion through the north of England undertaken by Rose, Hawksworth & Coppins (Rose et al., 1970) during which 66 localities were visited.

Fig. 5.1 Francis Rose (P. Loughran).

While such lichen survey work was proceeding apace, parallel research into the history of British woodlands using early documents and field archaeology techniques was providing complementary data on woodland management down the ages (Rackham, 1976; 1980; 1986). When the two are put together they yield deep insights into the ecology of our woodland lichen flora, the continuing interpretation of which has been one of the principal lichenological achievements of the last 20 years. Today we have some kind of answer to the basic questions we were then asking. Why is this wood or parkland better than another? Why is one tree better than its neighbour? We are now asking questions with a conservation bias, as we try to understand the functioning, dynamics, rates and methods of spread of our rarer epiphytes.

Factors affecting lichen diversity

Age and continuity

Ratcliffe (1968), in his masterly account of Atlantic bryophytes in the British Isles, was the first person to correlate woodland continuity with a rich lower plant flora. He gives the example of two superficially similar wooded glens in North Wales: Tyn-y-Groes with an outstanding bryophyte flora, and Torrent Walk, 6 km away, with a luxuriant growth of bryophytes, but composed almost entirely of common species. The explanation for the difference is that the rich woodland has never been completely cleared, so has always provided a refuge for drought-sensitive Atlantic species, whereas that at Torrent Walk was at one time clear-felled and replanted. During the period with no protective tree canopy, humidity dropped, the sensitive species were scorched out, and only those with vigorous powers of spread have returned.

In an important series of papers, Rose (1974; 1976; 1992) has developed the concept of 'ancient woodland indicator' lichens and produced lists which, in non-polluted areas, can be used to assess quantitatively the degree to which a woodland has had a long history of canopy continuity. His most popular scale, the Revised Index of Ecological Continuity (RIEC) (Rose, 1976), employs 30 lichens (Table 5.1) which, experience has shown, are faithful to documented ancient high forest or pasture woodland. Most belong to the *Lobarion pulmonariae* alliance (James *et al.*, 1977), which seems to have been the major epiphytic community at the time of the wildwood. If any 20 species on the list are present at a site this is an excellent indication of a woodland with ecological continuity stretching back to early medieval and quite possibly pre-Roman times. Direct connection with the wildwood is probably not required, but the forest or park, must have originated at a time when fragments of very ancient woodland existed near to the site. Outside the west of Scotland, most species on the list have very poor dispersal powers over any except the shortest of distances, but in earlier times of pure air and undrained well-wooded landscapes, dispersal must have been freer. Advice on how to use the RIEC and its more sensitive successor, the New Index of Ecological Continuity (NIEC) is provided in Appendix 3, but be warned, only a handful of professional or near-professional lichenologists can operate the latter.

Table 5.1 Lichen epiphytes used to calculate the Revised Index of Ecological Continuity (RIEC) (Rose, 1976).

Arthonia vinosa	*Pannaria conoplea*
Arthopyrenia ranunculospora	*Parmelia crinita*
Biatora sphaeroides	*P. reddenda*
Catillaria atropurpurea	*Parmeliella triptophylla* or *Degelia atlantica*
Dimerella lutea	*Peltigera collina*
Enterographa crassa	*P. horizontalis*
Lecanactis lyncea	*Porina leptalea*
L. premnea	*Pyrenula chlorospila* or *P. macrospora*
Lobaria amplissima	*Rinodina isidioides*
L. pulmonaria	*Schismatomma quercicola*
L. scrobiculata	*Stenocybe septata*
L. virens	*Sticta limbata*
Loxospora elatinum	*S. sylvatica*
Nephroma laevigatum	*Thelopsis rubella*
Pachyphiale carneola	*Thelotrema lepadinum*

Woodland structure

While the degree of continuity of humid forest conditions and ancient bark surfaces as measured by the RIEC explains, to a large extent, why some woods are better for lichens than others, their structure is nearly as important. Most lichens require reasonable light levels so thrive best in open structured woodland, with glades as a main feature. Sanderson (1998) has observed that the best sites include glades of various sizes accommodating the varied tolerances of different groups of lichens. Crustose species, often with *Trentepohlia* as the algal symbiont, are confined to well-lit, broken-canopied high forest, avoiding both large glades and closed canopy woodland. Characteristic species are closely adpressed to the bark and include *Agonimia octospora, Lobaria virens, Opegrapha fumosa, Phyllospora rosei, Pyrenula nitida* and *Zamenhofia hibernica*. Glade edges are preferred by foliose species that have quite high requirements for light, but also require shelter, i.e. *Lobaria amplissima*, *L. pulmonaria* and *Parmelia* species. A third group comprises lichens of dry bark that thrive on old trees standing in well-lit open woodland or parkland, such as *Enterographa sorediata*, *Lecanactis lyncea*, *L. premnea*, *Opegrapha prosodea* and many pinhead lichens (Caliciales).

There is currently much discussion regarding the structure of the original wildwood. The prevalent view is that it contained a well-developed glade structure imposed by herds of red deer (*Cervus elaphus*), aurochs (*Bos taurus primigenius*) and other large herbivores. So many woodland species of plant, bird and butterfly are associated with forest interface areas that this niche must have been continuously available throughout the postglacial forest period. The present New Forest woods, with their mosaic of 'lawns' of all sizes, are one of the best models we have. Grazing would also have kept down scrub, which is a major enemy of the lichen flora; for instance, in the New Forest, expanding holly is starting to shade out important tree bole communities.

Closed high forest, i.e. tall trees growing in close canopy, is nearly always the result of planting or the manipulation of pre-existing stands by, for example, the singling of coppice. Most are dense and dark, and regular felling cycles prevent any continuity of really old trees. When compared with irregular and gladed forests it becomes apparent that the epiphyte flora of closed high forest is second-rate, being composed of widespread species with good dispersal mechanisms.

Coppice woodland, while excellent for flowers, insects, birds and dormice, is only third-rate for lichens even where it has continuity with the wildwood. This results from intensive management. Two examples will suffice. An area of ancient coppice-with-standards at Bradfield Woods Site of Special Scientific Interest in Suffolk has only 15 epiphytic lichens in an area of 1 km^2, while old oak coppice in the southern Lake District supports fewer than 20 lichen species per 1 km^2; in the absence of coppicing over 100 could be expected.

Large old trees

Many lichen species depend on niches present on trees that a forester would regard as over-mature, so they are normally absent from managed woodland. Such large old trees, hollow and gradually falling apart, are magnets to lichenologists who have become expert at seeking them out (Fig. 5.2). More survive in Britain than elsewhere in Europe. Francis Rose tells the story of how he took the great Dutch authority on epiphytes, Prof. J. J. Barkman, to show him the internationally famous communities in the New Forest, but what impressed him more was the car journey across the Weald where he saw more ancient oaks than are present in the whole of The Netherlands.

Most ancient trees result from a type of land use known as wood pasture, now rarely practised, and we have to thank the historical researches of Oliver Rackham (Rackham, 1976) for illuminating its detail. It entailed grazing animals in association with pollarded trees so that crops of wood and meat could be obtained from the same ground. The practice was well established by the time of the Norman Conquest, being distinguished in the Domesday Book (AD 1086) as *silva pastilis*. Pasture woodlands were used either for hunting or grazing domestic stock. They are found:

1) on the site of royal, baronial or ecclesiastical hunting forests, for example much of the New Forest and the Forest of Dean;

Fig. 5.2 Lichenologists examining an old tree in the Forest of Dean (J.R. Laundon).

2) in medieval deer parks where relics of the wildwood were enclosed along with the deer, their structure now perpetuated by planting for conservation and landscape reasons;

3) as boundary trees along woodland edges, field margins and beside rivers, for example in the Lake District (Fig. 5.3) and Exmoor;

4) in woodland complexes associated with grazed coppices as at Cranborne Chase and Hatfield Forest; and

5) in the few remaining large wooded commons of pre-Enclosure Act age, Burnham Beeches in Buckinghamshire and Ebernoe Common in Sussex being fine examples. Such areas form only a tiny fraction of our woodland cover, but are highly significant for the survival of epiphytic lichen communities from the past.

Kilvert's Diary entry for 22 April 1876 (Plomer, 1938), contains a vivid account of the appearance and atmosphere of wood pasture in Moccas Park, Herefordshire:

> '.... we came upon the tallest stateliest ash I ever saw and what seemed at first in the dusk to be a great grey ruined tower, but which proved to be the vast ruin of the King Oak of Moccas Park, hollow and broken but still alive and vigorous in parts and actually pushing out new shoots and branches. I fear these grey old men of Moccas, those grey, gnarled, low-browed, knock-kneed, bowed, bent, huge, strange, long-armed, deformed, hunchbacked, misshapen oak men that stand waiting and watching century after century. No human hand set these oaks. They are "the trees which the Lord hath planted". They look as if they had been at the beginning and making of the world, and they will probably see its end' (Fig. 5.4).

Climate

On a European scale, many epiphytic lichens show a strong western distributional bias, being concentrated along the Atlantic coast where the climate is mild and humid. In both number of species, their luxuriance, and proportion of near endemics this 'Atlantic' element is more strongly represented in the British Isles than anywhere else in Europe. We are fortunate that at least 25 macrolichens, including *Degelia atlantica*, *Lobaria scrobiculata*, *L. virens*, *Parmelia sinuosa*, and *Pseudocyphellaria* and *Sticta* spp., which are strongly threatened on the European mainland, are still common in our western woods. That is why their flora is important on an international scale. As Britain falls entirely within the Atlantic climatic province, as defined by Troll (1925), some degree of subdivision is desirable. In a pioneering paper, Coppins (1976) tentatively divided the British epiphytic lichen flora into the following groupings:

1. Western: *Leptogium burgessii, Parmelia laevigata, P. sinuosa*.
2. Western and Southern: *Dimerella lutea, Ochrolechia inversa*.
3. Western and Northern: *Nephroma parile*.
4. Southern: *Arthonia impolita, Phlyctis agelaea, Usnea articulata*.
5. Eastern: *Anaptychia ciliaris, Caloplaca luteoalba, Parmelia acetabulum*.
6. Northern: *Cetraria pinastri, C. sepincola, Mycoblastus sanguinarius*.
7. Widespread: many species including *Arthonia radiata, Evernia prunastri, Pertusaria amara*.

These groupings represent a useful start in this area, though taking account of extra-British records would produce a more authoritative and desirable classification.

Today, phytogeographers use as their framework the concept of an oceanity-continentality gradient in which humidity and temperature, together with

Fig. 5.3 Ancient, recently pollarded ash at Watendlath, Borrowdale, showing the wide range of niches available to lichens.

insolation, radiation and windiness can be expressed as a single function. Figure 5.5, showing the oceanity gradient across Europe, immediately reveals the unique climate experienced by western Britain and the tip of Brittany. The Atlantic climatic zone favours the following four woodland types (after Sanderson, 1998):

1. Hyperoceanic Temperate Rainforest: the internationally important, very wet, mossy, oak/hazel woodlands of low ground in the west of Scotland and the west of Ireland, supporting rich and distinctive assemblages of luxuriant macrolichens, many of them very rare on the Continent; a slightly impoverished version of this flora is present in hyperoceanic areas of the Lake District, Wales and Southwest England.

2. Oceanic Temperate Woodlands: beyond the Atlantic fringe, where not affected by pollution, all but the most eastern areas carry an oceanic flora which lacks the rare specialists of the hyperoceanic regions, but is still rich in Atlantic lichens, with crustose species often more prominent and southern species making an appearance; standard Lobarion is the climax community on rough bark and the Graphidion on smooth bark.

3. Sub-oceanic Temperate Woodlands: present in the more continental parts of the country especially east Scotland, Northumberland and the Welsh Marches, holding an epiphyte flora without many oceanic species which are

replaced by others with a more continental distribution. The original extent of this flora, now obscured by the effects of pollution, probably included all of Eastern England apart from the more oceanic Weald.

4. Boreal Woodlands: high ground in the west supporting pine and birch woods, with the climax epiphytic community, the *Parmelietum laevigatae*, spreading to oak and alder where exposure to leaching increases bark acidity; at lower altitudes there is a transition to the Lobarion on oak, eastward to an acid bark community known as the *Pseudevernion furfuraceae*. Pinewoods have their own highly distinctive flora.

This four zone division of the British epiphyte flora on climatic grounds is an oversimplification of a multi-dimensional continuum, the definitive account of which has still to be written. From an international perspective, however, the division of the Atlantic Province of Europe into four zones makes for a manageable level of detail.

Other factors

While woodland management history, structure, macroclimate and the level of air pollution experienced have been identified as major influences on the distribution and ecology of epiphytic lichens, many additional factors control their occurrence, particularly at a site level. Some of these are large-scale, such as topography, which affects aspect and mesoclimate, and geology, which through the soil influences the types of tree present, their bark chemistry and the degree of waterlogging; humidity on the whole favours the development of profuse lichen growth.

At the level of the individual tree, whether bark is acid, neutral, basic or nutrient-enriched is very important in determining what lichens will be present; likewise, whether bark is rough or smooth, exposed or sheltered, wetted by rain or dry, or absent so that lignum is exposed. Competition with bryophytes is a significant but little researched topic. These subjects, and all others imaginable, are exhaustively discussed by Barkman in his monumental book, *Phytosociology of Ecology of Cryptogamic Epiphytes* (1958).

Fig. 5.4 ... misshapen oak men, The Old Deer Park, Chatsworth.

It would be somewhat tedious to supply methodical examples of these factors in operation. Instead, portraits follow of some of the richest and most exciting woodland sites in Britain with ecological detail highlighted.

The New Forest

Despite its name, the New Forest is anything but new; it has been described as the finest remnant of old forest remaining in Britain. This is an understatement; in the opinion of many experts it is the finest ancient forest remaining on the lowland plains of northwest Europe. This view is based on its extent, the very large number of old trees (Plate 5a), the relative lack of human interference, its invertebrate fauna, the amount of dead wood and its incomparably rich epiphytic lichen and bryophyte flora. What factors have conspired to create this jewel?

Fig. 5.5 Map showing the gradient of oceanity across Europe. The unique position of western areas of the British Isles can be clearly seen. Provided by A.B.G. Averis.

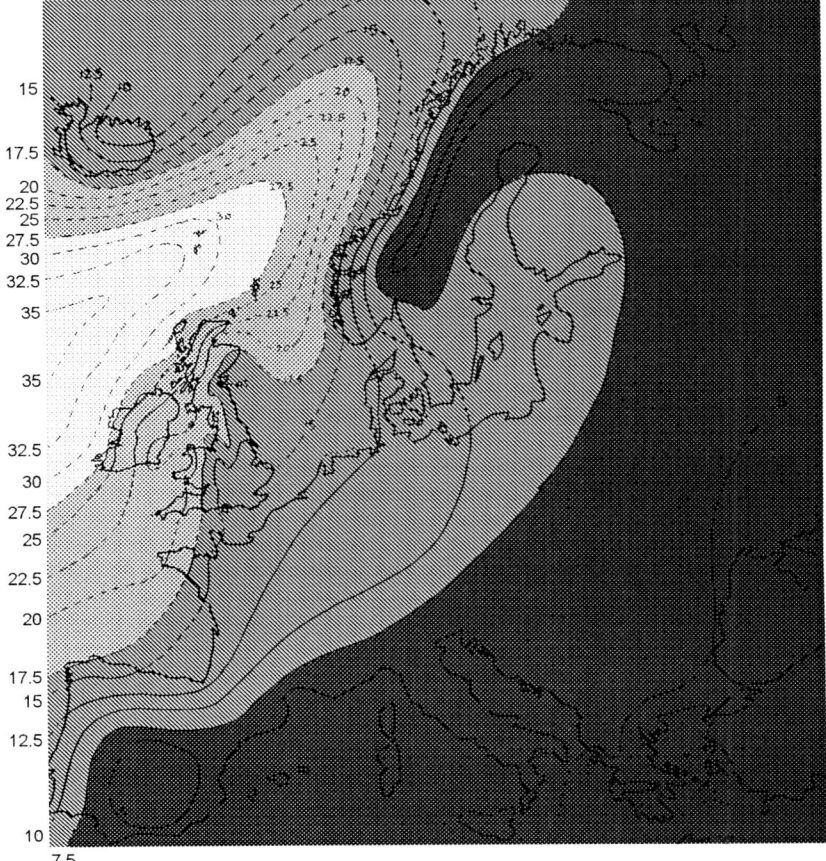

First came the passion for hunting shown by British royalty. In the twelfth century the area was made a vast Royal hunting forest to preserve the deer; this acted as a brake on its exploitation. The gladed forest was managed as pasture woodland with pollarding carried out for fodder, but there was no large-scale timber removal. From the sixteenth century the Crown allowed oak extraction to force down prices for the benefit of the Navy, but this was on a selective not a clear-felling system, and pollarding continued to provide an unbroken sequence of ancient trees (Fig. 5.6) until it was banned in 1698. In the eighteenth century an attempt to enclose the forest started, which would have led to more intensive management, but it was never completed due to resistance from the commoners. In 1851 deer forest status ended, many deer were removed, and a burst of natural regeneration converted the pasture woodland to high forest. Ecologists wishing to know more details of how 20,000 ha of semi-natural vegetation have survived to the present day, are referred to Tubbs (1968; 1986).

The last thousand years of land use has, by chance, preserved many of the features of the original wildwood; the forest is uneven-aged, has plenty of old trees, there is much variation in canopy density, the water table remains high, and the ancient flora can be seen almost intact with regard to content, though diminished in abundance. Standing among the large ancient trees the modern lichenologist could believe he had been transported back a thousand years, that car horns were the sound of hunting horns and that the next glade might contain King Arthur and his knights.

Fig. 5.6 A giant beech pollard dwarfs surrounding trees in the New Forest (T. Heathcote).

The New Forest has been thoroughly worked for lichens since 1967, first by Rose and James who published a benchmark paper in 1974 incorporating records submitted by Coppins and Davey. It catalogued the flora, paying particular attention to historical records, and then analysed the lichens with respect to biogeographical relationships, ecology, phytosociology, regeneration and the effects of air pollution. Subsequently Giavarini and Sandell have been active and more recently Sanderson has carried out conservation-related research undertaking specific studies of the rarest lichens, working on rates of recolonisation and investigating the effects of deer and pony grazing on the lichen flora of tree boles (Sanderson, 1998).

Fig. 5.7 Shaded beech trunks in the New Forest support a wide range of rare crustose lichens (T. Heathcote).

Progress in cataloguing and understanding the flora has been continuous. By 1974, 259 epiphytic taxa were known from the old woodlands, a number now increased to 344, of which ten are in the national Red Data List (Church *et al.*, 1997). Two features are of particular note. First there is a suite of lichens typical of grazed high forest, but rare in or absent from parkland as they require shade and humidity; these New Forest specialists include *Agonimia octospora, Arthonia astroidestra, Enterographa sorediata, Micarea pycnidiophora, Parmelia minarum, Pertusaria velata, Pyrenula nitida, Rinodina isidioides, Zamenhofia coralloidea, Z. hibernica* and *Z. rosei*. The Forest is also remarkable for the richness of beech (Fig. 5.7), which elsewhere in the UK is a rather poor tree for lichens; beech specialists are *Catillaria laureri, Enterographa elaborata, Parmelia minarum, Pertusaria velata* and *Pyrenula nitida*. The smooth bark of beech and holly also supports a community with numerous writing lichens known as the Graphidion.

A rich Lobarion has survived on the mossy boles of beech and oak, but is nowhere frequent. An enigma, given that *Lobaria pulmonaria* and *L. virens* occur abundantly on several hundred widely dispersed trees, is their apparent inability to spread to adjacent, seemingly quite suitable ones. Low levels of air pollution may be responsible. *Lobaria amplissima* is present on just two oaks and one beech, while *L. scrobiculata* has not been seen this century. Other Lobarion taxa present include abundant *Normandina pulchella, Parmelia crinita* and *P. reticulata*, with smaller amounts of *Collema subfurvum, Nephroma laevigatum, Pannaria mediterranea, P. pityrea, P. sampaiana* and *Sticta limbata*. Several of these are restricted to niches with more basic bark, such as the occasional ash, while *Catillaria laureri* is found on damaged pollards. Though fully developed Lobarion is rare, a pre-Lobarion association composed of demanding crustose lichens together with characteristic *Parmelia* species is widespread, succession to the climax being halted due to a failure of the larger taxa to establish. In many woods nearly every tree supports *Enterographa crassa* and *Thelotrema lepadinum*. Work in South Bentley Inclosure by Sanderson has shown how the old forest element relies on trees of large girth. Trees with a girth less than 2 m held very few such lichens, while a third of those over 3 m in girth supported RIEC species.

While the New Forest contains a galaxy of rich epiphytic communities that embrace acid bark, mesic bark, flushed bark (beech), rough bark, smooth bark, twigs and lignum, it is particularly notable for the *Lecanactidetum premneae* association of ancient dry bark. In the New Forest this is only found on very old oaks in well-illuminated situations. Faithful species are *Arthonia impolita*, *Lecanactis amylacea*, *L. lyncea*, *L. premnea*, *Schismatomma decolorans* and several Caliciales. They are characteristic of areas of undisturbed pasture woodland trapped in the inclosures. Recent studies on rates of recolonisation (Sanderson, *pers. comm.*) have revealed that in relatively undisturbed parts of the Forest most species can recolonise within 200–300 years, but the *Lecanactidetum* appears to require at least 400 years to recover fully from clearfelling, though it is resistant to partial clearance.

Over the last 30 years it has become apparent that deer grazing is essential for the maintenance of a rich epiphytic flora on tree boles. In one series of observations, heavy shading from holly growth, resulting from 30 years of enclosure, had reduced lichen diversity on the lower trunks by 50%, both total number and number of old woodland indicators being equally affected. Currently, volunteers are removing holly from around selected trees, for example those carrying populations of *Lobaria virens*.

A community rarely found in the old woodlands of the Forest is the Xanthorion. It is present in a fragmented form only on nutrient streaks below tree wounds and on branches used as bird roosts. This scarcity has led to the suggestion that it may be a feature of the cultural landscape, having either followed man to these islands or invaded from the coast once the wildwood had been replaced by agriculture.

What is there left to discover about the lichens of the New Forest? It is so large and complex a habitat that a few species will have escaped detection and some sterile crusts remain to be described, so there is infilling still to do. One neglected habitat is the tree canopy, as most lichenologists confine their attention to the lower 2.5 m of trunk and to spreading boughs. Francis Rose has been known to carry a light aluminium ladder that can be dismantled into sections; it was used to confirm *Lobaria amplissima* growing high on a tree. Evidence for other unusual species in the canopy has come from examining debris on the ground after gales; this has provided some of the only records of *Usnea articulata* in the Forest.

No woodland lichenologist has come of age until an apprenticeship has been spent in the New Forest, as the experience will form a benchmark against which other sites can be gauged. However, one note of warning: in common with other extensive areas, like Ben Lawers, a visit here by the uninitiated can be disappointing; they are advised to make their first excursion in the company of an expert able to guide them to the richer localities and help them to distinguish between the confusing variety of crusts.

Loch Sunart, Ardgour

Distinguished visitors to the Royal Botanic Garden, Edinburgh, who wish to experience the lichens of temperate rainforest, are taken to Loch Sunart, where a narrow road winds along the north shore through woods of oak and birch, fringed by rocky outcrops and profusions of golden whin (*Ulex europaeus*). At every bend the view changes as the shore's 30 bays come into view (Fig. 5.8). The first impression of the woods is how incredibly mossy they are; a deep blanket of bryophytes covers the floor, enveloping outcrops and dead wood, forming a thick skirt around tree boles, climbing inclined trunks and festooning horizontal boughs (Plate 5b), while in the canopy there is a progression to mats of purple *Frullania* and dark green cushions of *Ulota*. The moss carpet absorbs large amounts of rain which evaporate slowly over a period of days, stabilising humidity.

The trees, which are mostly even aged and of only medium size (Plate 5c), are surprisingly varied with respect to their lichen cover. Some are smothered in *Lobaria pulmonaria* (Plate 5b) colonies the size of dinner plates; they stand out from the trunk like brackets in response to the constantly high humidity. If the tree is very mossy, there may be heaps of *Lobaria* at its base where it has peeled off, so much of the thallus having been unattached. An adjacent tree, that has more bark exposed, will be covered with tufts and beards of *Usnea cornuta*, *U. flammea*, *U. fragilescens*, *U. rubicunda* and *U. subfloridana*. Another tree will support huge patches of white *Ochrolechia androgyna*, *O. tartarea*, *Pertusaria* species or *Thelotrema lepadinum*, perhaps with *Peltigera horizontalis* on its base. The large size that crustose species attain suggests they have a mechanism that prevents them from becoming overgrown by bryophytes. Competition with bryophytes is a major feature of lichen ecology in this region. All the best lichen woods are on sunny, south-facing slopes while those richest in rare Atlantic bryophytes are darker with a north aspect. This means that rich woodland bryophyte floras and rich woodland lichen floras do not often coincide.

The most luxuriant Lobarion is on trees near the edge of woodland, where sidelight falls on the trunks and bark pH is 5 or over (Plate 5d). At certain sites along Loch Sunart it is possible to find up to 26 (out of a possible total of 30)

Fig. 5.8 Some of the richest temperate rainforest in Europe is found along the north shore of Loch Sunart (W. Hunter).

RIEC species. All four *Lobaria* species are regularly present, together with up to five *Pannaria*, six *Leptogium*, four *Sticta*, three *Pseudocyphellaria* and *Peltigera collina*. Also conspicuous are crustose species that can overgrow moss, such as *Biatora sphaeroides*, *Catillaria atropurpurea* and *Dimerella lutea*. The species that have to be searched for are not the macrolichens, which are luxuriant and stand out, it is crustose species that require a moss-free, rough-bark niche such as *Loxospora elatinum* and *Pachyphiale carneola*.

The woods may have had a long history of grazing as occasional old pollards point to late medieval wood pasture. These trees, easily the oldest in the area, are hollow giants, covered in ferns and bilberry and notable for their rich flora of pinhead lichens. The current winter grazing is a mixed blessing. It prevents a shrub layer from developing and shading the boles, but at the same time almost eliminates regeneration and reduces the vascular plant interest. The answer may be temporary fencing or the use of tree shelters to foster young oaks, though it will be a long time before the issue becomes critical.

While the whole area is a lichenologist's Mecca, west of Scotland specialities are particularly abundant where stream gorges cut down through the woods. In one such locality, a secret ravine full of tangled vegetation, where fallen trees, huge mossy overarching hazels, a few giant ashes, elms (*Ulmus glabra*), bird cherries (*Prunus padus*) and willows provide variety from oak as a host tree, the flora is exceptional. The 'jewels' include *Parmentaria chilensis* which is locally frequent on the smooth bark of hazel (Plate 6a). This is its only site in Britain, though it is present in a few woods in Co. Kerry; it is accurately described in the Red Data List as having the appearance of blackberries floating in custard. Other national rarities are *Arthothelium reagens*, *Graphina ruiziana*, *Graphis alboscripta*, *Lecanora cinereofusca* (willow and rowan), *Leptogium hibernicum*, *Polychidium dendriscum*, *Pseudocyphellaria norvegica* and *Stenocybe bryophila*, together with several undescribed species. One hundred and ninety epiphytes are present. Humidity in the gorge is so consistently high that filmy ferns (*Hymenophyllum* sp.) have replaced bryophytes on outcropping rock which, as a bonus, locally supports sheets of the lichens *Peltigera leucophlebia* and *Placynthium flabellosum*.

On high ground, oak gives way to open, grazed, birch woodland. This supports a different, but equally distinctive, variant of the hyperoceanic rainforest community. The *Parmelion laevigatae*, typical of acid bark, is rich in silvery-grey *Hypogymnia physodes* (Plate 6b), *Parmelia endochlora*, *P. laevigata* and *P. taylorensis*, together with yellow *P. sinuosa*, *Usnea* and *Cladonia* species and luxuriant tufts of *Sphaerophorus globosus* which get nibbled by the deer. Scarcer species include *Catillaria pulverea*, *Cetrelia olivetorum*, *Lecidea doliformis*, *Micarea stipitata*, *Pertusaria opthalmiza* and *Mycoblastus caesium*. The low bark pH of 3.5 to 4.5 is a result of leaching by the high rainfall.

The history of the Loch Sunart oakwoods is only now coming to light (Kirby, 1998). Far from being fragments of the wildwood, parts were intensively managed for charcoal to feed a blast furnace on Loch Etive and for the local smelting of bog iron ore, they also provided bark for the Irish tanning industry and supplied 1,400 tons of wood each year to a bobbin mill at Salen. Exploitation ceased about 150 years ago when the coppice was singled up and some planting was undertaken to convert them to high forest. Hundreds of charcoal burners' platforms are still visible. This history of management makes the flora all the more remarkable. Under an extreme Atlantic climate it appears that

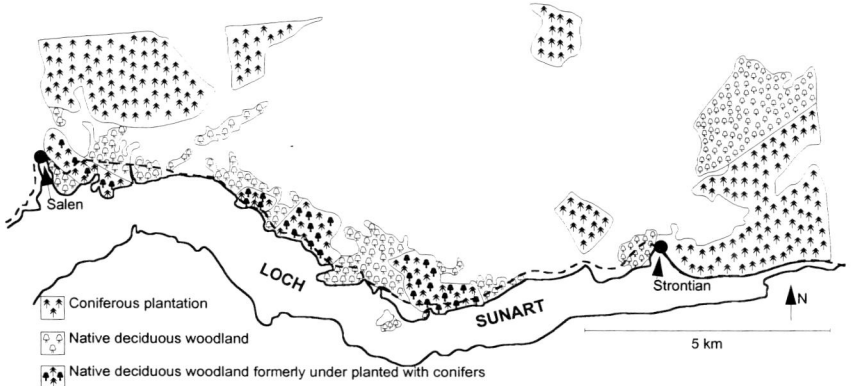

Fig. 5.9 Map showing plans for the rehabilitation of deciduous woodland along the north shore of Loch Sunart.

species normally regarded as old woodland indicator lichens can not only survive in highly fragmented woodland, but retain sufficient vigour for the speedy recolonisation of new sites and young trees.

In the 1970s the Forestry Commission thinned four large blocks of native oak woodland along the north shore and under-planted them with exotic conifers as the first step in converting them to commercial softwood plantations (Fig. 5.9). Now that the international importance of the area for lower plants has been recognised, Forest Enterprise are restoring the oakwoods by felling the conifers to waste and erecting many miles of deer/sheep-proof fencing to give the oak a chance to regenerate naturally. Some planting, using local material, may be carried out. This pioneering, positive management for lower plants shows every prospect of being successful due to the favourable climate and the reservoir of rich communities remaining in the area.

Scottish pinewoods

The 'Caledonian' pinewoods are composed of a subspecies of pine (*Pinus sylvestris* ssp. *scotica*) unknown outside Scotland. It has short needles and maintains a pyramidal form until late in life when it becomes round-headed. Rackham (1986) provides evidence that these woods were discovered commercially in about 1600, since when they have been systematically exploited, although, due to their ready ability to self-seed, the total area has not greatly changed. Sir T. D. Lauder, in about 1830, found giant decayed trees that had escaped felling in Glenmore. This is probably the last record of extensive wildwood dominated by pine, though it is rumoured that a few trees over 500 years old have recently come to light. There never was a 'Great Wood of Caledon', pine has always been patchily distributed; today it covers an area of *c.*10,500 ha, scattered in 30 major blocks, much of it very open.

Pinewoods received only casual attention from lichenologists until Brian Coppins moved to Scotland in 1974 and started looking at them in earnest.

Fig. 5.10 Dead pine standing among juniper, Glenmore Forest, Aviemore. Though scarce, dead pines support many of the rarest and most characteristic lichens of the Caledonian pinewoods.

Since then he has worked most areas preparing lists and making ecological observations. The woods, which also contain birch, rowan, juniper, etc., have so far yielded 371 epiphytic lichens, 167 of which are present on the bark and lignum of pine. Nineteen of these species are restricted to native pinewoods.

Such woods can be difficult to work, as a survey usually involves stumbling about in knee-deep heather or pushing through head-high juniper. At first sight, only common lichens such as *Bryoria fuscescens*, *Hypogymnia physodes*, *Hypocenomyce scalaris*, *Imshaugia aleurites*, *Mycoblastus sanguinarius*, *Parmeliopsis ambigua*, *P. hyperopta*, *Platismatia glauca*, *Ochrolechia tartarea* and *Pseudevernia furfuracea* appear to be present, though growing luxuriantly, with some hanging off the trees in great tassels. This is the classic acid bark assemblage known as the Physodion; it has a preferential pH range of 3.0 to 4.0. Peering up the trunks extensive sheets of *Chrysothrix flavovirens* may come into view, colouring the underside of major branches yellowish-green. After a while additional species become apparent, perhaps growing around the edge of bark scales, on horizontal boughs or in crevices. They might well include such pinewood specialists as *Bryoria capillaris*, *B. furcellata*, *Calicium parvum*, *Hypocenomyce friesii*, *Lecanora cadubriae*, *Lecidea turgidula*, *Ochrolechia microstictoides* and *Protoparmelia ochrococca*. Areas of lignum will be found to support a wide range of pinhead lichens and several species of *Xylographa*, their pale, narrow fruits aligned along the grain of the wood and scarcely resembling a lichen. On hardened resin, the honorary lichens *Sarea* (*Tromera*) *difformis* and *S. resinae* may be found. All these species can be seen during a walk through the Glenmore Forest Park, near Aviemore.

A major difference from deciduous forests is that a high proportion of the lichen interest is concentrated on stumps and standing dead trees (Fig. 5.10). These may be present at a very low density, but are of the utmost importance for preserving diversity as slowly decaying pine lignum is the preferred niche of many of the rarest and most characteristic pinewood species. Standing, partly decorticated dead 'pillars', perhaps burnt by a lightning strike or wildfire, are the place to search for *Hypocenomyce anthracophila, H. leucococca, H. sorophora, H. xanthococca* and *Micarea elachista*. The sheltered side of standing dead trunks and the underside of fallen ones, picturesquely known as the 'bones' of the forest, are rich in pin-head lichens, most of which need to be collected for accurate identification. The decay-resistant, often very hard lignum of dead trees supports *Cyphelium tigillare* and *C. inquinans*. The latter is better known from fence posts in the south of England, and it was a major surprise when its native habitat was discovered to be the Caledonian pinewoods. Many of the pinhead lichens employ passive means of spore dispersal (rather than explosive), which makes them unique among the lichenised fungi. It has been suggested that insects may play a significant role in their dissemination; the spores, being small and ornamented, readily adhere to anything that touches the fruit body. The Finnish word for Caliciales is 'nokinuppinen', which translates as 'smut-knobs' (Plate 6c).

Nine pinewood species are included in the Red Data List. Although two of these have not been seen for a while it would be unwise to consider them extinct as the habitat must be considered underworked by today's standards. The pinewood lichen everyone hopes to re-find is *Cetraria juniperina*, which forms small yellow rosettes on the trunks of pines and the understorey juniper. It was last collected in Rothiemurchus Forest in the nineteenth century, but despite special searches has eluded modern lichenologists. Tree bases support communities rich in unusual *Cladonia* species such as *C. carneola, C. cenotea, C. sulphurina* and the diminutive *Cladonia botrytes*. This small, distinctively coloured species with pale brown, grape-like clusters of apothecia, was recorded a dozen times on the stumps of felled pine in the 1960s and early 1970s, but had not been seen since. Its decline was believed to be linked to a preference for stumps left by Canadian wartime felling gangs which, for a time, provided optimum conditions, but have now decayed away. Recently, following a special search of sites where large Scots pine have been felled within the last 10 to 20 years, very small amounts have been found at four localities in Badenoch and Strathspey. *Bryoria implexa*, one of several brown 'beard-lichens' found on conifers, is known from four nineteenth-century collections, but has not been seen recently.

To complement the RIEC scale for deciduous woodland, Coppins generated a Pinewood Index. One difficulty to be overcome was the need to take into account the steep east-west climatic gradient across Scotland, which strongly influences the pinewood flora. This was dealt with by arranging the woods into three climatic/floristic groups: Western (W), Central (C) and Eastern (E). Accordingly, the Index comprises 50 species: 20 are generally distributed, for example *Lecidea hypopta, Mycoblastus affinis* and *Protoparmelia ochrococca*; 15 are confined to the W and C groups, for example *Cavernularia hultenii, Micarea adnata* and *Platismatia norvegica*; and 15 to the E and C groups, for example

Bryoria furcellata, Lecanora cadubriae and *Microcalicium disseminatum.* Using the Index, which is announced, but remains unpublished, the top ten pinewood sites in Britain have been identified (Table 5.2).

Table 5.2 The top ten native pinewoods identified using the Pinewood Index; the number of Index species present is shown in brackets. W = western group; C = central group; E = eastern group of woods. (Coppins, 1990).

1. Glen Strathfarrar (C) (36)	6. Rothiemurchus (E) (26)
2. Glen Affric (C) (32)	7. Achnashellach (W) (24)
3. Glen Guisachan (C) (30)	8. Ballochbuie (E) (23)
4. Coulin (W) (30)	9. Loch Maree (W) (23)
5. Black Wood of Rannoch (C) (29)	10. Abernethy (E) (22)

Coniferous plantations

Due to their recent origin and dark interiors, commercial softwood plantations detain few lichenologists. This is a misjudgement, as they support a specialised flora that is unlikely to be present in the surrounding countryside. A randomly selected 40-year-old plantation of Scots pine was investigated in the southern Pennines in an area where air pollution has greatly reduced the epiphytic flora. The richest habitat in the plantation was stumps left after thinning operations. These were in various states of decomposition, each stage having a characteristic lichen assemblage. Within a year or two of exposure most stumps are colonised by minute, lime-green dots which are the fruits of *Thelocarpon epibolum*; next to appear are *Fellhanera subtilis, Micarea nitschkeana* and *Placynthiella icmalea*. As the bark decays and becomes more water-retentive, sheets of *Micarea prasina* develop on the side of the stump, lasting until the bark falls off (Fig. 5.11). Older stumps are colonised by the moss *Orthotrichum lineare* together with a range of common *Cladonia, Placynthiella* and *Trapeliopsis* species,

Fig. 5.11 The distribution of lichens on a coniferous stump in a Forestry Commission plantation, Snake Pass, Derbyshire (M.J. Lindley).

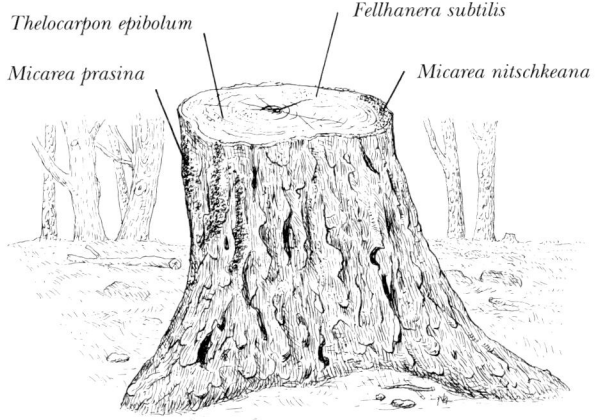

Fig. 5.12 Interior of an undisturbed hazel wood, Struidh, Eigg, Inner Hebrides. Such sites are exceptionally rich in strongly oceanic lichens.

though surprises like *Bacidia saxenii* occasionally turn up. By contrast, the living pine trunks carried only a grey-green crust of *Lecanora conizaeoides*.

The extensive pine plantations at Culbin on the east coast of Scotland date back to the 1920s. A recent survey (Coppins, *pers. comm.*) showed that after 60 years they had accumulated most of the commoner lichens of native pinewoods, including *Buellia schaereri, Lecanora aitema* and *Lecidea turgidula*, but only four Index Species were present. It is clear that within the area where pines are native, plantations diversify faster than elsewhere.

Western hazel woods

Recent research into the history of hazel suggests that exposed sites along the Atlantic seaboard of Britain, particularly where the soils are not too acid or waterlogged, have supported stands of hazel for thousands of years. The pollen record indicates a forest cover dominated by hazel with some elm, aspen, birch and rowan. This raises the possibility that certain remote hazel woods in places such as Skye, Eigg, Mull and at Seil, near Oban, are oceanic climax woodland of great antiquity.

On the small Hebridian Island of Eigg around 30 ha of hazel woodland survives on steep boulder-strewn slopes below basalt cliffs and climbs a little way up the faces on broad ledges (Gilbert, 1984b). The free-draining, brown-earth soil under the trees supports a high density of spring flowers such as wood anemone, woodruff, primrose and wood sanicle together with large pleurocarpous mosses which in themselves are evidence that the woodland is long established. The tiny, compact, remote, 3 ha wood at Struidh on the east coast of the island is distinctive for its giant hazel stools, the great girth of the mossy stems, its natural appearance, and for being an Aladdin's cave of lichens (Fig. 5.12).

The upper sides of leaning main trunks are thickly clothed with large foliose lichens including all British species of *Lobaria* and *Sticta*, *Nephroma laevigatum*, *N. parile*, *Pannaria conoplea*, *P. rubiginosa*, *Parmelia crinita* and *Peltigera collina*. These are present throughout the wood, the sheltered interior having in addition *Dimerella lutea*, *Collema furfuraceum*, *C. subflaccidum*, *Leptogium brebissonii*, *L. burgessii*, *L. cyanescens*, the beautiful *Pseudocyphellaria crocata* with its network of bright yellow soralia, *P. intricata* and *P. norvegica*. All are growing with exceptional luxuriance, many foliose species assuming a subfruticose style of growth; the wood is truly 'dripping' with lichens. This assemblage firmly places the community in the moss-dominated, euoceanic, basiphilous facies of the Lobarion that is probably endemic to Great Britain and southwest Ireland, but is not yet adequately described.

One is easily seduced by the glamour of the Lobarion; unless an effort is made, other communities can get overlooked. In hazel woods, however, it is no trouble to work along the arching mossy boughs to the ultimate twigs to see how assemblages replace each other as light, humidity and bark characteristics change. Less mossy, smaller branches carry a *Parmelia*-dominated community in which *Cetrelia olivetorum*, *Menegazzia terebrata*, *Parmelia glabratula*, *P. laevigata*, *P. perlata* and *Pertusaria hymenea* are conspicuous.

Intermingled with the *Parmelia*-dominated community, extending out onto the finer branches, and covering long, straight, shaded poles, there is a dense mosaic of crustose lichens that is in many ways as remarkable as the Lobarion, the only differences being that many of the species need to be collected in order to have their identity confirmed. This community, known as the Graphidion, forms well-integrated pink, white, grey, green and brown mosaics on smooth bark (Plate 6d, e); at Struidh the species involved include *Arthonia didyma*, *A. cinnabarina*, *Arthothelium reagens*, *Bactrospora homalotropa*, *Graphina anguina*, *Graphis scripta*, *Pyrenula chlorospila*, *P. laevigata*, *P. occidentalis*, *Thelotrema monospora* and *T. subtile*. This wood was one of the first places from which a strange *Graphis* with white fruit bodies was collected; it has since been found in five similar localities and described as a new species with the name *G. alboscripta*. It is endemic. Averis & A. M. Coppins (1998) believe that this rich, hyperoceanic facies of the Graphidion is a relic community that accompanied the northward spread of hazel, rowan and holly at the retreat of the last ice age. As it favours bark surfaces less than 25 years old, the species must be fairly mobile, given favourable conditions of light and humidity.

Other niches available to lichens in a hazel wood are well-lit twigs, which may carry *Parmelia sinuosa* and several species of *Usnea*. Sheltered recesses at the base of stools provide a habitat for *Arthonia elegans*, *Chaenotheca furfuracea* (Plate 6c) and *Dimerella diluta*. This 3 ha wood on Eigg yielded 118 epiphytic lichens on hazel, a total which increases to 130 if species from two old willows and an elm are included. The RIEC score of 95 is exceptional considering that values are normally calculated per 100 ha of habitat. In common with other western woods many of the demanding epiphytes are also present on shaded, mossy rocks beneath the tree canopy.

At Cleadale, on the west coast of Eigg, a 20 ha hazel wood has been exploited by a nearby crofting settlement. The oldest inhabitant on the island, Duncan McKay, described how each crofter used his own patch for fuel and timber, some using more than others. The last time it was utilised to any extent was in 1940–45 while coal was unavailable, when just the bigger stems were removed, starting at the lower margin. Small-scale extraction still goes on. Though elements of the Lobarion are present giving RIEC scores ranging from 30 to 55, few of the more demanding species seen at Struidh are present and the total number of epiphytes is only 80.

As explained above, old woodland indicator species have largely been selected because of their dependence on the continuously high humidity of the forest microclimate, often coupled with a poor colonising ability. It might be expected that their restriction to ancient woodland would break down in hyperoceanic areas where over 220 rain days (>1 mm in 24 hours) are coupled with equable temperatures. Why are any species restricted to older areas of woodland in such a climate? In the absence of experimental work, one can only speculate. Possible explanations are a requirement for ultra-high humidities, perhaps associated with the water-retentive properties of a luxuriant bryophyte carpet, or a dependence on throughfall for an adequate and uninterrupted nutrient supply. Many of the largest species, particularly those with a cyanobacterial partner, require a basiphilous environment if they are to function optimally, which is inconsistent with the rather low pH (4.8–5.1) of bark samples from Struidh, but this could be regularly raised by the canopy intercepting salt spray which is driven far inland by Atlantic storms.

The cropping ash of the Lake District

The heads of many valleys in the central Lake District are occupied by medieval landscapes in which regularly pollarded ash trees line field boundaries, tracks and watercourses (Fig. 5.13). Until recently these pollards were an

Fig. 5.13 The trunks of the 'cropping' ash, which have continuity with the wildwood, support a rich flora of ancient woodland indicator lichens; Watendlath, Borrowdale.

essential part of the farming economy, providing browse wood, fencing stakes, firewood, hurdle-making material and so on. As their use declined, the National Trust has taken over the pollarding, cropping them on a 20 to 30 year rotation, thus maintaining a tradition thought to have started a thousand years ago with the Norse settlers. The 'cropping ash', as they are called, are mostly hollow with a girth exceeding 4 m; the age of the oldest generation has been estimated at 500 years, making them among the oldest ash in the country. Maiden ash only last around 200 years. Pollarding greatly increases longevity provided it is continued, but if boughs are allowed to grow too heavy they split the tree open.

The ancient trunks are covered with cryptogamic communities that were investigated in side valleys off Borrowdale at Watendlath, Stonethwaite and Seathwaite. It soon became clear that the cropping ash can be divided into two groups with regard to their epiphytes. Certain rather acidic trees support a bryophyte cover dominated by *Hypnum cupressiforme*, which provides a substrate for *Cladonia coniocraea*, *Parmelia glabratula* and *P. saxatilis*, while their bark is host to *Lecanora chlarotera*, *Ochrolechia androgyna*, *Pertusaria albescens* var. *corallina*, *P. amara* and *P. pertusa*. Other trees have a more alkaline bark and carry elements of the Lobarion. These ashes can be recognised immediately from the bright, silky-green moss *Homalothecium sericeum* around their base which usually also bears large patches of *Peltigera*. In Borrowdale, acid-barked ash outnumber alkaline ones by about 3 to 1. Bark differences that cannot be obviously correlated with soil type, exposure, size, or age of tree, presumably have a genetic basis.

The exciting feature of this Lobarion is its richness; it is the finest Lobarion in England, outside Cornwall, and there is nothing approaching it in southern Scotland. Ivan Day (1989), who has made a detailed study of cropping ash in the Seathwaite Valley, tagging the best trees and making management recommendations, has found the most abundant Lobarion macrolichens to be *Collema subflaccidum*, *Lobaria virens*, *Nephroma laevigata*, *Pannaria conoplea*, *Parmeliella triptophylla*, *Peltigera horizontalis* and *Sticta sylvatica*, with smaller amounts of *Collema furfuraceum*, *Degelia plumbea*, *Nephroma parile*, *Sticta canariensis*, (both morphotypes), *S. fuliginosa* and *S. limbata*. *Leptogium burgessii*, *Lobaria amplissima* and *L. pulmonaria* are present in the valley, but prefer maiden ash.

The macrolichens are supported by a wide range of smaller species, the distribution of which depends on the niche structure of the individual pollard. The more massive trunks provide ridges (*Pertusaria* spp.), folds (*Thelopsis rubella*, *Thelotrema lepadinum*, *Pachyphiale carneola*), seams of dead wood (*Caloplaca obscurella*, *Micarea lignaria*), bosses that create underhangs (*Graphis* and *Opegrapha* spp.), nutrient streaks below wounds (*Candelariella xanthostigma*, *Wadea dendrographa*), mossy areas (*Biatora sphaeroides*, *Dimerella lutea*, *Lecidea vernalis*, *Normandina pulchella*) and smooth bark associated with callus (*Arthonia radiata*, *Lecanora chlarotera*, *Lecidella elaeochroma*, *Pyrenula macrospora*) (Plate 6e). The RIEC score for the Seatoller woods is 130, strong evidence that woodland in the valley has an unbroken history and represents a rare fragment of the original forest cover.

Parkland

In the fourteenth century there were over 1,900 deer parks in England and Wales, many of which still exist in a modified form. They were typically created by enclosing an area of low agricultural value that often contained elements of the wildwood. Though the trees were usually pollarded, many parks received little management until the eighteenth century when landscape architects such as Brown, Kent and Repton turned these pasture woods into picturesque landscape by the addition of lakes, temples, and supplementary planting in the form of belts, clumps and avenues. This conversion modified the habitat without destroying it.

Francis Rose, in the late 1960s, was first to discover the lichen interest of medieval parklands. During the following decade it became a fashionable day out for small groups of lichenologists to identify a suitable-looking parkland from the Ordnance Survey map, contact the estate office, then spend a most pleasant and rewarding time exploring these exclusive habitats, perhaps finishing up sharing a glass of sherry with the owner. Now that virtually all parks have been surveyed, many repeatedly, this pioneering phase is drawing to a close. Parks of medieval origin have become one of the most thoroughly worked habitat in the country (Rose, 1992; Hodgetts, 1996), and are appreciated as a distinctly British phenomenon, being hardly known on the Continent where they have been lost through conversion to formal landscape. Parks of eighteenth-century origin, fashioned from farmland, have no particular lichen interest.

The most significant factor in the ecology of old deer parks is the presence of numerous mature to over-mature trees that are dying on their feet; these provide a continuity of substrate dating back to the days of the wildwood. The trees are a mixture of neglected pollards and maidens, oak being most frequent, accompanied by ash and elm where the soil is more fertile, and with planted exotics such as sycamore, lime, walnut, tulip tree and horse chestnut nearer the house. There are often numerous decorticate fallen trunks which provide a lignicolous habitat for lichens and are also highly regarded for their invertebrates and fungi. Grazing, nowadays mostly by domestic stock, is beneficial in preventing scrub or ivy from shading the trunks, but creates a problem by eliminating natural regeneration.

Some of the best parklands in Britain are listed in Table 5.3, overpage, together with their species totals and RIEC scores. The bigger ones, like Arlington Park in Devon, Boconnoc Park in Cornwall, Melbury Park in Dorset, Parham and Eridge in Sussex, and Drummond Park on Tayside are of international significance; they are vast, complex and take several days to explore thoroughly. At a county level the significance of old parklands is that they will probably be the only habitat in which the Lobarion, the climax community of the wildwood, survives. In parks that experience low grade air pollution the Lobarion is often restricted to just a few of the oldest trees, preferring slightly leaning specimens standing in a sheltered position and with a bark pH >5. Giant ash trees at the foot of a slope or by a watercourse are a favourite refugium; they account for most of the 50 *Lobaria amplissima* trees in England.

A complementary parkland lichen community, confined to the sheltered side of well-lit trees over 250 years old, is the *Lecanactidetum premneae*, which comprises *Arthonia impolita, Lecanactis premnea, L. lyncea, L. amylacea, Opegrapha prosodea, Schismatomma decolorans* and *S. cretaceum* (James et al., 1977). It is found on dry, acid-barked oaks and is of international importance, occurring in England, Wales and Ireland, but rarely on the Continent where there are few trees of sufficient age. While the Lobarion requires light, humidity and a continuity of trees, the Lecanactidetum, tolerant of a wider range of climatic conditions, needs only dry ancient bark surfaces.

Table 5.3 Relatively intact medieval deer parks with (a) the total number of epiphytic/lignicolous lichens at each and (b) scores on the Revised Index of Ecological Continuity (from Rose, 1992).

Name	(a)	(b)
Boconnoc Park, Cornwall	191	145
Trebartha Park, Cornwall	162	130
Arlington Park, Devon	213	130
Dunsland Park, Devon	163	115
Widdon Park, Devon	163	100
Mells Park, Somerset	142	80
Melbury Park, Dorset	218	110
Lulworth Park, Dorset	150	75
Longleat Park, Wiltshire	159	90
Parham Park, Sussex	190	65
Eridge Park, Sussex,	185	95
Ashburnham Park, Sussex	172	80
Brampton Bryan Park, Hereford	176	65
Dynevor Park, Dyfed	137	85
Domelynllyn Park, Gwynedd	157	125
Inverary Park, Strathclyde	134	90
Drummond Park, nr Crieff, Tayside	162	65
Cawdor Castle, nr Nairn, Highland	131	60

The best parks for lichens remain 'unimproved', the trees standing over bracken or acid grassland. Such areas are often known to the estate as 'the old deer park'. Unfortunately, the lichen interest in the majority of parks has been greatly reduced by the regular application of agrochemicals or ploughing and reseeding to increase stocking rates. A small amount of eutrophication from fertiliser or farm stock leads to conditions suitable for the colourful Xanthorion, which in the east of the country can be very rich, containing species of *Anaptychia, Buellia, Caloplaca, Parmelia, Physcia, Physconia, Ramalina, Teloschistes* and *Xanthoria*. This community is characteristic of dry, open, sunny woodland in southern Europe and may have spread to Britain as a result of human activity. Levels of eutrophication that become too high are disastrous, the lichens developing a coat of green algae and dying. Over most of Britain the Xanthorion is in heavy decline, but with levels of fertiliser use generally lower in parkland than in the surrounding countryside this is another community the best examples of which are becoming restricted to parkland and track sides (see p. 166).

The top parks in each county are mostly Sites of Special Scientific Interest (SSSIs) on account of their lichen/invertebrate importance (Harding & Rose, 1986). They are safeguarded by regulations that control the input of agrochemicals, encourage replacement planting, maintain light levels around trunks, and promote transplant work to perpetuate lichen populations. Regional levels of air pollution, however, cannot be controlled and have

Fig. 5.14 Hulk of a giant oak lying among bracken, the Old Deer Park, Chatsworth.

greatly reduced the lichen interest of parks in the Midlands and the east of the country (see Chapter 4). What has survived is sometimes surprising as the following description of the old deer park at Chatsworth shows.

Chatsworth Park, Derbyshire

This extensive park, still with herds of red and fallow deer (*Dama dama*), is an SSSI, but has largely been stripped of its former lichen flora by a long history of air pollution. Wide-ranging grassland improvement which leaves white skirts of powdered fertiliser around tree bases has further reduced its interest. Yet in one remote valley, 500-year-old oaks standing over acid grassland (Fig. 5.4) hold relic lichen communities that survived the industrial period either deep in bark crevices or on fallen trunks lying among bracken. These slow-grown oak hulks, when exposed to full sunlight, take hundreds of years to decay (Fig. 5.14). Their surfaces, which appear to be as unchanging, hard and acidic as the gritstone boulders among which they lie, support such locally scarce lichens as *Bryoria fuscescens*, *Cladonia digitata*, *C. parasitica*, *Hypocenomyce caradocensis*, *Imshaugia aleurites*, *Lecanora piniperda*, *Lecidea hypopta*, *Parmeliopsis ambigua* and *Protoparmelia oleagina* together with other commoner species. For several this is their only station in the county.

The living oaks at Chatsworth are mainly bare of lichens, with their bark acidified to pH 3.0 by decades of air pollution. The upper side of inclined trunks carry a grey-green crust of *Lecanora conizaeoides*, deep bark crevices provide a habitat for occasional populations of *Chaenotheca ferruginea*, *C. stemonea*, *C. trichialis*, *Chaenothecopsis nigra*, *Chrysothrix candelaris* and *Cyphelium inquinans*, while root buttresses are white with *Ochrolechia turneri*. The lesson to be learnt from Chatsworth is that even seriously polluted parkland can provide the unexpected; the lichen flora of the old hulks shows links to that of the Caledonian pinewoods.

6

Acid Rock

Sir Halford Mackinder (1902) first observed that if a line is drawn from the mouth of the River Tees to the mouth of the River Exe it divides the country into Highland Britain and Lowland Britain. Large and continuous areas of Highland Britain are more than 305 m above sea level, rock outcrops are frequent, and steep slopes, thin soils and a high rainfall make farming difficult so that there is much uncultivated moorland. By contrast, Lowland Britain is flat or composed of lines of undulating hills reaching upwards to around 200 m, the greater part of it cultivated land favoured by the deep soils, low rainfall and high sunshine. A further difference is that Highland Britain is composed of hard rocks, older than the Coal Measures, whereas Lowland Britain is underlain by softer rocks younger than the Coal Measures (< 280 million years) and often calcareous.

This division of the British Isles into two roughly equal halves has great significance for lichenologists. Those living in Highland Britain have ample opportunity to study acid rock habitats, preferably as natural outcrop, but also in quarries and on stone walls. The deprived lichenologists in the lowlands, unless they live in one or two very special areas, have to make do with studying acid substrates in secondary habitats such as churchyards or the lichenologically neglected military buildings (castles) and country houses where no expense was spared in transporting suitably hard stone over a great distance.

This chapter concentrates on acid rock lichens found below 610 m; lichens of higher level outcrops are dealt with in the chapter on mountains. The varied geology of the British Isles provides a wide range of acid rocks in this area, for example sedimentary grits, sandstones, mudstones and tuffs; metamorphic quartzites, slates and schists; and almost the entire sequence of igneous rocks including granites, rhyolites, diorites, andesites, gabbros, basalts and serpentine. Rocks influence the development of lichen communities through differences in texture, hardness, aspect and especially chemical composition, which controls acidity.

Acidity can be viewed from three perspectives. To chemists and soil scientists it is straightforward. A solution may be acid, neutral or alkaline (basic) depending on the proportion of hydrogen ions (H^+) and hydroxyl ions (OH^-) present. If hydrogen ions are more abundant the solution is acid, if the reverse is true the solution is alkaline and if the proportion is equal it is neutral. The relationship is expressed as a pH measurement which under field conditions can range from below 3 to above 12 (Table 6.1); neutrality is 7. Each unit step on the pH scale represents a tenfold difference in hydrogen ion concentration, so a solution of pH 4 has ten times the concentration of hydrogen ions of one of pH 5, a hundred times as many as a solution of pH 6 and a thousand times more than a solution of pH 7. A dry rock surface does not have a pH, but once it is wet the film of water possesses one which may be measured accurately with a flat electrode or more approximately using narrow range pH papers that turn different colours.

A botanist's view of pH is similar to that of a chemist, but the critical pH is that which separates calcicoles from calcifuges: this is not neutrality, but somewhere between 6 and 7. Evidence from higher plant ecology suggests that pH can influence plant growth directly through the toxic effect of a high concentration of hydrogen or hydroxyl ions, or indirectly through its influence on the solubility of certain elements. At low pHs, aluminium (Al^{3+}), which is potentially toxic, becomes soluble; it seems probable that calcicole lichens are sensitive to this ion while calcifuge species can tolerate it. However, this is supposition; lack of its confirmation is one of the outstanding gaps in our knowledge of lichen ecology. In addition, essential nutrients such as phosphorus become less available at the two ends of the pH spectrum, and at low pHs nitrification slows down until below pH $c.$ 5.5, the lichen photobiont *Nostoc* fails to function. Whatever the mechanism responsible, many lichens that grow in intimate contact with their substratum are so sensitive to these pH effects that they can be used to estimate the pH of rock surfaces (Table 6.1).

Table 6.1. The pH range of commonly occurring rocks and their derived soils; lichens characteristic of the different pHs are listed.

	Reaction	**pH**	**Substratum**	**Lichen species**
Acid	Very strong	3	Sulphide-rich rocks, Dry acid peat, Quartzite	*Cladonia coccifera* *Fuscidea kockiana* *Micarea lignaria*
	Strong	4	Gritstone, Granite, Slate	*Rhizocarpon geographicum*
	Moderate	5	Red sandstone, Tuff, Mudstone, Basalt	*Acaropara fuscata* *Ochrolechia parella* *Parmelia loxodes*
	Slightly	6	Serpentine, Mica-schist, Epidorite, Calcareous sandstone	*Caloplaca flavovirescens* *Porpidia speirea* *Rhizocarpon concentricum*
Neutral		7	Dolomite, Ironstone	*Caloplaca isidiigera* *Lecania erysibe*
Alkaline	Slightly			*Protoblastenia rupestris*
	Moderate	8	Chalk, Limestone Oolite, Marble, Seaspray	*Caloplaca lactea* *Clauzadea immersa* *Verrucaria baldensis*
	Strong	9		
		10	Evaporites	*Fulgensia fulgens* *Toninia sedifolia* *Squamarina lentigera*
	Very strong	11		

The geologist's classification of igneous rocks is based on the proportion of silica present, since it forms acidic compounds and is the dominant component in most of them. The most siliceous igneous rocks are known to geologists as acid rocks (granite, rhyolite) and those with progressively less silica as intermediate (diorite, andesite), basic (basalt, gabbro) and ultrabasic (serpentine) respectively. Artificially defined silica percentages differentiate the classes, for example, the acid/intermediate rock boundary is 66% silica (SiO_2). Though still retained in a general sense, the choice of terms, which was originated by nineteenth-century geologists, is now regarded as unfortunate, as even ultrabasic rocks rarely give rise to soils with a pH over 5.5.

Acid rock in lowland Britain

The High Weald

One of the few areas in southeast England where acid rocks outcrop is the High Weald of the Kent-Sussex border. Here the soft, poorly consolidated Hastings Beds form low rounded outcrops on well-wooded valley sides (Fig. 6.1). Borrer (1805) and Forster (1816) published lichen lists from these rocks and they have recently been resurveyed by Pentecost & Rose (1985) so are a classic locality for studying change over nearly 200 years. A total of 90 saxicolous lichens have been recorded of which 18 can not now be re-found, and 11 others have seriously declined. The present flora is dominated by *Baeomyces*

Fig. 6.1 Wealden sandstone rocks at Eridge with a light sprinkling of snow. One hundred and fifty years ago they supported *Lobaria pulmonaria* (A. Pentecost).

rufus, *Hypogymnia physodes*, *Lecidea fuscoatra*, *Lepraria incana*, *Micarea lignaria*, *Ochrolechia androgyna*, *Parmelia saxatilis*, *Psilolechia lucida* and *Trapeliopsis granulosa*. Many of these species are also widespread in secondary habitats, but there is also an element that in southeast England is almost exclusively confined to these sandstone rocks; it comprises *Cladonia incrassata*, *C. subcervicornis*, *Coriscium viride*, *Cystocoleus ebeneus*, *Micarea botryoides*, *Pertusaria aspergilla*, *P. corallina*, *Sphaerophorus globosus* and *S. melanocarpus*. Many lichenologists are surprised that such species are present in the southeast as they are more typical of the oceanic western parts of Britain. The reason lies in the porous, water-holding nature of the soft sandstone and the microclimate of the steep ravines in which many of the outcrops occur. The 'Atlantic' nature of the flora extends to the bryophytes and ferns with the Tunbridge filmy fern (*Hymenophyllum tunbridgense*) and hay-scented buckler fern (*Dryopteris aemula*) present at many sites. No comparable habitat is known east of Exmoor.

A number of large foliose lichens have disappeared from these rocks since the early 1800s, for example *Lobaria pulmonaria*, *L. scrobiculata*, *Nephroma laevigatum*, *N. parile*, *Parmelia conspersa* and *P. crinita*, with severe declines shown by *Bryoria fuscescens* (four sites down to one), *Cladonia caespiticia* (five down to one), *Coriscium viride* (four down to one), *Spaerophorus globosus* (five down to two) and *S. melanocarpus* (five down to three). These changes are thought to be due to over-collection, air pollution, rock climbing and the spread of densely shading woody vegetation, particularly *Rhododendron ponticum*. The richest sites today are High Rocks, Paddockhurst, Penns and Uckfield, each with between 28 and 36 saxicolous species.

Charnwood Forest

The Charnwood Forest inlier of Precambrian rock rises unexpectedly from the Midland plain 10 km north of Leicester. It gives rise to a compact area of hills with craggy tor-like summits and bracken-covered slopes. These hard ancient rocks are varied in composition so every hill has a different lichen flora. At Altar Stones, by Exit 22 off the M1 motorway, the outcrops are of a silicified volcanic ash known as hornstone. Here, among graffiti and broken glass, the largest population of *Parmelia disjuncta* in England occurs on sloping rock surfaces. It can be separated from other brown *Parmelia* species by its small neat rosettes of very narrow lobes bearing tiny isidia and their C- reaction. Associated lichens on the still smoke-blackened rock are abundant *Acarospora fuscata*, *Buellia aethelea*, *Lecanora rupicola*, *Lecanora soralifera* and *Miriquidica leucophaea*, lesser amounts of *Candelariella coralliza* and *Parmelia incurva*, and on walls *Lecanora epanora*. There are also well-developed communities of terricolous lichens.

Five hundred metres south of Altar Stones, at Markfield, a trigonometric point (220 m) indicates the summit of a granite outcrop. This has largely been quarried away, but the few natural exposures left carry dense swards of *Umbilicaria deusta* and smaller amounts of *U. polyphylla* growing with *Acarospora fuscata*, *Trapelia involuta* and *T. obtegens*. The normally acid granite has been slightly eutrophicated by quarry dust. A few kilometres further east, near Old John Tower in Bradgate Country Park, outcrops of slate occur (Fig. 6.2). These again are smoke-blackened, but carry a rich lichen flora that includes sheets of *Lasallia pustulata*. Precambrian outcrops elsewhere in Charnwood Forest support *Ochrolechia tartarea*, *Ophioparma ventosum* (Plate 7a) and *Porpidia speirea*.

Fig. 6.2 Hill-top outcrops of Pre-Cambrian slate at Bradgate Park, Leicestershire, support several northern species.

The acid rock flora has experienced its share of extinctions. Species known to the Rev. A. Bloxam (1801–78) and others, but which have not been seen recently, include *Alectoria fuscescens*, *Parmelia conspersa*, *P. omphalodes*, *Sphaerophorus globosus* and *Stereocaulon dactylophyllum* (Hawksworth, 1971).

Sarsen stones

In the vicinity of the Marlborough Downs, Wiltshire, the Lambourn Downs, Berkshire, and Portisham, Dorset, there are several groups of sarsen stones lying in grassland. These rounded sandstone boulders are composed of hardened quartzite and are believed to be the remains of a former sandstone deposit of Tertiary age that overlay the chalk. Soil movement has carried most of them down slope so they now occur as boulder trains in valley bottoms. They can be huge, weighing up to 60 metric tons, but most are less than a metre high. If any lichen habitat can be described as charismatic the sarsens (dark strangers) qualify due to their singular nature and exceptional interest. Their lichen flora, remarkable in possessing both upland and maritime species, has been studied by Laundon (1976), O'Dare & Laundon (1986) and A. M. Coppins (1996). The lichen interest includes those sarsens used to construct stone circles thousands of years old at Avebury and Stonehenge (Fig. 6.3); in fact, hollows in the Slaughter Stone at the latter support the rare nitrophilous species *Phaeophyscia sciastra*.

The maritime element, which is widespread on the sarsens, comprises *Anaptychia runcinata*, *Aspicilia leprosescens*, *Buellia subdisciformis*, *Caloplaca ceracea*, *Lecanora fugiens*, *Ramalina siliquosa*, *Rinodina atrocinerea* and *R. confragosa*. It has never been satisfactorily explained why these species are present so far from the coast. If the sarsens were on hilltops they might occasionally catch salt spray blown over 100 km on westerly gales, but they are in valleys. Glider pilots report that on many days sea breezes reach the Berkshire Downs eliminating

Fig. 6.3 Stonehenge; it has never been satisfactorily explained why these ancient stones support lichens with maritime affinities. The surveyors are Peter James and Francis Rose (D. Mansell).

thermals; perhaps they carry sufficient salt to these inland sites. Alternatively, a tolerance of fluctuating pH levels may be part of the explanation; acid coastal rocks get drenched in alkaline sea spray and the acidic sarsens may periodically be flushed with wind-blown alkaline material and detritus from birds, but this is mere speculation.

The sarsen concentrations at Avebury, Fyfield Down, Lockeridge Dene and Piggle Dene lie at between 150 and 210 m above sea level (Fig. 6.4). Despite this low altitude they support a number of species more typical of the highland zone such as *Aspicilia epiglypta, Fuscidea cyathoides, Lecanora gangaleoides, L. rupicola, Parmelia conspersa, P. omphalodes* and *Rhizocarpon geographicum.* These occur alongside the general sarsen flora of *Aspicilia caesiocinerea, Buellia stellulata, Candelariella coralliza, Haematomma ochroleucum, Lecanora polytropa, Lecidella carpathica, Parmelia loxodes, P. mougeotii, P. pastillifera, Pertusaria pseudocorallina* and *Porpidia soredizodes.* One lichen, *Buellia saxorum,* appears to be confined in Britain to sarsens where it is very common and can be recognised by its superficial apothecia on a pale mosaic-forming thallus that is C+ red. It is one of the few acid rock lichens to have its headquarters in lowland Britain.

In the past, sarsens were removed in great numbers and broken up to provide building materials. Today, the largest concentrations are protected as NNRs or SSSIs, but at a number of sites increasingly intensive land use in the form of fertiliser and herbicide applications, which drift onto the stones, has all but eliminated the lichen flora. A monitoring programme has been initiated to quantify the damage.

The Malvern Hills

The 'English Alps' of Celia Fiennes rise to a height of 425 m from the Severn and Herefordshire Plains. The main ridge, 13 km long, is fashioned from Precambrian schists and gneisses heavily intruded by plutonic and volcanic rocks. The lichens of the area were thoroughly investigated in the middle of the last century (Lees, 1852; 1868) when 260 species were recorded; since then little organised field work has been undertaken until the area was visited in connection with preparations for this book. The outstanding feature of the rocks in the north Malverns is the abundance of *Lasallia pustulata* (Plate 7b), which penetrates the town to within 200 m of Woolworths. This species covers entire outcrops with close-set thalli, up to 10 cm in diameter, only *Lepraria lobificans* and *Leproloma membranaceum* occurring under the shade of its lobes. It shows a preference for sloping or near-vertical faces which, during heavy rain, are flushed by water that has drained through mineral soil.

Associated saxicolous communities on the massive diorites belong to the *Parmelion conspersae*, an alliance of well-lit, slightly to markedly nutrient-enriched siliceous substrata, the enrichment in this instance being a result of rock composition. Characteristic species of the alliance in the Malverns include *Acarospora fuscata*, *Aspicilia caesiocinerea*, *Diploschistes scruposus*, *Lecanora gangaleoides*, *L. rupicola*, *Lecidea fuscoatra*, *Lecidella scabra*, *Miriquidica leucophaea*, *Parmelia loxodes*, *P. mougeotii*, *Pertusaria aspergilla*, *Protoparmelia badia*, *Schaereria fuscocinerea*, *Scoliciosporum umbrinum* and *Trapelia* spp. Less frequently encountered are *Aspicilia cinerea*, *Caloplaca crenularia*, *Parmelia conspersa*, *P. revoluta* and *Rhizocarpon viridiatrum*. Despite lying close to the Exe-Tees line, the Malvern list is reminiscent of other hard rock sites in the lowlands, any upland influence on the composition of the lichen flora being minimal.

On the western slopes of Broad Down an exposure of the Warren House Volcanics takes the form of basaltic pillow lavas crisscrossed with veins of

Fig. 6.4 Sarsen stones line a valley at Fyfield on the Marlborough Downs.

secondary minerals such as calcite and epidote, which introduce a weakly basic influence. This does not obviously affect the lichen flora which is still a facies of the *Parmelion conspersae*, here dominated by *Lecanora soralifera* with a few additional species such as *L. sulphurea*, *Rhizocarpon lavatum* and *Tephromela grumosa*, the sorediate counterpart to *T. atra*. Prominences used as bird-perching stones support the 'wild' form of *Lecanora muralis*, which is smaller and browner than the aggressively colonising pale green, white-margined race seen in urban areas. Among the outcrops terricolous species are frequent with abundant *Cladonia foliacea*.

A comparison of the current lichen flora of the Malvern Ridge with that recorded in the middle of the last century shows that the most interesting species have all been lost. Lees (Lees, 1852) describes finding the golden hair-lichen (*Teloschistes flavicans*) on the craggy rocks of North Hill, together with *Bryoria fuscescens*, *Lobaria scrobiculata*, *Parmelia omphalodes*, *Sphaerophorus globosus* and *S. melanocarpus*, 'all growing in a luxuriant and beautiful manner truly gratifying to lovers of nature'. None of these have been seen recently, while *Ochrolechia parella*, formerly abundant throughout the chain, is now scarce. Thanks to the thoroughness of the Victorian recorders, the Malverns provide an excellent demonstration of the extent to which the acid saxicolous lichen flora of exposed sites has been depressed by industrial pollution. The upland element, here at the edge of its range, has experienced the greatest decline. There is still scope for further survey work, especially at the northern end of the range.

Other sites

The above four areas are the most important in Lowland Britain for acid saxicolous lichens. Sites such as the North York Moors, the Quantock Hills and the ancient rocks of Shropshire lie close to the Exe-Tees line and more properly belong in Highland Britain. While most Tertiary rocks are so soft that they can be dug using one's hands, there are hardened horizons that like the sarsens would make a worthwhile study. For example, the flint-quartzite conglomerate known as Hertfordshire Puddingstone, the ferricretes of the Reading area, the Goldstone of Brighton, or the saccharoidal quartzites that occur near High Wycombe. While much of this material has been broken up and incorporated into buildings, particularly churches, large lumps were used as mounting blocks, corner guards or to mark fords. Other places to search for these rocks are along roadsides and field margins, on village greens and beside gravel pits. On a smaller scale the study of pebbles is still in its infancy.

Acid rock in upland Britain

Due to uneven coverage by lichenologists, acid rocks in the uplands will be dealt with according to rock type rather than by region.

Millstone grit

Millstone grit underlies the Pennine moorlands of Derbyshire, Staffordshire, Yorkshire and Lancashire; it is also found around the South Wales coalfield and in the Midland Valley of Scotland. Another massive grit of Carboniferous age, the fell sandstone of Northumberland, is included here. All these gritstones are well exposed in scarps, often several kilometres long, below which boulder scree accumulates.

In recent years the gritstone outcrops of the southern Pennines have been intensively studied. The early workers did not pay a great deal of attention to this habitat so we will never know the full extent of its lichen flora prior to the Industrial Revolution. From the luxuriance of the few herbarium specimens that have survived from that time it can be conjectured that the outcrops formerly carried a dense shaggy coat composed of many species among which shrubby thalli of *Bryoria*, *Sphaerophorus* and *Usnea* would have been conspicuous. Today, high-level outcrops catch pollution particularly badly; those on the Crowden Moors, downwind of Manchester for instance, are scrubbed bare by the pollution-laden wind. Conversely, the richest sites for millstone grit lichens tend to be outcrops in sheltered valleys east of the main watershed as these are furthest from the major emission sources in Lancashire. The presence of a light woodland cover helps filter out pollution. Some of the best sites so far located are the Snake Pass, Padley Gorge, Chatsworth Edge, Burbage and Gardom's Edge; all easily accessible from Sheffield, the pollution from which is dispersed eastwards.

The above sites each contain between 45 and 55 saxicolous species including large amounts of the common sandstone lichens *Mycoblastus sanguinarius*, *Parmelia saxatilis*, *P. omphalodes*, *Pertusaria corallina*, *Platismatia glauca*, *Protoparmelia badia*, *Pseudevernia furfuracea* and *Rhizocarpon geographicum*. Twenty years ago there was a lot of *Lecanora conizaeoides*, but this is being replaced by swards of *Trapelia involuta*. Other species that appear to have increased as pollution levels fall include *Lecanora epanora*, *L. subaurea*, *Lepraria caesioalba* and *Parmelia incurva*. There are also small populations of the pollution-sensitive lichens *Bryoria fuscescens*, *Ochrolechia tartarea*, *Umbilicaria deusta*, *U. polyrrhiza*, *U. torrefacta*, *Usnea hirta* and *U. wasmuthii*. In addition, these exposures have recently produced a crop of remarkable records, which has led to a reappraisal of their interest. Particularly notable northern rarities that have been discovered include *Mycoblastus alpinus* on the vertical side of shaded sandstone boulders at two sites, *Cornicularia normoerica* at 490 m on the Derwent watershed and *Lecidea commaculans* locally frequent on Burbage Moor (its next nearest site is the Isle of Mull), together with *Fuscidea austera*, *Rhizocarpon lecanorinum*, *R. subgeminatum* and *Schaereria cinereofusca*. A dark, usually sterile crust common on many of the sandstone edges has been identified as *Protoparmelia picea*, which may have its British headquarters here. The first UK specimen of *Lecidea pernigra* came from Cracken Edge, near Chinley, Derbyshire, in 1979. Since then it has been found at a dozen places in the area. It is considered that many of these species must have survived the industrial period as small populations in obscure microhabitats and are now expanding onto more conspicuous parts of the outcrops.

The Northumberland-Cumberland border is the only place in England where a fully developed gritstone lichen flora can still be studied and an attempt made to reconstruct the attenuated assemblages that occur further south. Echo Crags is a small, north-facing, hill-top exposure of fell sandstone at 460 m. Chief among the lichens it supports is the sulphur-yellow *Alectoria sarmentosa*, which locally forms a dense covering on slightly sheltered rock faces. Exposed buttresses carry luxuriant stands of *Usnea subfloridana*, while near ground level large patches of *Arthonia arthonioides* colour the dry underhangs pink. All three species of *Sphaerophorus* are abundant in sheltered crevices. Other species of interest include *Bryoria bicolor*, *B. chalybeiformis*, *B. fuscescens*,

Fig. 6.5 Muckle Samuel's Crags, Northumberland, carry a fully developed sandstone lichen flora that has affinities with cloud zone assemblages.

Imshaugia aleurites, Mycoblastus affinis, Ochrolechia frigida, Parmelia mougeotii and *Umbilicaria cylindrica* at one of its few sites on sandstone in England. Thirteen kilometres west of Echo Crags lies the Kielderstone, one of the best known landmarks on the English-Scottish border. It is an isolated block of sandstone, as large as a house, where border parties met to settle their differences. It deserves to be equally well known for its lichens, as one side is densely covered with fertile *Alectoria sarmentosa*, first discovered there in 1969 (Gilbert, 1980a), though the Berwickshire Naturalists' Club record of '*Usnea barbata* on the Kielderstone' (Anon, 1905) almost certainly refers to this species. It has been suggested that it formerly occurred in native pine woodland in the borders but, now the pines have gone, survives as a relic, safe from moorland fires, on these sheltered sandstone cliffs.

Another small but impressive crag in the same general area is Muckle Samuel's, sited on a watershed at 335 m (Fig. 6.5). It was here in 1977 that Michael Crawley, a non-lichenologist, collected a large *Usnea* which he passed on to Mark Seaward. The latter noticed an unusual *Bryoria* mixed in with the *Usnea*, which was subsequently identified as *B. nadvornikiana*, new to the British Isles. A visit by the author revealed that much of the crag is dominated by this lichen, which is intermixed with large specimens of *Usnea filipendula*. They occur below bilberry-covered ledges, and together form a most striking spectacle. The environmental conditions on these remote watersheds – pure air, high atmospheric humidity, frequent low cloud, a persistent breeziness – suggest that these prodigious communities of lichens with a fruticose growth form have an affinity with fog-induced and cloud zone assemblages (Fig. 6.6).

South of the Tyne Gap, upland sandstone crags, some of them in Co. Durham, are equally interesting, with *Alectoria nigricans, Allantoparmelia alpicola, Pseudephebe pubescens* and formerly *Cetraria hepatizon*. How far down the Pennines these species formerly occurred is an open question, but they are still present on the millstone grit summit of Cross Fell (893 m), a site covered in Chapter 11.

Quartzite

Quartzites are tough, highly siliceous sandstones composed of pure quartz sand cemented by a clear quartz cement. Due to their lack of iron they form dazzling white cliffs and tors, which are well displayed on the Stipperstones Ridge, Shropshire, and the Lickey Hills of Worcestershire. Exposures in the neighbourhood of Cranberry Rock on the Stipperstones are almost entirely covered with a mosaic of *Fuscidea cyathoides* and *F. kockiana*. Nothing like this is known on the millstone grit. Block scree surrounding the tors is somewhat richer than the *in situ* rock, probably a result of surfaces being less deeply leached.

To study the most extensive quartzite landscapes in Britain it is necessary to travel to northwest Scotland where, in Sutherland, there are entire vistas dominated by shining white hills. The landowner is the Duke of Westminster who named his legendary racehorses Arkle and Foinaven after quartzite peaks on his estate. The lichen flora of the Cambrian quartzite, which occurs from sea level to 908 m, has been examined on the remote mountain of Foinaven (Gilbert & Fox, 1986). The hardness, acidity and lack of nutrients in the rock result in a lichen flora so specialised that it must rank among the poorest of any substrate in Britain (Fig. 6.7). In many places the lichen cover, even of stable blocks, is around 1%. The commonest species on the frost-shattered rock, at around 500 m, are *Fuscidea intercincta*, *Porpidia tuberculosa*, *Ionaspis suaveolens* and *Rhizocarpon geographicum*. Other species of importance are *Fuscidea austera*, *F. cyathoides*, *F. gothoburgensis*, *F. lygaea-tenebrica*, *Lecidea lactea*, *Rhizocarpon lecanorinum* and *Schaereria tenebrosa*. Most of these lichens appear untypical as there is a massive development of dark prothallus on which scattered groups of areoles

Fig. 6.6 The prodigious lichen growth on Muckle Samuel's Crags includes the shrubby species *Bryoria fuscescens*, *B. nadvornikiana* and *Usnea filipendula*.

Fig. 6.7 The quartzite screes of Foinaven carry highly specialised lichen communities; they rank among the most species poor in Britain.

occur. For this reason, thin layer chromatography is useful in confirming identifications. A black form of *Tremolechia atrata* is present, which has reduced areoles impregnated with a blackish pigment that masks the normal rust-red colour.

This spectacular countryside is full of surprises. The quartzite screes are never silent, on still days creaking audibly as they constantly adjust to changing temperatures and to gravity. It was regularly observed that on the quartzite occasional thalli of normally pale crustaceous lichens were purple (Plate 7c). Usually the entire thallus was stained, making it stand out from adjacent colonies. The explanation appears to be that the colour originates from the droppings of ptarmigan (*Lagopus mutus*) that have been eating crowberries or bilberries, the stain being taken up from the guano and concentrated by the lichen. Its spread throughout the surface of a contaminated lichen suggests lateral transport. Nothing is known regarding the persistence of the effect, though shaking for five hours in rainwater failed to dull the colour, which was concentrated in the upper cortex.

Basalt

Basalt is used here as a generic term to cover all dark lavas with a fine-grained texture and a basic composition. In Tertiary times the north of Britain was on the edge of a volcanic province that produced a vast basaltic plateau the rocks of which are well exposed on the Inner Hebridean islands of Arran, Mull, Eigg and Skye, on the mainland at Ardnamurchan and in northeast Ireland. Some flows show a remarkably regular columnar jointing which has made them world famous, notably the Giant's Causeway in Antrim and Fingal's Cave, Staffa. Elsewhere, pre-Tertiary basalts, mainly of Carboniferous age, form the

Whin Sill which runs diagonally from Teesdale to the Farne Islands, the toadstones of Derbyshire, and a number of abrupt hills in the Lothians; it is to these we turn for an introduction to their lichen flora. But first a note of warning; the rock splinters under the impact of a hammer and chisel, so wise lichenologists, unless bespectacled, wear protective goggles.

The conical hill, North Berwick Law, East Lothian (187 m), rises sharply from the cornfields of the coastal plain, its sides showing numerous dark outcrops of basalt. The dominant species on these exposures are *Acarospora fuscata, Lecanora intricata, L. polytropa, L. rupicola* and *Rhizocarpon geographicum*, while sheltered faces support *Diploschistes scruposus, Dirina massiliensis* f. *sorediata, Enterographa zonata, Lecidea orosthea, Leproloma membranaceum* and *Ramalina siliquosa*. A characteristic nutrient-enhanced facies of the basalt flora develops on slabs near ground level that are strewn with rabbit and sheep droppings. Here *Anaptychia runcinata, Aspicilia caesiocinerea, Candelariella coralliza, C. vitellina, Lecanora gangaleoides, L. muralis, Parmelia conspersa, P. glabratula, P. loxodes, Pertusaria pseudocorallina, Rinodina atrocinerea* and *Tephromela grumosa* dominate the rock surfaces. On faces flushed by runoff the rare *Ramalina polymorpha* locally forms dense swards often accompanied by *Xanthoria candelaria* (Plate 7d).

Eight kilometres further inland, Traprain Law (221 m) forms a superb viewpoint for the other Lothian igneous hills – Garlton, Pencraig, Bass Rock and to the north Castle rock and Arthur's Seat in Edinburgh. The latter is one of the best places to see *Lecanora andrewii*, a member of the *L. dispersa* group distinguished by its chemical spot tests of Pd+ orange, C+ orange-red, its coarsely crenulate thalline margin and penchant for sheltered basalt. Traprain Law, the richest of the hills so far explored, has produced some remarkable lichens. On the north side, intermittently irrigated slabs of black basalt, down which water drains after rain, are home to a wide range of cyanophilic species (Plate 7e), including *Collema glebulentum, Nephroma parile, Placynthium flabellosum, Porocyphus coccodes, Pyrenopsis grumulifera* and *P. impolita*. Some of these species are inconspicuous and easily missed unless the slabs are examined while drying out after rain. Associated lichens include a number of mild calcicoles such as *Agonimia tristicula, Catapyrenium lachneum, Dermatocarpon miniatum, Leptogium gelatinosum* and *Polysporina lapponica;* the source of the alkalinity is bands of secondary minerals which weather to release alkalis. Recesses in large north-facing underhangs are lined with *Dirina, Haematomma ochroleucum, Lecanora orosthea* and the wrinkled white thalli of the rare *Lecanora swartzii*. The latter is only associated with underhangs sufficiently large to provide shelter for three or four people in a rainstorm.

The sunny, south facing side of Traprain Law is home to *Carbonea assimilis, C. vorticosa, Lecanora andrewii, L. caesiosora, Lecidea furvella, Parmelia disjuncta, Rimularia intercedens* and *Umbilicaria deusta* together with a wide range of ubiquitous calcifuge lichens. These frequently form closed mosaics with little or no uncolonised rock between the crustose thalli (Fig. 6.8). This situation has been likened to a battlefield where species struggle to overcome their neighbours with various results. In one study over a 13 year period (Hawksworth & Chater, 1979), no growth took place at the junction between *Ochrolechia parella* and *Lecanora gangaleoides*, a mound of dark tissue building up where the thalli were in contact; equilibrium had been reached. *O. parella*, however, had little difficulty advancing over colonies of *Rhizocarpon obscuratum*, though *L. gangaleoides* was in equilibrium where it came into contact with this species. Small colonies

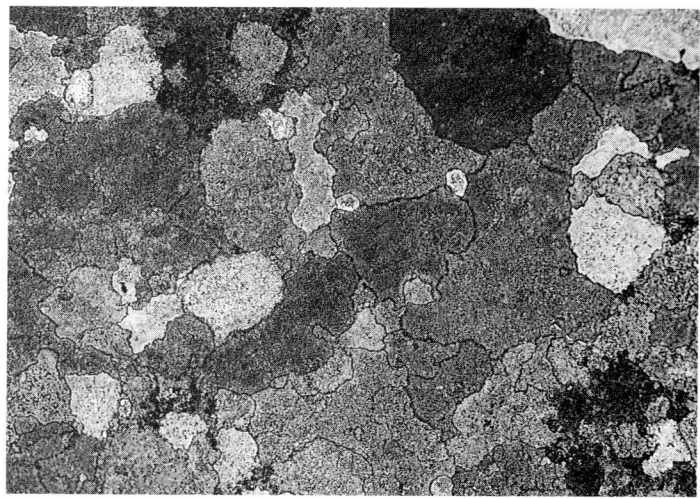

Fig. 6.8 Lichen mosaics on acid rock have been compared to a battlefield (P.A. Gainey).

of *L. gangaleoides* were seen to coalesce into a single large thallus, an ability that *Rhizocarpon geographicum* does not have; it lays down a black 'no man's land', which is why it is called the map lichen. When a fast-growing, short-lived foliose lichen becomes established on top of a crustose mosaic it passes over it like a temporary storm, the mosaic reforming in its wake. These little studied interactions are important as they control competition, succession and stability in mosaics of crustose species.

Other choice basalt sites in southern Scotland are Minto Craigs and Broad Law, Ancrum, both in Roxburghshire, and Hume Castle, Berwickshire. They add extra species such as *Caloplaca arenaria* and *Immersaria athroocarpa* and hold additional populations of the normally coastal lichens *Anaptychia runcinata*, *Ramalina polymorpha*, *R. siliquosa* and *R. subfarinacea*.

The Whin Sill (dolerite) forms a series of bold escarpments that span northeast England. Most outcrops are remote and surrounded by rough grazing, so have escaped the worst effects of both air and agricultural pollution (Fig. 6.9). Their lichen flora, especially where low grassy outcrops are concerned, is similar to that of North Berwick Law, but with *Aspicilia cinerea*, *Lecanora rupicola* f. *sorediata*, *Parmelia verruculifera*, *P. revoluta*, *Physcia dubia*, *Porpidia speirea* and *Scoliciosporum umbrinum* more fully represented. The extensive north-facing outcrops along the Roman Wall, set in moorland, support a more calcifuge flora in which *Lecidea lithophila*, *Ophioparma ventosum* and *Tephromela aglaea* are prominent. Dry underhangs bring in *Lecanora subcarnea* and *Tylothallia biformigera* (coastal). The high metal content of whinstone locally favours the development of the *Acarosporion sinopicae* alliance. Characteristic species, many of which were first found by J. F. Skinner, include *Acarospora atrata*, *A. smaragdula*, *Lecanora epanora*, *L. subaurea*, *Stereocaulon evolutum*, *S. pileatum* and *Tremolechia atrata*. A further direction of variation is the occurrence of seepage tracks which, on the dip slope of the Whin Sill near Crag Lough in Northumberland, support *Amygdalaria pelobotryon*, *Aspicilia laevata*, *Ephebe lanata*, *Peltigera britannica*, *Placopsis gelida*, *Polychidium muscicola* and *Umbilicaria deusta*. Wet outcrops in Upper Teesdale have recently added *Dermatocarpon intestiniforme* to the Whin Sill flora. A detailed study along its 110 km outcrop has still to be undertaken.

There are a number of isolated basalt hills in the Welsh Border country, for example Roundton Hill in Montgomeryshire and Stanner rocks in Radnorshire, some of which are famous for their higher plants. Their lichen flora fits in well with that of similar volcanic outcrops, with *Lasallia pustulata*, *Lecidea fuliginosa*, *Leprocaulon microscopicum* and *Rhizocarpon viridiatrum* well developed. The Derbyshire toadstone outcrops vary slightly in composition, but attempts by Brian Fox to match the flora of individual exposures to their mineralogy have met with little success. Instead, the lichen flora appears to be controlled by the depth of the brownish-red oxidation crust which can readily be seen when a rock is broken open; it takes about 50 years to develop a depth of 1 mm. Around 0.8 mm of weathering is needed before the first lichens (*Trapelia* species) start to grow; then a lichen succession occurs as weathering gradually penetrates to 10–15 mm deep. There appears to be a simple relationship; the deeper the oxidation crust the richer the lichen community.

Little has been written on the basalts of the Tertiary province, though those on Mull (James, 1978), Arran (Giavarini, 1986) and Skye have been surveyed. They appear to support a northern and western element that includes *Buellia ocellata*, *Caloplaca arnoldii*, *C. obliterans*, *Coccotrema citrinescens*, *Fuscidea lygaea*, *Micarea submoestula*, *Pertusaria chiodectonoides*, *P. flavicans*, *P. pseudocorallina*, *Pilophorus strumaticus* and *Polyblastia theleodes*. These occur together with the anticipated mildly calcicolous element comprising species such as *Caloplaca flavovirescens*, *Porpidia speirea*, *Rhizocarpon concentricum* and *R. umbilicatum*.

Fig. 6.9 Basaltic outcrops along the Whin Sill have been largely ignored by lichenologists; Peel Crag, Northumberland.

Damp basalt on the Trotternish Ridge, Skye, provided considerable excitement for members on a British Lichen Society meeting (1990) by yielding a genus new to Britain. This was the olive-brown, rosette-forming *Vestergrenopsis elaeina* which Alan Fryday has subsequently found in a similar habitat in Glencoe. When mossy, or influenced by sea spray, basalts lose much of their distinctiveness. In Ireland many of the best exposures are coastal. There, inland, deeply weathered boulders, bluffs and crags of basalt have yielded *Aspicilia epiglypta, Bacidia viridifarinosa, Caloplaca arenaria, C. arnoldii, C. crenularia, Micarea coppinsii,* a sorediate morph of *Ochrolechia parella, Pertusaria amarascens* and *Polysporina lapponica* (Coppins & Coppins, *pers. comm.*).

The flora of the Tertiary basalts is sufficiently different from those further south to suggest a regional distinctiveness. The definitive monograph on the lichen flora of British basalt has yet to be written.

Granite

Granite is widespread in upland Britain, its resistance to weathering giving rise to distinctive scenery on Bodmin Moor, Dartmoor, in the Lake District, Cheviot Hills, Rhinns of Kells and parts of the Scottish Highlands and Islands (Fig. 6.10). It is a coarse-grained acid rock in which huge pink and white feldspars are set in a glassy quartz-biotite matrix. No lichen appears to be confined to granite in Britain, though a number show a strong preference for it. These include the calcifuge species *Clauzadeana macula, Lecidea auriculata, L. sarcogynoides, Parmelia mougeotii, Pertusaria monogona, P. excludens, Sarcogyne clavus, S. privigna* and, when growing on rock, *Teloschistes flavicans*. Collecting off rounded granite boulders is a quick way to blunt a chisel; anyone wishing to become acquainted with the lichen flora of granite is advised to carry a 1.5 kg lump hammer and at least three freshly sharpened chisels.

Fig. 6.10 The granite habitat; Hound Tor, Dartmoor (V.J. Giavarini).

A detailed ecological study of the Dartmoor granite has been made by Giavarini (1990) who, carrying the above equipment, surveyed 25 large tors including the famous Haytor and Hound Tor. He recorded 273 lichens and concluded that the size of the tor and its surrounding 'clitter' of boulders, was less important in determining its lichen richness than habitat diversity. The number of taxa per granite tor ranged from 59 (Heltor) to 96 (Black Tor). A tor can be broken down into the following habitats; vertical faces, flat ledges, sloping ledges, sheltered rock faces, nutrient-poor rock, nutrient-enriched rock and 'clitter'. This classification, of course, also stands for other types of outcrop.

Vertical faces are dominated almost exclusively by brown to grey coloured mosaics of *Fuscidea cyathoides*, *F. kockiana* and *F. lygaea*, joined on slightly sloping surfaces by *Lecanora gangaleoides*, *Ophioparma ventosa* and *Pertusaria corallina*. Where rainwater tracks run through vegetation and then down a rock face, this dramatically alters the lichen flora which is replaced by a monoculture of *Lasallia pustulata*, a phenomenon well displayed on Great Staple Tor. Wide, damp vertical joints at the foot of walls may be lined with *Pilophorus strumaticus*, *Porina curnowii* and *P. lectissima*. Dry sheltered recesses are not present on all tors, but where they occur yielded *Lepraria*, *Micarea*, *Opegrapha* and *Porina* species. Deep crevices may contain the luminous moss *Schistostegia pennata*, which floods dark corners with a beautiful green light.

Flat ledges on which a little peat has accumulated are rich in *Cladonia* species together with *Omphalina ericetorum*, *Trapeliopsis gelatinosa*, *T. glaucolepidea* and *T. pseudogranulosa*. Sloping ledges are often seepage-flushed with the lichens arranged in distinct zones, *Ephebe lanata* in the wettest part grading successively into *Cladonia subcervicornis*, *Lecidea fuscoatra*, *Massalongia carnosa*, *Pertusaria excludens* and *Stereocaulon evolutum*.

Summit outcrops that are little grazed or trampled such as Yes Tor, Fur Tor and Great Mis Tor are now the best sites for siliceous lichens on Dartmoor. On the highest tors a montane element composed of *Cetraria hepatizon*, *Cornicularia normoerica*, *Pseudephebe pubescens* and *Umbilicaria torrefacta* occurs. There is a slight mystery concerning *U. cylindrica*. The last definite Dartmoor sighting was on Crockern Tor near Two Bridges in 1829. There is, however, a note of it being 'very rare' in an unpublished report based on a British Lichen Society field meeting in 1969. Its rediscovery would be most exciting. Of all the siliceous rock communities on Dartmoor, those typical of nutrient-enriched substrates are now the most widespread and are thought to be increasing as a result of higher stocking densities of sheep and cattle. These colourful communities are dominated by *Candelariella coralliza*, *Lecidea sulphurea*, *Parmelia conspersa*, *P. loxodes* and *Rinodina atrocinerea* with a wide diversity of other species.

The ecology of the lichen flora of the Dartmoor granite can largely be explained in terms of an upland effect relating to altitude and rainfall, tor morphology, a maritime influence at the margins, and increased stocking rates causing eutrophication. This last effect is alarming as it is threatening the strongly acidic communities which are now well developed in only a few remote sites. Calcifuge species such as *Bryoria bicolor* and *B. fuscescens* have declined sharply in abundance, size and condition of the individual thalli since the mid-1970s. A similar trend has been reported from North Wales where analogous granite habitats occur.

Old red and new red sandstones

Red ploughed fields indicate areas underlain by the desert sands that accumulated during the Devonian (old red) and Permo-Triassic (new red) periods. They are cemented by ferruginous or a mixture of ferruginous and calcareous matter, so that certain beds are quite hard and give rise to undulating topography. Between them these formations are widespread in England, Scotland and Wales, with particularly fine developments in Caithness, the Penrith District and the Brecon Beacons. Many inland exposures are overhanging river cliffs in sheltered ravines. For this reason they have remained unpopular and little is known regarding their lichen flora. Two features, however, stand out. First, the cliffs support a range of species more normally thought of as epiphytic such as *Bacidea arceutina*, *Coniocybe furfuracea*, *Opegrapha ochrocheila*, *Pertusaria amara*, *P. hymenea*, *P. pertusa*, *Ramalina canariensis*, *R. pollinaria* and *Usnea filipendula*. The porous, water-holding properties of the rock are probably responsible for these examples of substrate switch. The other feature is the presence of a range of calcicolous species that occur in response to the locally calcareous cement that binds the red, iron-encrusted sand grains. In the Black Mountains of Herefordshire these include *Agonimia tristicula*, *Aspicilia contorta*, *Bacidia sabuletorum*, *Caloplaca flavovirescens*, *Collema cristatum*, *Gyalecta jenensis*, *G. ulmi*, *Pertusaria albescens* and *Psora lurida*.

Schist, gneiss, slate and tuff below 610 m

It is in the upland areas of Wales, the Lake District and Scotland, where hard, metamorphosed Pre-Cambrian and Lower Palaeozoic rocks outcrop, that acid saxicolous lichens reach their greatest abundance and diversity. These areas provide a severe testing ground for lichenologists who must learn to distinguish mosaics of species that in the field look similar and even under the microscope may need a well-prepared and stained ascus apex to separate out the genera. Only half a dozen people in Britain are completely at home in this habitat. For a long time the British expert on the habitat was the well-known amateur Scottish botanist Ursula Duncan who lived at Arbroath (Fig. 6.11). The genera mostly frequently met with are *Fuscidea*, *Lecidea*, *Micarea*,

Fig. 6.11 Ursula Duncan, for many years the British expert on upland saxicolous lichens; Glen Isla, 1969 (C.D. Bird).

Miriquidica, Porpidia, Rhizocarpon, Rimularia and *Stereocaulon*. Often material needs to be collected and taken back for close examination or thin layer chromatography, which means that progress in the field is slow and gets slower as the weight of specimens increases. It is more rewarding to work a single boulder for an hour or a corrie for a day than to cover a lot of ground superficially; a party will always record more comprehensively than an individual. Macrolichens provide light relief.

Exposed rock, present as isolated boulders, block scree or ridge-line outcrop, experiences full illumination and sharp wetting/drying cycles, which favour crustose lichens. Easily identifiable species favouring this type of habitat are *Lecidea lithophila, Lepraria caesioalba, Lecanora polytropa, Ophioparma ventosa, Pertusaria corallina* and *Tremolechia atrata*. Only three yellow *Rhizocarpon* species are likely to be met with, the other five being montane. The map lichen, *R. geographicum*, so-called because it forms mosaics that look like continents with the countries marked out by black lines, is distinguished by its sharply angled areoles, while *R. lecanorinum* has crescent-shaped areoles that form a false margin round the black fruits. *R. viridiatrum* occurs as small scattered patches among other lichens, especially *Aspicilia caecsocinerea*, on which it is parasitic when young. If there are rock basins where temporary pools of water collect, their margins should be examined for the reddish-brown areoles of *R. geminatum*. The genus *Fuscidea* is well represented on dry acid rock. With experience and a set of chemicals, members of this genus can usually be named in the field. *F. cyathoides* is the only species that is Pd+ red, *F. praeruptorum* has C+ red soralia, *F. gothobergensis* has white areoles on a black hypothallus and 'snail-grazed' soredia, while other species possess distinctive fruits with or without margins (Fig. 6.12). Several *Rimularia* species were virtually overlooked in Britain until recently, but are now known to be widespread in the uplands. *R. intercedens* and *R. illita* are both C+ red, but the former has a distinctive nipple-like protuberance in the centre of each areole. *R. mullensis* has a brownish, verrucose thallus that is K+ yellow to red, the same reaction as the grey thallus of *Lecidea lactea*. So, by paying attention to small morphological differences, chemical reaction and ecological preference, many of these difficult

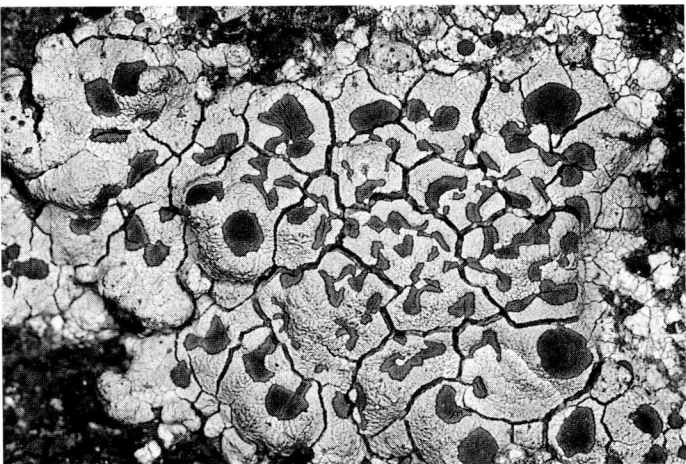

Fig. 6.12 The genus *Fuscidea* is well represented on dry acid rock; *F. kochiana* is distinctive for its irregularly rounded, sunken fruit bodies (J.M. Gray).

mosaic-forming crusts can, with practice, be sorted out. A battery operated UV lamp is useful for certain separations.

The macrolichens of exposed outcrops and boulders include *Bryoria fuscescens*, *Cetraria chlorophylla*, *Parmelia omphalodes*, *Platismatia glauca*, *Pseudevernia furfuracea* and *Sphaerophorus globosus*, often accompanied by thick crusts of *Ochrolechia tartarea*. This is the community that crofters collecting the dye lichens crottle and cudbear, must have become expert at locating. *Umbilicaria* species, which can also be used for dyeing, have a habit of being dominant on one or two boulders, but absent from many apparently similar ones. This suggests an ability for very local dispersal, but difficulties over longer distances. This would be a suitable subject to investigate by transplant experiments.

Rock faces and corrie walls provide a more sheltered, dirtier and damper habitat for siliceous lichens. Such habitats can be very acid; five samples of soil and moss cushions from the Lake District gave pH readings in distilled water of below 3. In such situations *Porpidia* species reach their greatest abundance. Most of these polymorphic taxa can only be separated by thin layer chromatography. Among the calcifuge, non-sorediate *Porpidia* species, spot tests are only helpful for identifying *P. platycarpoides*, which contains norstictic acid (K+ yellow to red). The lichen flora of cliffs in high rainfall areas can be rather dull, the conditions there favouring *Arthrorhaphis citrinella*, *Baeomyces* spp., *Cladonia subcervicornis* (abundant), *Lecanora intricata*, *Miriquidica leucophaea*, *Pertusaria lactea*, *Porpidia tuberculosa*, *Stereocaulon* and *Trapelia* spp. At the front edge of ledges a mixture of wet peat, bryophytes and algae provide a substrate for lichens such as *Bryophagus gloeocapsa*, *Micarea leprosula*, *M. lignaria*, *Omphalina* spp. and *Trapeliopsis* spp. To lengthen the list of species obtained from such a cliff it is necessary to investigate the microhabitats.

Vertical and overhanging faces should be searched for *Diploschistes scruposus*, *Haematomma ochroleucum* and *Lithographa tesserata* together with trentepohlioid species in the genera *Lecanactis*, *Opegrapha* and *Porina*. A black, felt-like covering on the rock will be one of two very curious lichens, *Cystocoleus ebeneus* and *Racodium rupestre*. Neither produce fruit bodies, conidia or lichen substances; Smith (1921a) separated them on the basis of *Racodium* having *Cladophora* as the algal partner, but they are now both considered to contain *Trentepohlia* and are distinguished on hyphal morphology (Fig. 6.13). The deeper crevices, into

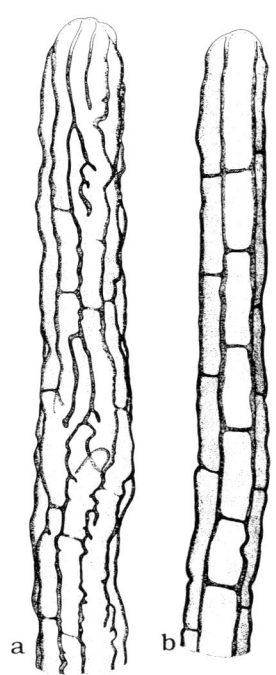

Fig. 6.13 Highly magnified filaments of (a) *Cystocoleus ebeneus* and (b) *Racodium rupestre* showing the difference in hyphal morphology (A. Orange).

which it is often difficult to see, are usually lined with the thin thalli and obscure fruits of *Micarea bauschiana, M. botryoides, M. lutulata, M. myriocarpa, M. pseudomarginata, M. sylvicola* or *M. tuberculata*. These need to be collected for determination under a microscope using the copiously illustrated European monograph on the genus by Coppins (1983). Drier crevices may be coloured bright yellow-green by sheets of the finely granular *Psilolechia lucida* with its tiny green fruits. On many occasions I have searched unsuccessfully for its parasite, *Microcalicium arenarium*, a pinhead lichen. The only specimen I have seen was on a shaded pebble protruding from the side of a ditch. In this instance the *P. lucida* population contained *Stichococcus* as the algal partner, rather than the more usual *Trebouxia*-like alga.

A further direction of variation on acid cliffs is areas that are intermittently irrigated by seepage water. Such sites support moss cushions amongst which grow *Amygdalaria pelobotryon, Cladonia cyathomorpha, Claurouxia chalybeioides, Ephebe lanata, Massalongia carnosa, Polychidium muscicola, Toninia thiospora* and *Stereocaulon vesuvianum*, to mention just a few. Damp pebbles on ledges attract a different range of species such as *Lecidea phaeops, Trapelia coarctata, T. placodioides* and *T. obtegens*.

Acid rock below 610 m is capable of producing surprises. This will be illustrated by reference to the genus *Umbilicaria*. In June 1889 J. A. Martindale of Stavely discovered in Langdale what is still the most famous lichen of Lakeland, *U. crustulosa*. This large species is locally dominant on Raven Crag and ascends the hillside above almost to its summit. Although now known from several other sites in the Lake District it occurs nowhere else in the country. The cliff, Carreg Wastad in the Llanberis Pass, Snowdonia, has a large population of the rare *U. hirsuta* that was discovered by a climber queuing to ascend Crackstone Rib. At neither site is climbing threatening the lichen populations. *U. spodochroa*, only known from acid boulders by Loch Eribol, Sutherland, was reputedly discovered by Peter James when attending to a call of nature; it has not been re-found.

Fortuitous discoveries of this kind are among the most pleasing as they involve little effort. One year Chris Hitch took the Kindrogan field course to survey the church at Old Struan. While eating lunch among some low outcrops he was called over to look at an unfamiliar glossy brown lichen which proved to be *Protoparmelia ariseda*, new to Britain. The trouble with finding a lichen new to Britain is that unless it is very distinctive it can remain unnamed for many months. This was not the case with the discovery of *Miriquidica garovaglii*. Brian Coppins and Ray Woods were attending a meeting of the Botanical Society of Scotland. It was a hot day and the rest of the party were pressing on with unreasonable haste into the hills north of Loch Maree. Having been hopelessly 'dropped' they turned their attention to some cliffs near the loch where they discovered an unfamiliar lichen with convex brown areoles that they quickly realised was new to Britain; it was the find of the meeting.

Serpentine and gabbro

To a higher plant botanist mention of serpentine conjures up pictures of the alpine catchfly (*Viscaria alpina*), the forked spleenwort (*Asplenium septentrionale*) and other choice rarities; are the lichens equally special? As a group the ultrabasic serpentine rocks have a rather varied chemical composition with magnesium, iron, chromium and nickel present in widely fluctuating amounts.

Fig. 6.14 Outcrops of serpentine in a valley leading down to the sea, The Lizard, Cornwall. Several lichens occur nowhere else in Britain.

Consequently, each serpentine area has a characteristic lichen flora resulting from the interplay of toxic ions, nutrient availability and climatic conditions. The largest area of serpentine in Great Britain is the Lizard Peninsula in Cornwall. It was here, during the British Lichen Society's spring field meeting of 1986, that five lichens new to Britain were discovered in a single morning (Gilbert & James, 1987). Most of the Lizard serpentine is covered in heathland, but small red and green cliffs outcrop on the side of valleys at Kynance, Gew-graze, Mullion and Dowas (Fig. 6.14). The mossy boulder screes at their foot locally carries large populations of *Heterodermia leucomelos, H. obscurata, H. isidiophora* (the only site in Europe) and *Nephroma tangeriense*. Sloping outcrops above the scree support the distinctive grey-green, richly fertile *Solenospora liparina*, which is associated with *Collema latzelii*, both of these Mediterranean species confined in Britain to the Lizard serpentine. Other rarities include *Arthonia atlantica, Parmelia tinctina* and, on a rocky bank, *Cladonia mediterranea*; these species together with others too numerous to mention, combine to make these valleys leading down to the coast among the lichenologically most exciting places in Britain.

Serpentine elsewhere in the UK has an impoverished and specialised lichen flora. On Unst in the Shetlands, the Keen of Hamer National Nature Reserve is underlain by nutrient-deficient harzburgite and dunite. These are sparsely colonised by a few ubiquitous species which show a mix of calcicole and calcifuge tendencies. A small increase in nutrient status towards the summit of outcrops, associated with bird perches, results in a marked increase in species richness. Species recorded here, but not on the Lizard, include *Fuscidea mollis* and *Pertusaria chiodectonoides*. Serpentine on the Island of Rhum (Gilbert, 1983) is even poorer; neither the dark rough rock nor the yellow sandy soil between boulders are a favourable substratum. The only species found with any regularity on the extensive area of 'barrens' are isolated thalli of *Baeomyces placophyllus, B. rufus, B. roseus, Micarea lignaria, M. leprosula, Peltigera* species, *Stereocaulon vesuvianum* and *Trapelia* species, which rarely afford more than a

5% cover. A markedly basiphilous element is missing, as are nationally rare lichens. So serpentine is not always rewarding.

Chemically, gabbros are the equivalent of basalts, but due to their slow cooling history they are coarsely crystalline. Gabbro occupies 25 km^2 on the east side of the Lizard where it outcrops inland as an ancient sea cliff while Crousa Down is littered with rounded blocks of the rock, known locally as 'crusairs'. Their origin is obscure, but ecologically they have much in common with sarsen stones. Their slightly basic nature favours *Ochrolechia parella* and *Pertusaria* species while discouraging *Lecidea* and *Porpidia;* over 70 saxicolous species have been recorded on them (Gilbert & James, 1987). The old sea cliff on the Lizard has not produced any notable lichens, neither have Tertiary gabbro outcrops on the islands of Arran, Mull, Rhum and Skye. The coarse texture of the rock, together with its high content of ferro-magnesium minerals, is believed to be responsible for the impoverished lichen flora.

Siliceous rock supports more lichen species than any other habitat in the British Isles, which is why this chapter has been so loaded with Latin names. Perhaps for this reason, and the fact that many belong to difficult genera, they are not particularly well known and so provide ample scope for research and original investigation. Much could be done by amateurs who, by specialising in a defined habitat, such as basalt, old and new red sandstones, quartzite, slate, Cheviot andesite, Welsh gabbro or Bunter pebbles, would soon become expert in that area. I have found that an advantage of working one habitat for a number of years is that eventually one becomes so familiar with the species that identification ceases to be a problem and it is possible to pay full attention to the ecology of the communities.

7

Heaths and Moors

More classifications have been proposed for moors, heaths and mires than for any other type of British vegetation. I have chosen to be guided by Rackham (1986), who, following a thousand years of English usage, separates the heaths of the lowlands from the moors of the Highland Zone. He maintains that there is a real distinction between the two, though whether this is evident in their lichen flora remains to be seen. Heaths occur in dry parts of the country, are subject to periodic droughts, and are usually found on light mineral soils. They are clearly the product of human activities and need management; if neglected they turn into woodland. Moors occur in high rainfall areas and are underlain by wet, acid peat; they are not so evidently an artefact and are more stable.

The vegetation of heathland is dominated by sub-shrubs, in particular heather (*Calluna vulgaris*) and sometimes gorse (*Ulex*) or other ericoids such as bell heather (*Erica cinerea*) in a range of combinations. Moorland, our most extensive type of natural vegetation, covers vast areas in the north of England and over half of Scotland. It, too, may be covered by heather, but the vegetation is more varied; possible dominants include bilberry, crowberry (*Empetrum nigrum*), cross-leaved heath (*Erica tetralix*), bog myrtle (*Myrica gale*), cottongrass (*Eriophorum*), grasses (*Agrostis, Molinia, Nardus*) and *Sphagnum* moss. Typically moors are underlain by waterlogged peat that varies from a few centimetres to a few metres thick; the latter is known as blanket bog. Montane heaths, maritime heaths and grey dunes are dealt with in Chapters 11 and 13.

Heather-dominated heaths and moors are found only in strongly oceanic regions of west Europe (Fig. 7.1), where they have been in marked decline over the last two centuries due to the demands of agriculture, forestry, urban development and sheer neglect which results in woodland invasion. England now has less than a tenth of the lowland heath of two centuries ago (Rackham, 1986), while losses on the Continent have been even greater. Moorland, in contrast, is less threatened and still dominates many northern landscapes. Although originally the result of progressive podsolization following forest destruction and, therefore, man-made, both communities have had several thousand years in which to develop a rich and distinctive lichen flora. Rackham (1986) provides evidence that by the end of the Iron Age all our larger moors existed, and that many southern heathlands are even older, dating from the Bronze Age. Their decline set in at the end of the seventeenth century as a result of innovations in farming and was chronicled by, among others, Gilbert White, John Clare, George Borrow and Thomas Hardy. More survives in Britain than elsewhere, giving us a special responsibility to study and conserve them.

The lichen flora

The lichen flora of heaths and moors have so much in common that they will be described together before their differences are considered. A walk across a stretch of heathland, with eyes trained on the ground, is likely to yield between ten and 25 terricolous lichens. Most of these will belong to the large genus *Cladonia*, which contains some of the most conspicuous lichens in our flora; several have English names and have attracted the attention of poets. There is a group with red fruits, known popularly as Bengal matches or British soldiers. Wordsworth wrote of them:

> 'Ah me what lovely tints are these,
> Of olive, green, and scarlet bright!
> In spikes and branches and in stars,
> Green, red and partly white'.

A kind with trumpet or goblet-shaped fruit bodies are called cup-lichens or pixie-cups and inspired Felicia Hemans to compose the following verse:

> 'Oh, green is the turf where my brothers play,
> Through the long bright hours of the summer day;
> They find the red-cup moss where they climb,
> And they chase the bee o'er the scented thyme.'

Fig. 7.1 Hatching indicates the main area in which heather-dominated heaths occur, while the continuous thick line shows the world distribution of one of their characteristic lichens – *Cladonia portentosa*.

A third group is intricately branched like coral and known as reindeer lichen; Crabbe described it as:

'The wiry moss that whitens all the hill'.

Beginners are strongly attracted to this diverse genus, as at first the species appear to be clearly separated by shape and colour and, growing on the ground, they are easy to collect. Greater experience brings with it an appreciation of the plasticity of many of the species, until, as still more are encountered, the differences become less clear cut and a grievous uncertainty prevails.

Fig. 7.2 (above) The common pixie-cup lichens *Cladonia chlorophaea* (left), *C. pyxidata*, (centre) and *C. fimbriata*, (right) compared. *(below)* Three specimens of *C. ramulosa* to show the range of variation (M.J. Lindley).

Eventually, with the help of a lens, a set of chemicals, a UV lamp and several years experience, order is restored and the majority can be named in the field.

They are best studied by lying prone in the heather when each lichen patch is seen to be composed of several species growing in distinct clumps, like a small garden. Most eye-catching are the scarlet-fruited *Cladonia coccifera*, *C. floerkeana*, *C. macilenta* and *C. polydactyla* (Plates 8a and 8b). All have a wide ecological amplitude, but the first two are less shade-tolerant and the last prefers sites with a high humidity. Brown-fruited cup-lichens may also be present, these can be separated from one another by the shape of the podetia and the size of the granules within the cups. *C. fimbriata* has elegant, wine-glass shaped podetia that are farinose-sorediate (grains the size of flour) throughout. *C. pyxidata* has funnel-shaped cups that gradually widen upwards from the base, their upper parts carrying dark corticate granules (the size of granulated sugar); while the cups of *C. chlorophaea* gradually taper to a basal stalk and are granular-sorediate (with granules the size of castor sugar). They are illustrated in Figure 7.2. Five chemical races of *C. chlorophaea* are known in Britain; although sometimes recognised at the species level, their status remains unclear as there is evidence of interbreeding.

The reindeer lichens, also known as *Cladinas* as they belong to the subgenus *Cladina*, have erect, richly branched podetia that lack small scales and cups. Most can be named in the field, but the chemical paraphenylenediamine is useful for confirming aberrant material. The very common *C. portentosa* has unorientated terminal branches and is Pd- (Plate 8c), the more local *C. arbuscula* has branch apices strongly orientated in one direction (Fig. 7.3) and is Pd+ red, while *C. ciliata* has orientated podetia that are purple-brown tinged, Pd+ red, and is unique in having apices that are dichotomously branched. The European endemic *C. portentosa* is particularly characteristic of lowland heaths (Fig. 7.1).

Other *Cladonia* species that might be expected under heather are *C. coniocraea* with simple pointed podetia, and *C. glauca* and *C. subulata*, which have branched antler-like ones. The mavericks, *C. crispa*, *C. furcata*, *C. ramulosa* and *C. squamosa* can be confusingly polymorphic. Some of the variation shown by *C. ramulosa*, the structure of which is based on proliferating asymmetrical cups is illustrated in Figure 7.2. More distinct are *C. gracilis*, a well-named, tall, elegant species, usually with slender cups, and *C. uncialis*, which has erect, swollen, dichotomously branched podetia with brown tips. A further dozen *Cladonia* species may be present, but they have a more restricted distribution.

The plasticity shown by many *Cladonia* species may have an adaptive significance when they are growing in a dynamic habitat such as heathland. Unlike animals, plants cannot move when conditions alter, but if they are able to respond to factors like increasing shade or humidity by changing their habit to more elongated, squamulose, branched or sorediate, or becoming less pigmented, this may allow them to survive through a heather cycle. This could also help to explain why many *Cladonia* species are virtually ubiquitous on moorland.

Though heathland lichen communities are dominated by the genus *Cladonia*, other terricolous species are usually present. Most conspicuous are richly branched, glossy brown tufts of *Coelocaulon aculeatum* and *C. muricatum*. Do not worry if you have a problem separating the pair; lichenologists cannot agree on their status, having at various times classified them as species, sub-

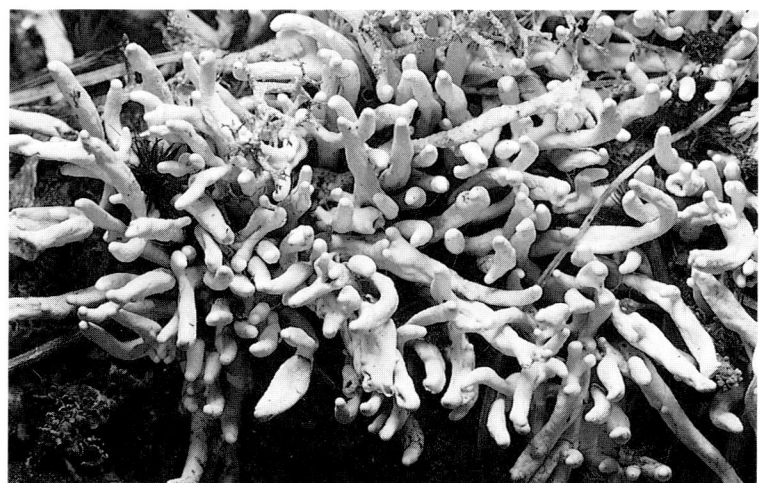

Fig. 7.3 Siphula ceratites (above) a rare lichen of wet moorland in northwest Scotland (I.C. Munro). Stout, orientated tips of the reindeer lichen *Cladonia arbuscula* (below) (P.A. Ardron).

species or mere varieties. A curious little lichen, relatively recently removed from *Cladonia* into a genus of its own is *Pycnothelia papillaria*, which has a whitish, swollen, hollow erect system, like babies' molar teeth. Crustose lichens to look out for include *Baeomyces roseus*, *B. rufus*, *Icmadophila ericetorum*, three species of *Placynthiella*, *Trapeliopsis granulosa*, *T. pseudogranulosa* and small lichenised agarics belonging to the genus *Omphalina*.

The succession following heather burning

Many heather moors are managed by rotational burning to promote the dominance of heather. On grouse moors best practice is to burn in long narrow strips, each covering between 0.5 and 1 ha, so the area becomes a mosaic of even-aged stands that can be picked out by their height difference and aged by counting growth rings in the heather stems. Unmanaged heathland is also regularly swept by accidental fires started by picnickers or vandals.

Ecologically the regenerating heather can be classified into four phases. The 'pioneer phase', which lasts up to seven years, sees a discontinuous heather cover, up to 24 cm high, gradually re-established. The 'building phase' lasts until the plants are around 14 years old, 50 cm high, and have achieved maximum canopy density. It is followed by the 'mature phase', when the heather plants reach their maximum height of 60–70 cm; at this time there is a tendency for the oldest branches to spread apart, permitting more light to reach the ground. After around 23 years the death of the central branches heralds the 'degenerate phase', large gaps appear in the canopy and the stems lie flat on the ground like the spokes of a wheel. Few heather plants survive past 33 years old. On unmanaged moors a similar cyclical process takes place around the life of individual heather bushes (Watt, 1955), but is less easy to study.

Coppins & Shimwell (1971) have studied the succession of lichens following burning on lowland heath at Skipwith Common, East Yorkshire (Fig. 7.4). After a fire the bare peat of the 'pioneer phase' is rapidly colonised by two brown-black crustose species, *Placynthiella icmalea* and *P. uliginosa*, which can be difficult to detect until they have been pointed out. They are accompanied by pale grey-green patches of *Trapeliopsis granulosa*, which may be sorediate and mostly sterile, or non-sorediate and richly fertile, with fruit bodies varying from pale pink, to reddish-brown, to white, to dark grey-green. This variability was responsible for its earlier name of *Lecidea quadricolor*. *Cladonia* squamules soon appear, but can rarely be identified to species level until they produce podetia; among the earliest colonisers are *C. coccifera* and *C. floerkeana*. The 'building phase' sees the entry or expansion of a wide range of *Cladonia* species such as *C. chlorophaea*, *C. crispata*, *C. fimbriata*, *C. macilenta* and *C. ramulosa*, which reached a peak after 10–13 years at Skipwith. During the 'mature phase' the cover of lichens declines as they become shaded out or partly suppressed by bryophytes, although if the heather is parted, it is surprising how many can be found. The onset of the 'degenerate phase' is marked by an abrupt rise in the quantity and variety of lichens as light levels increase, and is notable for the appearance of reindeer lichens. Finally, following a fire, or death of the heather due to extreme old age, the succession starts again with another 'pioneer phase'.

The fire succession has also been studied on heather-dominated heaths in Surrey by Fritsch & Salisbury (1915) and on a heather-bearberry (*Calluna-Arctostaphylos*) heath in Aberdeenshire (Ward, 1970). In both instances its course was similar to that outlined for Yorkshire. Ward reported the first appearance of *Cladonia* squamules after two to three years, of thalli with podetia after three to four years, and maximum development in the tenth year when the lichen cover reached 66% with a mean of 8 lichen species per 4 m^2. *C. gracilis*, *C. portentosa* and *Hypogymnia physodes* were characteristic of the later stages. Studies such as these make good student projects as they promote an understanding of dispersal, establishment, growth rates and competition.

Persistent *Cladonia* patches in closed heathland

As recounted above, *Cladonia* species increase to a maximum about ten years after a fire, then start to decline as the dwarf-shrub canopy closes. However, it

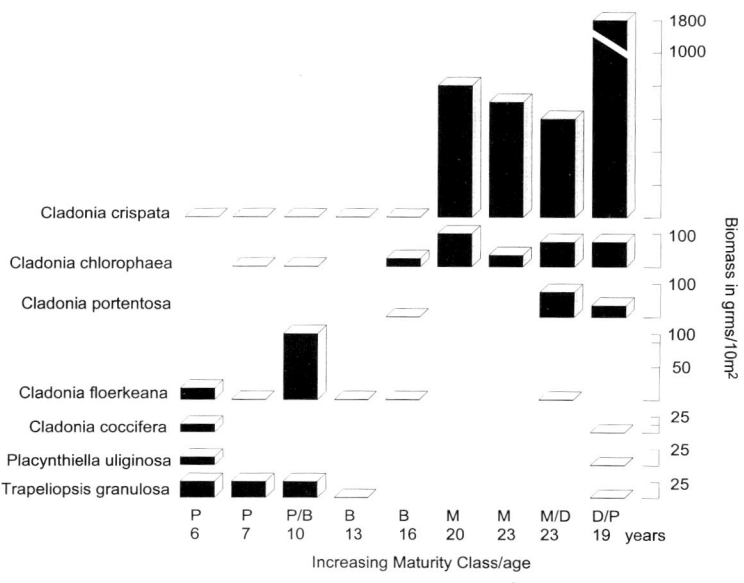

Fig. 7.4 The biomass of lichens in burnt heather moorland of different ages. P = pioneer; B = building; M = mature; and D = degenerate phase. From Coppins & Shimwell, 1971.

is noticeable that patches of mixed *Cladonia* species, about 50 cm across, can persist in an otherwise closed stand. They disappear only after being overtopped and shaded out by the surrounding bushes. Close observation reveals that the lichen patches are devoid of seedlings of either ericaceous or herbaceous species. This phenomenon has been investigated by Hobbs (1985) working on the Muir of Dinnet, Aberdeenshire. He discovered that aqueous extracts from detached podetia, and leachates from transplanted *Cladonia* turfs, had an inhibitory action on the seeds of wild oat (*Avena fatua*), heather and wavy hair-grass (*Deschampsia flexuosa*) in the laboratory. Field experiments confirmed that mixed *Cladonia* patches can inhibit the germination and establishment of vascular plants and thus maintain themselves in otherwise closed heathland stands.

Other workers have reported similar allelopathic effects. For example, Pyatt (1967) showed that *Peltigera* had an inhibitory influence on the germination and subsequent growth of sand-dune grasses, while Ramaut & Corvisier (1975) found that extracts from *Cladonia portentosa, C. gracilis* and *Coelocaulon muricatum* inhibited the germination of the seeds of Scots pine indirectly by acting on the mycorrhizae. Lawrey (1977) extended the work by showing that certain lichen compounds, such as squamatic acid, commonly found in *Cladonia* species, inhibited spore germination in three common mosses under laboratory conditions. If these observations can be confirmed in the field they have important implications for lichen ecology as they contradict the traditional view of lichens as pioneers paving the way for other species. The investigations would be suitable for amateurs using home-made equipment.

The difference between heaths and moors

The broad distinction between heaths and moors made at the start of this chapter is recognised throughout Europe. Gimingham (1972) refers to them as 'Dry heaths' and 'Humid and wet heaths'. The majority of descriptions, including those in the National Vegetation Classification (Rodwell, 1991), do not cover lichens in sufficient detail to clarify whether the two types are lichenologically distinct. As most of the 30 or so lichens already mentioned have wide ecological tolerances, growing equally well on mineral and peaty soils, and show no particular sensitivity to waterlogging, they are of little use as differential species. There are, however, a small number of terricolous lichens that show a degree of restriction to northern moorlands. These include *Cladonia cornuta*, *C. sulphurina*, *Icmadophila ericetorum*, *Omphalina hudsoniana*, *O. luteovitellina*, *Pycnothelia papillaria*, *Trapeliopsis gelatinosa* and *T. glaucolepidea*. Several of these are striking plants. The tall, lacerated podetia of *C. sulphurina*, which look as if they have been freshly dipped in powdered sulphur, can often be seen on peat faces along forestry roads, while the large, flat, coral-pink fruits of *Icmadophila ericetorum* are sufficient to stop even non-botanists in their tracks. *Omphalina* is the only lichen genus in Britain to have an agaric as the fruit body. The thallus of *O. luteovitellina* is a crust of dark green granular globules (the *Botrydina*-type), which give rise to small chrome-yellow agarics. *O. hudsoniana* has a thallus of green circular squamules with raised margins (the *Coriscium*-type), which also produce small yellow agarics. The regular association of the *Coriscium*-type thallus with the toadstool was first noticed relatively recently (Gams, 1962); before then they were regarded as separate entities. Decaying *Coriscium* squamules should be inspected with a lens for the tiny yellow, pimple-like fruit bodies of the lichen *Thelocarpon epibolum*.

A moorland lichen worth making a pilgrimage to see is *Siphula ceratites*. The thallus is composed of compact, ivory-white tufts of stout cylindrical branches up to 7 cm tall (Fig. 7.3). It forms swards in wet peaty hollows in northwest Scotland. A convenient place to see it is by the path up Suilven from Lochinver where it is locally abundant on blanket bog at an altitude of 120 m.

Other terricolous lichens that show a preference for moorland over heaths are *Cetraria islandica*, *Cladonia cervicornis* subsp. *verticillata*, *C. strepsilis* and *Micarea lignaria*. All have isolated stations in the lowlands, some of them regarded as relic, for example *Cetraria islandica* on the outskirts of King's Lynn and at Linwood Warren, Lincolnshire; other examples are *Cladonia zopfii* and *M. lignaria* in Surrey. *C. strepsilis* is frequent on wet heaths in the New Forest where sorediate specimens with stunted podetia and conspicuous yellow-green soralia have recently been collected. At first they were erroneously assigned to *C. brevis*, a species new to Britain, but they are now considered to be an undescribed subspecies of *C. strepsilis*. To sum up, most heathland lichens in Britain are widespread; none appear to be preferential for lowland heaths, but a small group is characteristic of moorlands.

The effect of grazing

Farmers down the ages have valued heather as nutritious winter grazing for cattle and sheep. Domestic stock, together with rabbits and deer, find heather most palatable when it is in the pioneer and building phases. It can be maintained in this state by a combination of heavy grazing and sporadic burning, which provide the semi-open conditions favoured by many lichens. So grazing, as it prevents the heather from becoming too tall and leggy, is beneficial to the maintenance of a rich terricolous lichen flora. Only the *Cladinas*, which are typical of the later stages of the heather cycle, are disadvantaged, and they usually find a niche on banks, by tracks or in association with acid grassland. It is a feature of heathland that the richest lichen communities are frequently associated with slopes, quarries, stream corridors, rocky outcrops, rabbit warrens and old peat cuttings, rather than standard flat areas.

The trampling associated with grazing breaks up lichen mats, reducing their cover and biomass, but not necessarily their diversity. This is strikingly demonstrated on Little Dun Fell in the northern Pennines where a series of grazing exclosures was erected in 1954. They enclose either heather-cottongrass (*Calluna-Eriophorum*) blanket bog or a drier bilberry-wavy hair-grass (*Vaccinium myrtillus-Deschampsia flexuosa*) moorland. Inside the exclosures *Cetraria islandica*, *Cladonia arbuscula* and *C. portentosa* have a high cover and are forming cushions up to 30 cm across and 8 cm deep (Fig. 7.5), while, outside, the same species are represented by only broken fragments of thallus. By 1994 the terricolous lichen cover inside the fences had increased to 20% compared to 5% outside and the difference in biomass was a hundred-fold. Such luxuriance is rarely seen on our overgrazed hills.

Roderick Corner observed a colony of *Cladonia rangiferina* on Great Calva in the Lake District for 12 years. Under increasingly intensive grazing it disappeared from the open hillside and is now confined to a rocky area which

Fig. 7.5 When protected from grazing, Iceland moss (*Cetraria islandica*) forms deep carpets; Cross Fell exclosure (A.M. Fryday).

affords it some protection from sheep. The effects of differential grazing can frequently be observed along fence lines enclosing forestry or, on lowland heaths, in the vicinity of rabbit warrens. The paradox is that moderate sheep and cattle grazing on heathlands encourages lichen diversity, but at the same time the biomass of the larger species suffers.

One of the great lichen experiences of Britain is to obtain permission to visit the Suffolk Wildlife Trust Reserve of Wangford Warren in Breckland to see the vast untrampled carpets of *Cladonia* species (Plate 8d). They lie 10 cm deep, creating a multicoloured blanket composed of white (*Cladina* spp.), yellow (*C. uncialis*) and dark brown (*C. furcata*, *C. gracilis*, *Coelocaulon* spp.). The carpet resists invasion by bryophytes and higher plants so the only associates are sparse sand sedge (*Carex arenaria*), a few annuals and the moss *Polytrichum piliferum*. Small banks and scrapes support other lichens, bringing the total terricolous lichen flora to around 25. This remarkable site is one of the few active dune systems left in Breckland and a remnant of the travelling sands described by John Evelyn in 1667 (Evelyn, 1818).

An ecological role for thick *Cladina* carpets in boreal forests, including perhaps Culbin Forest in Aberdeenshire, has recently been discovered (Tikkanen & Niemela, 1995). Temperature measurements in the upper layers of the soil under lichen-rich vegetation, and under heavily grazed lichen-poor plots, showed that the lichen layer acts as an insulating mulch. This is important as early autumn frosts, occurring before there has been any snowfall, are sufficiently severe to kill the roots of pine trees that are not protected by a lichen cover. An additional benefit of the lichen cover is to increase microbial activity, and therefore decomposition rates, on the forest floor, a result of more stable temperature and moisture conditions under the lichen mat.

Epiphytes on dwarf shrubs

In the west of Britain, particularly at windswept coastal sites, the dwarf-shrub canopy supports conspicuous tufts of epiphytes, including *Ramalina farinacea*, *R. pollinaria*, *Teloschistes flavicans*, *Usnea flammea*, *U. hirta*, *U. subscabrosa* and a range of *Parmelia* species. These are accompanied by a number of unusual crustose lichens that should be sought in the humid depths. This little known community, tolerant of deep shade, is described under maritime heath in Chapter 13.

On inland heather moors most lichenologists record just the conspicuous epiphytes such as *Hypogymnia physodes*, *H. tubulosa* and *Platismatia glauca*. However, as Brian Coppins was the first to point out, a range of inconspicuous additional species are frequently present. For example, *Micarea denigrata* and *M. nitschkeana* are widespread on heather twigs and litter in eastern areas, while *M. lignaria*, *M. peliocarpa*, *M. prasina* and *Scoliciosporum chlorococcum* are also regularly present on the living stems of heather, and sometimes also on heaths (*Erica*), gorse and bilberry. Heather in the degenerate phase of its life cycle may become colonised by *Fuscidea lightfootii*, *Lecanora carpinea*, *L. symmicta* and in polluted areas by a thick crust of *L. conizaeoides*. A list of 63 lichens epiphytic on heather in Ireland is provided by McCarthy *et al* (1985). A paper detailing lichens epiphytic on bilberry in Poland (Miadlikowska, 1997) has alerted lichenologists to a further neglected habitat. Seven species were present including *Dimerella pineti*, *Fellhanera myrtillicola* and *F. subtilis*. Bilberry in Britain has not been subjected to such close scrutiny, but is known to support

F. subtilis, and in 1966 Ursula Duncan found a stand in Glen Shee carrying a large colony of *Cetraria pinastri*.

Throughout Britain the smooth woody stems of bog myrtle carry the inconspicuous black perithecia of *Mycoglaena myricae*. It is so regularly present that those unfamiliar with the species might start their search in the higher plant herbarium.

Pebbles on the heath

A further habitat for lichens on heathland is acid pebbles, which may be of flint, granite, sandstone, pieces of brick or some other material. As in chalk grassland, the better colonised pebbles are the larger ones, and those embedded in the ground hold richer communities than ones lying loose. Good sport can be had walking over heathland with a knife or chisel levering pebbles out of the ground for inspection. It is rare for an individual pebble to support more than two or three lichens, but gradually an impressive list can be built up.

In the New Forest heathland *Micarea erratica, Rhizocarpon obscuratum, Trapelia coarctata, T. obtegens* and *Verrucaria dolosa* are among the commoner species on flint pebbles. From Godlingstone heath in Dorset, Vince Giavarini has recorded a pebble flora of a dozen species that include *Catillaria atomarioides, Micarea erratica, M. lignaria* and *M. lithinella*, while pieces of iron pan that had weathered out of the soil produced *Rhizocarpon oederi*. Several of these were first county records. Hardy's 'Egdon Heath' at Winfrith yielded *Micarea peliocarpa* on flint. The Pennine moors have a somewhat different flora on their coarse gritstone pebbles, the early colonisers being *Placynthiella icmalea, Trapelia involuta, T. obtegens* and *Trapeliopsis granulosa*.

More work needs to be undertaken to typify the lichen flora of heathland pebbles in different parts of the country and to work out the succession of species following a fire. This habitat is particularly important in the south where highly acidic saxicolous habitat is rare. On northern moorlands the pebble flora is less impressive as most of the species also occur on adjacent outcrops or on scree.

Conservation

Most of the top heathland sites in Britain, and indeed in Europe, for example Bardsey Island, Culbin Sands, Cuthill Links, Lizard Heathlands, Nare Head and Studland Heath are examples of maritime heath which is dealt with in Chapter 13. The finest inland heaths are those of the New Forest, which extend to 14,000 ha. They form part of a series, known as the 'Anglo-Norman' heaths, that stretch from Normandy to Wiltshire, Hampshire and the High Weald of Sussex. As they are greatly depleted in France the British examples, which represent more than 50% of the remaining area, are considered to be of international importance.

With the publication of the National Vegetation Classification (Rodwell, 1991), the heather-dwarf gorse (*Calluna-Ulex minor*) heaths of the New Forest and Sussex can be seen to occupy a central position in a national series of climatic types that extend from the heather-sheep's fescue (*Calluna-Festuca ovina*) heaths of continental eastern England, to the oceanic heather-western gorse (*Calluna-U. gallii*) heaths of southwest Britain. Within this series there is local variability related to soil wetness, soil base-status and management. Further

Plate 1

(a) The lichenologist Ray Woods wearing a jersey knitted from lichen-dyed wool (J. Woods).

(b) Lichen dyes. Cudbear (top) produces red dyes and crottle shades of brown (D.J. Hill).

(c) A modern lichen dyer with dye-pot, lichens and treated wool, Strontian.

(e) Canons Ashby Church, Northamptonshire, showing how lichens enhance architectural detail (D.J. Hill).

(d) Central panel of a well-dressing tableau made from petals and several types of lichen, Derbyshire.

Plate 2

(a) The lichen alga *Trebouxia* growing in culture (The Natural History Museum, London).

(b) Jelly lichens contain the cyanobacterial partner *Nostoc* (J.M. Gray).

(c) Sticta canariensis 'green algal morph' and *S. canariensis* 'cyanobacterial morph' contain the same fungus (J.M. Gray).

(d) The fruticose growth form exhibited by a beard lichen (*Usnea articulata*) pendant to 60 cm (J.M. Gray).

Plate 3

(a) Nest of a long-tailed tit (*Aegithales caudatus*) decorated with lichen to aid concealment by light reflection (B.J. Hatchwell).

(b) Caterpillar of the light crimson underwing moth (*Catocala promissa*) mimicking a lichen-covered surface (J. Porter).

(c) Caterpillar of the dotted carpet moth (*Alcis jubata*) feeding on *Usnea* (P.A. Ardron).

(d) Autumn green carpet moth (*Chloroclysta miata*) at rest on a lichen-covered tree trunk (R.W. Barnes).

Plate 4

(a) Thick sward of the pollution-tolerant lichen *Lecanora conizaeoides* on larch.

(b) Usnea florida, a beard lichen that is highly sensitive to several forms of pollution (J.M. Gray).

(c) Lichens on beech that have been killed by airborne fluorides, Invergordon, Scotland.

Plate 5

(a) Old beech woodland, New Forest, Hampshire (T. Heathcote).

(b) *Lobaria pulmonaria* festooning mossy Atlantic rainforest, Western Scotland (J.M. Gray).

(c) Atlantic oak woodland above Loch Sunart, Ardgour.

(d) A rich Lobarion community covering a bough at Loch Sunart; the lichen with golden soralia is *Pseudocyphellaria crocata* (F. Rose).

Plate 6

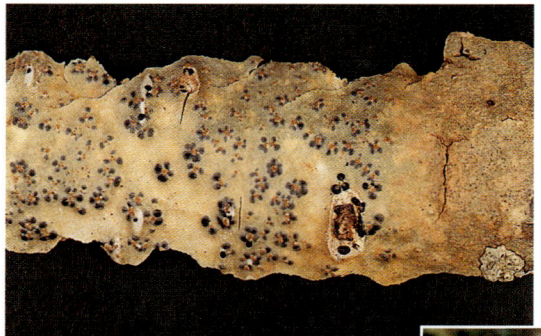

(a) Parmentaria chilensis, a strongly oceanic species known in Britain from one hazel wood at Loch Sunart (A.M. Coppins).

(b) Hypogymnia physodes, a common species of birch woods in the Highlands (J.M. Gray).

(c) A pin-head lichen, *Chaenotheca furfuracea* (J.M. Gray).

(d) A writing lichen, *Graphis scripta*, characteristic of smooth bark (J.M. Gray).

(e) Well-lit, smooth bark on many deciduous trees supports a mosaic of small crustose lichens (J.M. Gray).

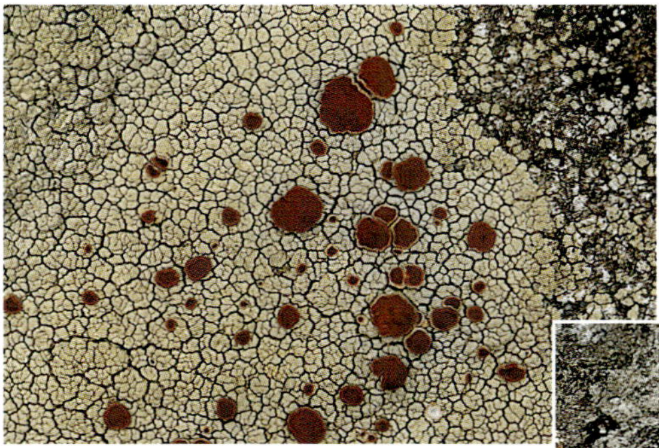

Plate 7

(a) Ophioparma ventosum, a lichen of acid rocks (I.C. Munro).

(b) Lasallia pustulata, a gregarious lichen (I.C. Munro).

(c) Purple-stained lichen on quartzite, Foinavon (D. Miller).

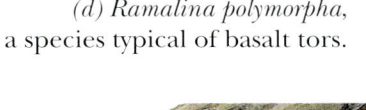

(d) Ramalina polymorpha, a species typical of basalt tors.

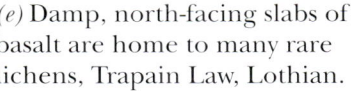

(e) Damp, north-facing slabs of basalt are home to many rare lichens, Trapain Law, Lothian.

Plate 8

(a) Cladonia coccifera, abundant on acid soils (J.M. Gray).

(b) Cladonia floerkeana, the 'Bengal match lichen', abundant in acid habitats (J.M. Gray).

(c) Cladonia portentosa, the commonest of the 'Reindeer lichens' (P.A. Ardron).

(d) Close-up of the lichen carpet at Wangford Warren, Breckland.

Plate 9

(a) Lichen-rich chalk grassland has developed where the surface was scraped off in 1940 to form a shooting butt, Martin Down.

(b) Lichenologists inspecting a path on the chalk downs, Butser Hill.

(c) Lichenologists at work on a limestone pavement, Ingleborough.

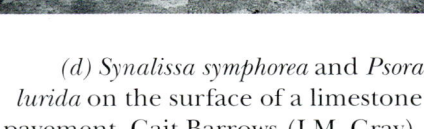

(d) Synalissa symphorea and *Psora lurida* on the surface of a limestone pavement, Gait Barrows (J.M. Gray).

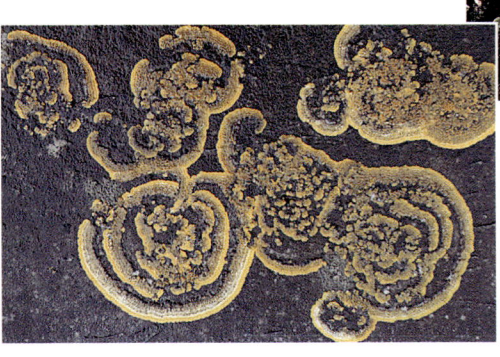

(e) Caloplaca aurantia, a species characteristic of Jurassic limestones (T.W. Chester).

Plate 10

(a) Lecanora polytropa growing on iron railings.

(b) Lecanora rupicola thickly encrusting a sandstone tombstone (T.W. Chester).

(c) Rhizocarpon geographicum on a slate tombstone.

(d) Timber-boarded Sussex barn carrying what is believed to be a unique assemblage of rare lichens, Parham Park.

Plate 11

(a) The international community dominated by *Lecanora dispersa* (white) that is present on concrete (I.C. Munro).

(b) Old open cast workings rich in heavy metals, Parys Mountain, Anglesey (O.W. Purvis).

(c) Baeomyces roseus only fruits regularly at acid mine sites (P.W. James).

(d) Baeomyces rufus is widespread at most acid mine sites (I.C. Munro).

(e) The normally brown *Acarospora smaragdula* becomes green when growing on copper-rich rocks (O.W. Purvis).

Plate 12

(a) Lower part of the Ben Alder buttress coloured yellow with *Fulgensia bracteata*.

(b) The rare alpine *Lecanora epibryon* growing with *Salix reticulata*, Ben Alder (A.M. Fryday).

(c) Margaret's Coffin almost devoid of snow, September 1996.

Plate 13

(a) Solorina crocea has a thallus with an orange underside (I.C. Munro).

(b) The rare alpine *Pertusaria glomerata*, Ben Lawers range (I.C. Munro).

(c) Catolechia wahlenbergii (Goblin lights), a rare lichen centred on the Ben Nevis range (A.M. Fryday).

Plate 14

(a) The River Jelly Lichen Steering Group at work by the River Eden, Cumbria (A.M. Coppins).

(b) Sites where shelving beds of rock flank a river are usually rich in aquatic lichens, South Tyne above Hexham.

(c) A mountain tarn with *Lecanora achariana* on the marginal boulders, Snowdonia.

Plate 15

(a) *Endocarpon adscendens* on mossy lakeside boulders, Windermere, Cumbria.

(b) *Dermatocarpon intestiniforme* dominates a high zone around Bassenthwaite Lake, Cumbria.

(c) Lichen zonation on a sea stack showing the black, orange and grey bands, North Cornish coast.

(d) The tiny, dot-like fruits of *Pyrenocollema halodytes* growing on barnacles (J.M. Gray).

Plate 16

(a) Colourful lichen assemblage of the grey zone (R.W. Barnes).

(b) The rare arctic-maritime species *Lecanora straminea* growing on a bird cliff, Flannan Isles.

(c) Teloschistes flavicans (Golden-hair lichen) grows on a few exposed cliff tops in southwest England and west Wales (P.A. Gainey).

directions of variation are increasing altitude or latitude, which are marked by an expansion in the proportion of grasses, cotton-grasses and bilberry at the expense of gorse. The lichens of heathlands are insufficiently known for the identification of species characteristic of these gradients, if any exist. Most calcifuge terricolous lichens have catholic tastes and are ubiquitous within the heathland area. They are not climatically limited in the UK, so any regional differences that exist are likely to be quantitative rather than qualitative.

Thus, despite their abundance, it is doubtful whether lichens will be of much help in identifying heathland sites of conservation importance unless recorded over a period of many years. Taking them into account would not greatly improve site assessments made without lichen data as was done, for example, in the Nature Conservation Review (Ratcliffe, 1977). Lichens are a fluctuating component, their coming and going being dependent on the phases of the heather cycle; there do not appear to be any that characterise continuity in the way they do for woodland.

Twenty-five terricolous lichens mark a heathland site out as lichen-rich. The New Forest heaths have 33; Ambersham-Heyshott Common in Sussex, which is believed to support the finest *Cladonia* communities in southeast England, has 20 *Cladonia* species; Ashdown Forest in Sussex has 24 terricolous lichens, and Lavington Common in Sussex, Chobham, Thursley and Hankley Commons in Surrey, Wangford Warren in Suffolk, Cawton-Marsham Heaths in Norfolk, Hartland Moor and Arne Heaths in Dorset, Stiperstones in Shropshire, Risby Warren in Lincolnshire and Skipwith Common in Yorkshire, also have well-developed terricolous lichen floras. Some southern heathlands are now poorer in lichens than they were 20 years ago due to the mature heather phase having taken over. For example, in 1975, 24 *Cladonia* species, including *C. rei* and *C. sulphurina*, were recorded from Lavington Common in West Sussex, by 1991 only eight could be found. Moorlands are less threatened than lowland heaths and so merit a lower priority for conservation.

Species with a distribution restricted by specialised habitat requirements are *Cladonia strepsilis, Icmadophila ericetorum, Pycnothelia papillaria* and *Trapeliopsis pseudogranulosa*, all show a preference for wet peat. The unique C+ emerald-green reaction of *C. strepsilis* is diagnostic, though beginners unfamiliar with the colour have been known to mistake the chlorophyll-green of a wetted thallus for the real thing. A few heathlands in the east of Britain support *Stereocaulon condensatum*, which is more typical of sites on the Continent; it was seen at Wangford Warren in the 1970s and is widespread at Culbin Sands in Aberdeenshire. *Cladonia arbuscula*, being sensitive to disturbance, including fire, is characteristic of high quality heathland. It has been over-recorded for *C. tenuis*, which is also Pd+; observing the branching pattern of the tips is the surest means of separating the two. In the north and west, acid, often grassy banks and tracksides where mineral soil comes to the surface may support *Baeomyces roseus, B, rufus, Cladonia cervicornis, C. callosa (C. fragilissima), C. cariosa, C. luteoalba* and *C. subcervicornis*.

At the present time the main threat to our lowland heaths and their lichen flora is neglect. In the past they were regularly managed as part of the farming system, used for small-scale peat and mineral extraction, and provided a home and livelihood for numerous country people. After a long gap this is now being taken over by Wildlife Trusts, County Councils and other conservation bodies who are implementing scrub clearance followed by carefully controlled grazing, mowing and burning schemes to restore them to their former condition. It must not be forgotten that terricolous lichens are very dependent on periodic disturbance. Tracks, small-scale sand and gravel excavation, stone quarries, ball clay pits, peat cuttings, road verges and military activity all favour acid terricolous lichens through bringing mineral soil to the surface, which diversifies the habitat mosaic.

The total lichen resource of inland heaths in Britain below 450 m, excluding the rarest and montane species, is around 50 terricolous and 15 epiphytic taxa, together with an unassessed element on pebbles. A national survey to characterise these three components of the lichen flora on a regional basis is needed.

8

Chalk and Limestone

Britain is well off for limestone; there are few counties in England that do not contain exposures, though in Scotland and Wales they are more scattered. Large areas of southern England are formed from the chalk, a soft limestone that radiates out from Salisbury Plain with long thin limbs spreading eastwards to create the North and South Downs, a narrow belt extending into Dorset and the Isle of Wight, and the longest arm forming the Chilterns then extending north to the Lincolnshire and Yorkshire Wolds (Fig. 8.1). These chalk landscapes are distinctive through their wide expanses of short, heavily grazed turf and the smooth rounded curves of the hills, their sides scarred by the white gashes of rabbit warrens and pale tracks. Ploughed land and beech woods help to create an evocative landscape, one that Huxley thought suggestive of mutton and pleasantness. There are few more attractive habitats in which to botanise; the lichenologist, on hands and knees, will experience particularly close acquaintance with the thyme-scented turf.

For most people, limestone scenery conjures up pictures of the Carboniferous or Mountain limestone of Yorkshire where massive, horizontal beds of hard, white limestone, 350 million years old, outcrop on hillsides forming parallel lines of scars with scree at their base. On terraces between the scars wide expanses of bare, grey limestone pavement add to the rockiness. Soils are thin, so cultivation is rarely practised, the land being used as sheep walk. This type of scenery, minus the pavements, is also found in the Mendips, the Bristol area, around the edge of the South Wales Coalfield, locally in North Wales, in the White Peak of Derbyshire, at the head of Morecambe Bay and in parts of Ireland. The classic area for this habitat, however, and the region in which calcicole lichens reach their maximum development, is the Craven District of Yorkshire; north of the Stainmoor Gap the limestones start to become less pure and subordinate to sandstones and shale, so that in Northumberland there are only half a dozen respectable exposures.

The Magnesian limestone, of Permian age, forms a well-wooded belt of ground about 8 km wide; it more or less follows the line of the Great North Road from Nottingham to the mouth of the River Tyne, a distance of 250 km (Fig. 8.1). It is a yellow-brown, dolomitic limestone, cut here and there by attractive, steep-sided ravines in which shaded exposures occur that hold a specialised lichen flora.

Running parallel to the west-facing chalk ridge that strikes diagonally across England, and separated from it by a series of clay vales, are the scarplands. This broad belt of countryside extends from Portland in Dorset to North Yorkshire and contains oolitic limestones of Jurassic age; these are around 180 million years old. The scarps are many and varied and give rise successively to the Cotswold Hills, the Northamptonshire Uplands, the Lincoln Edge and the Cleveland Hills. The honey-coloured limestone, which weathers to rich tints of yellow and brown, is mostly visible in drystone walls, quarries and buildings. In

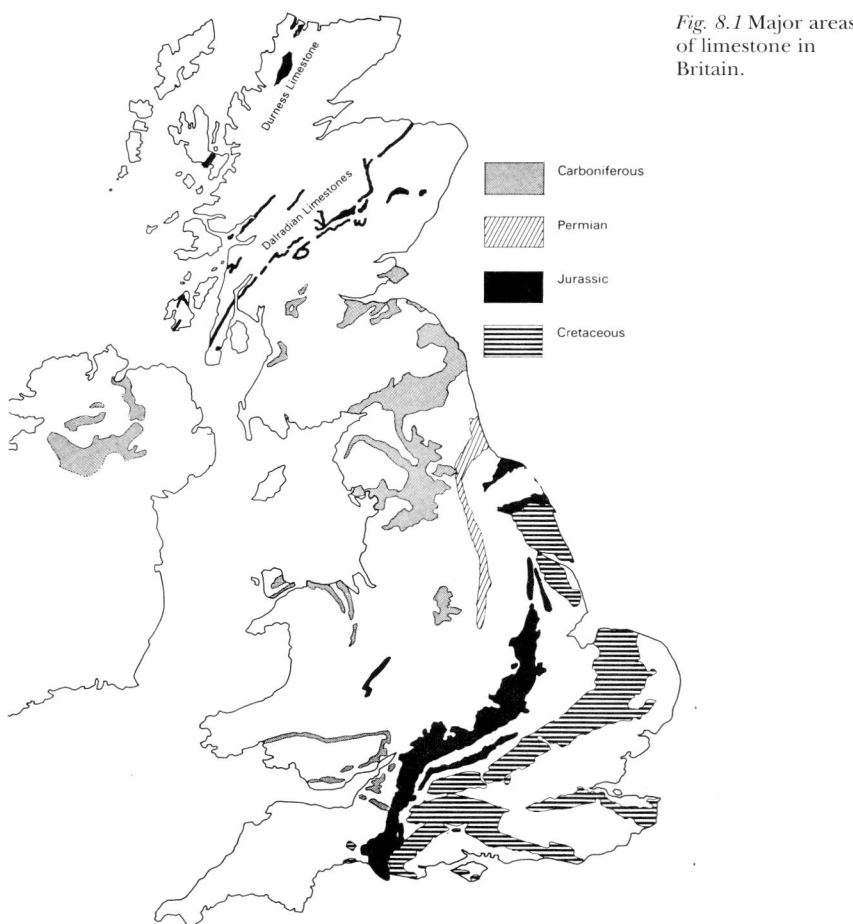

Fig. 8.1 Major areas of limestone in Britain.

Cleveland, limestones are subordinate to sandstones and shales, but a rich calcareous lichen flora is still locally present.

The only other significant limestone in England is the well-known double Silurian limestone scarp of Wenlock Edge and Aymestry Edge in Shropshire; its lichens are hardly known. Scotland has some lichenologically exciting bands of Dalradian limestone while in the far north, the Durness limestone (Cambrian) outcrop, though narrow, it is sufficiently extensive to give rise to wild karst scenery; it has been only partially explored by lichenologists.

The chalk

A survey carried out in 1966 revealed that only 3.3% of the English chalk, amounting to 43,500 ha, carried unsown grassland. Since then there have been many further losses to ploughing, tree planting and scrub invasion. The survival of the remaining fragments depends on the occurrence of escarpments too steep to cultivate and the use of certain areas for military training. Most of the richest sites, such as Tennyson Down on the Isle of Wight, Butser Hill in Hampshire and Box Hill in Surrey, are managed as country parks or, in the case of Lakenheath Warren and Heyshott Down, as nature reserves. The largest and most important area of chalk grassland in England, and probably in the world, is the Porton Ranges, owned by the Ministry of Defence.

Chalk grassland is created and maintained by sheep grazing. Landscape historians believe that this practice was established in southern England by the third century AD, so the chalk grassland habitat has existed for at least 1,700 years. It probably reached its maximum extent around 1600 AD, after which economic factors resulted in much conversion to arable. Due to rotational farming and changes in demand, it is rare for any individual area of chalk grassland to be more than 150 years old. Subsidiary habitats such as flint mines, chalk pits, hill forts, long barrows, tracks and linear earthworks have often been left alone by farmers so are more ancient.

The lichen flora of chalk grassland can be divided into three components which will be treated separately – those occurring on flints, those on chalk pebbles and the terricolous component growing on soil and dead vegetation.

Terricolous habitats

Grassland rich in terricolous lichens is very restricted on the chalk, but the pleasures of lying on the short turf and examining one's first good site are not easily forgotten. It is possible to walk over many kilometres of downland and find nothing more than a few scattered lichen thalli where the tightly grazed continuous turf is broken by sheep tracks or terracettes. For the development of terricolous lichens, a degree of stress or disturbance is required to reduce the vigour of the grassland. Appropriate stress factors are usually associated with the activities of man. For example, on the 300 ha Martin Down National Nature Reserve in Hampshire, only three sites support significant lichens. The richest of these are two areas where the surface of the downland was scraped off and piled up to form shooting butts in 1940 (Plate 9a). The flat, bared areas adjacent to the butts carry a short, open, herb-dominated sward containing abundant lichens, including *Catapyrenium squamulosum*, *Diploschistes muscorum* and *Verrucaria bryoctona*. The third area is the side of a sunken lane or hollow way where the commonest lichens of chalk grassland can be seen. These are *Bacidia muscorum*, *B. sabuletorum*, *Cladonia furcata*, *C. pocillum*. *C. rangiformis*, *Collema auriforme*, *C. tenax*, *Leptogium gelatinosum*, *L. schraderi* and *Polyblastia gelatinosum*. With diligent searching these species can be found at most sites, but are more likely to be beside tracks, associated with earthworks, or around chalk pits than on the open downland.

Paths, especially broad, long-established ones created by people (not sheep) moving up and down spurs, can also produce conditions suitable for terricolous lichens (Plate 9b). It was one such path on Watlington Hill in the Chilterns that first alerted me to the potential richness of this habitat for lichens. In the centre of the path the grassland was worn to a height of only 4 mm and was fairly open. It supported a lichen cover of 15–35% composed of seven species including large patches of *Diploschistes muscorum* and *Sarcosagium campestre* (Fig. 8.2). A similar phenomenon can be seen on Box Hill in Surrey, Ivinghoe Beacon and at other beauty spots. An advantage of trampling by shoes is that unlike the hooves of sheep, which tend to disrupt the sward, shoes compact it and most terricolous lichens favour a compact rather than a loose substratum.

Very rich stands of lichen vegetation are associated with exhaustive arable cropping, with old rabbit warrens, and particularly with sites where there has been a combination of the two. The outstanding lichen-rich grassland now found on the Porton Ranges was subject to intensive agriculture in mid-Victorian times, then from $c.1890$ until 1916 it was used as a commercial rabbit warren, which lowered the nutrient content of the soil through removal of carcasses. In places *Cladonia rangiformis* covers the ground like hoarfrost, attaining an 80% cover and leaving little space for other lichens. Elsewhere, particularly in stony, flint-strewn grassland, uncommon lichens such as *Bacidia herbarum*, *Catapyrenium pilosellum*, *Leptogium intermedium*, *L. teretiusculum*, *L. tenuissimum*, *Psora decipiens* and *Toninia sedifolia* can be found; all are strict and demanding calcicoles. There appears to have been a decline in the abundance and extent of terricolous lichens at Porton since the communities were first described in the early 1970s (Wells *et al.*, 1976); nitrogen deposition associated with acid rain may be involved.

An insight into the way in which over-steepening can disrupt a grazed sward allowing the entry of lichens was gained by examining the rounded brow of a road cutting on the South Downs. Well before the steep slope was reached the almost level grass/herb sward became less dense and a 2 m wide band of terricolous lichens containing up to five species per 20 x 20 cm quadrat developed. Where the slope steepened to over 15° the substrate became too unstable to support terricolous lichens. The opening up of the tight sward appeared to be due to drought. Over-steepening caused by natural factors is rare, but

Fig. 8.2 Transect across a path through chalk grassland showing how terricolous lichens are restricted to the area where the vegetation is shorter and the soil more compacted.

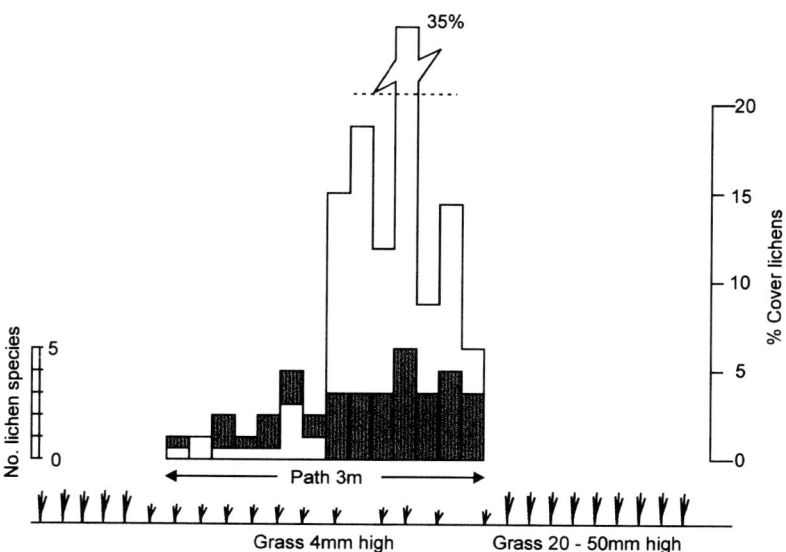

such localities are prolific for lichens. Small landslip features along the cliff top at Tennyson Down, Isle of Wight have produced a series of low, whale-backed mounds which support the richest maritime chalk grassland lichen communities in the country. The colourful assemblage is dominated by the scrambled egg-coloured *Fulgensia fulgens*, here at its only site on the chalk; it shares dominance with the cream-coloured *Squamarina cartiliginea*. Other rarities include *Megaspora verrucosa* at only its second site on the chalk, and *Toninia lobulata* (Fig. 8.3). A similar but less rich community is present on landslip areas at Box Hill, high above the River Mole.

Breckland is famous for its terricolous, calcicole lichens but, as with the rest of the chalk, good localities are scarce and have mostly experienced recent disturbance. In this district they are associated with places where surplus military equipment was buried at the end of the Second World War (Lakenheath East, Little Heath, Barnham), ground disturbed by road construction (Thetford Heath, Barnham Military Training area) or where defence work was carried out in 1940 (Weeting Heath). The Breckland rarities are *Buellia asterella*, *Fulgensia fulgens*, *Squamarina cartilaginea* and *S. lentigera*. They are members of the scarce *Fulgensietum fulgentis* community recorded only from Tennyson Down, Lakenheath Warren and (formerly) two places east of Newhaven where it was last seen by J. Hemmings in 1868. The community has also experienced a serious decline in Breckland. This set in between 1984 and 1988 when thalli started to become discoloured, became detached from the soil and were found lying upside down. *Fulgensia* became extinct in Breckland in 1995, while localities with *Buellia* declined from four to one. The latter, now exceedingly scarce at its sole remaining site (Lakenheath Warren), must be regarded as one of

Fig. 8.3 Over-steepening associated with natural features, such as minor landslips, produces conditions that are particularly favourable for terricolous lichens. The *Fulgensia* site, Tennyson Down, Isle of Wight.

Britain's most threatened lichens. This rapid, synchronised decline, has tentatively been linked to episodes of enhanced nitrogen deposition; a species recovery programme has been initiated.

In a survey of 83 chalk grassland sites (Gilbert, 1993; 1995) the number of terricolous lichens found per site varied from zero in the thick swards dominated by *Carex humilis* that occur in parts of west Wiltshire and Dorset, to over twenty. The highest numbers of species were all from localities in the south of England such as Deep Dene, Sussex (18), Lakenheath Warren (19), Tennyson Down (21), Ballard Down (21), Watlington Hill, Oxfordshire (23), Butser Hill, Hampshire (29), and, way ahead of the rest, Porton Down (42). The richest sites include small areas of chalk heath with *Cladonia portentosa* and *Coelocaulon aculeatum*, but where superficial deposits, such as the clay-with-flints, are well developed, red-fruited *Cladonia* species appear and a transition to true heathland occurs.

Chalk pebbles

Chalk pebbles are brought to the surface by rabbits burrowing and scraping. On steep slopes sheep grazing can also be sufficiently heavy to disrupt the turf and unearth them. Around large warrens extensive fans of loose nodules are present, but these are mostly too mobile to support lichens. It is in the vicinity of abandoned warrens that the chalk pebble lichen community reaches its maximum development. Excavated pebbles collect in declivities, in folds in the hillside and at the foot of slopes, forming a fine scree. These are the richest sites, but old scrapings, no more than 10 cm across and containing only a dozen or so nodules, also repay examination. Eventually all these sites become overgrown and return to grassland. Pebbles associated with molehills, sheep tracks and terracettes are less interesting.

The pebble community is dominated by a rather small number of lichens, among which *Verrucaria hochstetteri*, *V. muralis*, *V. murina* and *V. nigrescens* are the main species, although a persistent search will often reveal *Caloplaca citrina*, *Lecidea lichenicola*, *Polyblastia dermatodes*, *Protoblastenia rupestris*, *Sarcogyne regularis*, *Staurothele hymenogonia* and *Thelidium incavatum*. Frequently small rounded pebbles are completely covered by a single species, their fruits covering every surface (Fig. 8.4). This suggests considerable mobility within a limited area.

There is a group of species with a curiously patchy distribution; it includes *Caloplaca lactea*, *Clauzadea immersa*, *C. metzleri*, *Petractis clausa* and *Verrucaria baldensis*. Initially it was believed to represent a relic community marking areas of grassland that had remained open throughout the post-myxomatosis-induced flush of coarse vegetation (1955–65) that still covers much of the chalk. Later it was realised that the group is strongly linked to sites where the chalk has been hardened by faulting (Durdle Door, Ballard Down) and by the weight of overburden (Yorkshire Wolds), or else it is picking out the narrow line of outcrop of the compact, nodular, Chalk Rock marker horizon that separates the Middle and Upper Chalk. This relationship is evident on the chalk north of Watlington Hill in the Chilterns where the group is found in a narrow zone halfway up the escarpment. This corresponds exactly to the position of the Chalk Rock on the geology map. The association is even clearer on Compton Down, Isle of Wight, where the Chalk Rock can be identified forming a buttress in a road cutting; when its line is followed across adjacent downland, the distinctive community is again encountered. Other species that belong in this 'hard chalk' assemblage are *Acarospora heppii, Caloplaca aurantia, Catillaria aphana, Hymenelia prevostii, Polyblastia albida* and *Rinodina bischoffii*.

Chalk pebbles in somewhat shaded situations such as the entrance to rabbit burrows, on steep north-facing slopes or under bushes support an array of lichens in which the normally dominant *Verrucaria* species are replaced by *Thelidium zwackhii* with lesser amounts of *Pyrenocollema monense, P. saxicola, Thelidium incavatum* and *T. minutulum*. On Heyshott Down, West Sussex, a remarkable community occurs on shaded chalk pebbles near the foot of a very steep north-facing slope in a long disused quarry; in addition to the expected species *Gyalidea lecideopsis, Leptogium cretaceum* and *Placynthium tantaleum* are locally frequent. This is the only site on the chalk where the first two have been recorded.

A long 'tail' of unexpected species has been found on chalk pebbles, which suggests it is something of a habitat for opportunist lichens. These include mostly single records of *Caloplaca luteoalba, Sarcopyrenia gibba, Staurothele caesia, S. rugulosa, S. rupicola* and *Thelocarpon pallida*. A total of 65 species has been recorded from this microhabitat; most are small, with immersed fruit bodies, and need a microscope to confirm their identification. The number found per site varies from four to around 20, with particularly high numbers from the maritime chalk in Dorset, for example Ballard Down (23) and Durdle Door (27). The large, well-studied area of chalk grassland at Porton Down yielded only 14 species as the ranges are underlain by the soft Upper Chalk.

Little is known regarding lichen succession on chalk pebbles. It appears that the first lichen to colonise remains dominant. The rate of colonisation of freshly exposed pebbles was investigated by examining the depressions created when trees blew down in the gale of October 1987. Chalk fragments associated with three craters on Box Hill were examined in December 1990 and found to be well colonised by seven different lichens with thallus diameters up to 16

Fig. 8.4 Lichen-covered chalk pebbles.

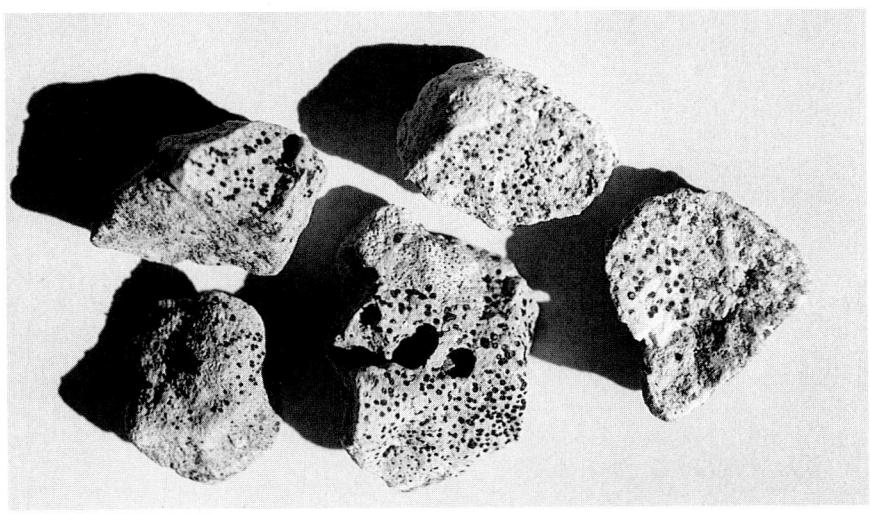

mm. This pattern of very rapid initial colonisation was confirmed at other sites in southeast England and on Freshwater Down on the Isle of Wight, where the cliff top grassland is littered with angular chalk fragments blown off the cliff face by the gale.

Flints

Geologists divide the chalk into three zones. The Lower Chalk is thin and marly; the Middle Chalk or Chalk Without Flints is purer and harder and also rather thin; the Upper Chalk is 400 m thick and contains abundant seams and tabular masses of flint. These are composed of hard, microcrystalline quartz with a porous 'rind or 'cortex'; their surface becomes acid after weathering. Flints are distributed very unevenly over the downs, being generally absent at localities on the Lower and Middle Chalk. Even on the Upper Chalk where they are invariably present in the bedrock, few may be visible at the surface. However, at certain sites such as Watlington Hill in the Chilterns, Ballard Down in Dorset and on the Porton Ranges in Wiltshire, they are locally so abundant that they form an open stony grassland; such areas are often the result of ploughing within the last hundred years. Occasionally, as at Thixendale in the Yorkshire Wolds, huge irregular flints are present at the foot of slopes. These were probably dragged into these positions and represent the clearance of flatter areas for agriculture. As a result of variation in the availability of the substratum, the number of species recorded from flint ranges from none at certain chalk grassland National Nature Reserves in Wiltshire to 41 at Porton Down. Sites with 25 species are exceptional. Large stable flints firmly embedded in the turf tend to be the richest, while coastal localities, on the whole, have a higher diversity than inland sites.

Flints, the commonest pebble in southern England, provide an acidic habitat in the generally alkaline chalk environment. When first exposed, the 'rind' retains traces of calcium carbonate so has an alkaline surface reaction. This can be demonstrated by applying cold dilute acid and watching the fizz of CO_2 production. With time, leaching produces flints with a mesotrophic surface (similar to basalt or calcareous sandstone), then eventually they become acidic. This pH range is why flints, with 75 species recorded during the survey of 83 sites, are richer in lichens than either chalk pebbles or terricolous substrata.

A dozen species represent the basic flora, the dominants being weak calcicoles, for example *Protoblastenia rupestris*, *Verrucaria dolosa*, *V. muralis* and *V. nigrescens*. Calcifuge species tend to be concentrated on the higher, more exposed and strongly leached parts of the larger flints and include *Micarea erratica*, *Porpidia crustulata*, *P. tuberculosa* and *Rhizocarpon obscuratum*. Irregularities often hold *Lecanora dispersa*, whereas damper and more sheltered areas near the soil surface carry *Lecania erysibe* and *Lecidella scabra*. Alkaline surfaces support *Aspicilia calcarea*, *A. contorta*, *Caloplaca flavescens* and *Verrucaria glaucina*, which do not occur on chalk pebbles.

Numerically the largest group is characteristic of acid substrata and includes species surprising to find on generally alkaline downland. For example, *Acarospora smaragdula*, *Buellia aethelea*, *B. ocellata*, *Fuscidea cyathoides*, *Ochrolechia parella*, *Parmelia mougeotii* and *Trapelia* spp. If the grass around the base of large flints is parted the damp, sheltered zone thus exposed may be covered with the small white pycnidia of *Bacidia arnoldiana*. Where recent ploughing of older grassland has occurred it is not uncommon to find piles of large flints that have been cleared and deposited along field boundaries; these usually appear bare of lichens, but a close examination will sometimes reveal abundant tiny yellow fruit bodies of *Thelocarpon intermediellum* or *T. pallidum*.

Three rare and declining species have been recorded from flints on the downs. The most recent collection of *Caloplaca atroflava* in the UK was made on Butser Hill, Hampshire by Brian Coppins in 1973; prior to that it was known only from isolated nineteenth century records on the Sussex Downs, Thetford Warren and on a few flint pebble beaches in the south. It resembles *Caloplaca holocarpa*, but has a black thallus. *Aspicilia tuberculosa* was last collected *c.* 1830; all specimens are on flint, the majority coming from the type locality, which is given by Borrer as the Sussex Downs. It is a poorly known species, so there is a possibility that it still survives. Re-finding this British endemic is a challenge to lichenologists. I would personally start by surveying Compton Down on the Isle of Wight, which is particularly rich in flints covered with *Aspicilia* spp. The third rare lichen of flint is *Rinodina aspersa*, known to Borrer from pebbles on the east Sussex Downs. It is apparently now confined to stabilised shingle beaches (Laundon, 1986).

Nodules of iron pyrites up to 15 cm across known as thunderbolts are an occasional feature of sites on the Lower Chalk. Their surface weathers to provide a highly acid microhabitat on which small thalli of species such as *Micarea erratica* and *Scoliciosporum umbrinum* can be found.

General ecology

The majority of downland lichens are widespread, so appear to be outside regional climatic control on the English chalk. Exceptions to this are *Buellia asterella*, *Fulgensia fulgens* and *Squamarina lentigera*, which are restricted to the warmest, driest and sunniest sites. Their main area of distribution outside Britain is the Mediterranean. A further example is the presence of *Gyalidea lecideopsis* and *Leptogium cretaceum* at Heyshott Down, where they occur alongside bryophytes of northern affinity. The particularly high rainfall of the western South Downs and the steep north-facing nature of the site are thought to facilitate their presence.

Edaphic factors can be important. For example, the hardness of the chalk is fundamental in determining the type of flora that develops on chalk pebbles. Also, the development of a rich terricolous lichen flora is favoured by the presence of a small wind borne fraction in the soil; the resulting red-brown silty loam being a more favourable substratum than the strongly granular, organic-rich upper layers of a typical chalk rendzina. Man's activities, particularly soil-stripping, are required to produce the best terricolous communities. Such activities probably reached their peak during Neolithic and Bronze Age times when huge earthworks were constructed and a pattern of shifting cultivation was practised. Today, Porton Down (Fig. 8.5) is the best analogue we have of that type of landscape, but even there the chalk does not yield up its secrets easily.

Following discussions with continental lichenologists it is considered that Porton Down, Tennyson Down, Ballard Down-Ulwell Gap and Butser Hill are sites of international importance and unequalled on the Continent. The chalk outcrops in Belgium, Denmark, France, Germany, Holland and southern Sweden are all rather small and overgrown. Reduced grazing and high levels

Fig. 8.5 Lichen-rich chalk grassland formerly under intensive agriculture followed by warrening; Porton Down (M. Reed).

of nitrates in the rain are the chief threats to our downland lichens together with lack of disturbance. It takes some readjustment to realise that a community of slow-growing lichens requires intermittent disturbance for its survival. Historical factors largely determine the pattern of distribution at any site.

Carboniferous limestone

Carboniferous limestone covers more of the British Isles than any other calcareous formation, so is well known to most people. The grey, bare-looking limestone scars that are such a feature of the Derbyshire and Yorkshire Dales are anything but bare. If the surface of the limestone is examined closely it will be seen to be covered with a mosaic of pale-coloured lichens. Many have a thallus which is immersed in the rock. This can be demonstrated by scratching with a fingernail until the green algal layer is exposed a little distance below the surface. Alternatively, if a piece of lichen-covered limestone is dissolved in dilute acid, one is left with a felt-like mat of hyphae that extends up to 16 mm into the rock. These immersed lichens provide few characters for field identification as only dark fruit bodies, usually $c.1$ mm across, are visible at the surface. For a firm identification a small sample needs to be collected, using hammer and chisel, so that at home the spores can be examined under a high-power microscope, whereupon most are rather easily keyed out. It may be the abundance of small species that has led to a neglect of these communities, for, despite the clear attractions of working in easily accessible, often spectacular surroundings, Carboniferous limestone must be regarded as a neglected habitat as far as lichens are concerned. For this reason it is not possible to provide as complete an overview of the habitat as it was for the chalk.

A typical area of limestone karst

It is believed that regional variation in the lichen flora of the Carboniferous limestone is not great. Most outcrops, whether on the Gower Peninsula, in the Avon Gorge or in the north of England, support a broadly similar flora, so the following description of an area of karst at the foot of Ingleborough in northwest Yorkshire (Fig. 8.6; Plate 9c), will serve as an introduction to the majority of sites.

The most widespread habitat around the skirt of Ingleborough is limestone pavement. The smooth surface of the clints present severe problems to lichens, as they are very freely draining, severely nutrient-deficient and snail-grazed. Rock analysis reveals that the limestone is 99.8% calcium carbonate, 0.1% silica and 0.1% water. This is the reason why the community is species poor and composed of specialist calcicoles. The scarcity of nitrogen is particularly limiting and encourages the growth of a thin, grey to black film of cyanophilic algae, which can fix their own nitrogen from the atmosphere, the dominants being species of *Gloeocapsa* and *Phormidium*. The most consistently abundant lichen is *Verrucaria baldensis*, which has very small, closely-set perithecia with fissures radiating from the central pore which are just visible through a x20 hand lens. It is supported by additional pyrenocarpous species with immersed thalli such as *Polyblastia albida*, *P. deminuta*, *Staurothele rupifraga*, *Thelidium decipiens*, *T. incavatum*, *Verrucaria dufourii* and *V. hochstetteri*. A number of other typical lichens have rounded apothecia and can be identified in the field; these include *Clauzadea immersa*, *Farnoldia jurana*, *Hymenelia prevostii*, *Lecanora dispersa*, *Protoblastenia calva* and *P. incrustans*. There are also occasional patches of larger, foliose, cyanophilic jelly lichens, for example *Collema auriforme*, *C. cristatum* and *Leptogium gelatinosum*. Shallow depressions where water collects are often completely lined with *Petractis clausa*, which has pale, closely-set, crater-like fruit bodies. A difficulty when attempting to obtain a complete species list is that the clint surfaces may carry a 50% cover of what is best described as nondescript growth. This is composed of sterile thalli, thalli with empty or incompletely developed fruit bodies, and cyanophilic algae. Diversity is particularly low in the driest stands; along cracks, fissures and other irregularities which are more mesic, a broader range of species can be found.

Fig. 8.6 Classic Carboniferous limestone habitat at Scar Close, Ingleborough.

The grykes, which are narrow, solution-widened joint planes one to two metres deep, provide a cool, humid and deeply shaded habitat. Their lichen flora is difficult to sample and, consequently, has rarely been surveyed. It is virtually impossible to collect off the sides of the grykes, but ten limestone fragments that had become jammed at depth were recovered and provided the following list. Lichens were only present on surfaces that received diffuse light and they had to compete with the orange-coloured free-living green alga *Trentepohlia*. The commonest lichens were species typical of damp, shaded limestone, for example *Acrocordia conoidea, Agonimia tristicula, Catillaria lenticularis, Collema mulipartitum, Gyalecta jenensis, Porina linearis* and *Staurothele rupifraga*. There were also three unexpected lichens, *Lempholemma myriococcum, Leptogium plicatile* and *Staurothele bacilligera*. The latter, present on three of the fragments, was first discovered in the United Kingdom on Ingleborough by W. Watson in 1934 and is still considered to be a very rare species. All three are also found associated with the upper levels of limestone streams, which emphasises the dampness of the deep gryke habitat.

What is often thought of as a typical limestone flora is rather restricted around the skirts of Ingleborough, being limited to isolated boulders, wall tops, cliff edges and places where sheep rub up against the rock. The characteristic species are robust and colourful calcicoles such as *Aspicilia calcarea, A. contorta, Caloplaca aurantia, C. flavescens, C. variabilis, Clauzadea immersa, Lecanora dispersa, Protoblastenia rupestris, Rhizocarpon umbilicatum* and *Verrucaria nigrescens*. Most of the clint species are also present in small amounts together with the vigorous bryophytes *Ctenidium molluscum, Homalothecium lutescens, H. sericeum* and *Schistidium apocarpum*. This community requires high exposure to both wind and light, and there is usually some evidence of bird droppings or other form of eutrophication such as a dusty farm track. Where the enrichment is concentrated, *Phaeophyscia orbicularis, Physcia caesia, P. adscendens, P. tenella* and *Xanthoria* spp. are likely to be present.

A suitable limestone habitat for beginners to examine is the vertical face of the scars, as many of the ledges and joint planes are covered with soil or filled with detritus that supports a mixture of mosses and larger lichens. One of the most attractive is the apple-green *Solorina saccata*, with its dark sunken fruit bodies, the source of its popular name – the socket lichen. Elsewhere squamulose species like *Catapyrenium squamulosum, Psora lurida, Squamarina cartilaginea, Toninia lobulata, T. sedifolia* and a range of Collemataceae occur. These latter species show considerable phenotypic plasticity and are not always easy to name. Moss cushions support *Agonimia tristicula, Bacidia sabuletorum, Cladonia pocillum* and *Peltigera* spp. Where water trickles down a rock face after rain there may be swards of the robust, button-like *Dermatocarpon miniatum* flanked by *Gyalecta jenensis, Placynthium nigrum, Thelidium papulare* and *Verrucaria caerulea*. Large overhanging faces tend to be bare of lichens. The dark sooty streaks on Malham Cove, for instance, which gave Charles Kingsley the idea for his story *The Water Babies*, are formed by chemical deposits, not lichens.

Solution hollows

One of the most fascinating habitats in this karst landscape are the solution cups that develop on the surface of the pavements. These shallow depressions, or troughs, up to 50 cm across and 10 cm deep, hold water for several days

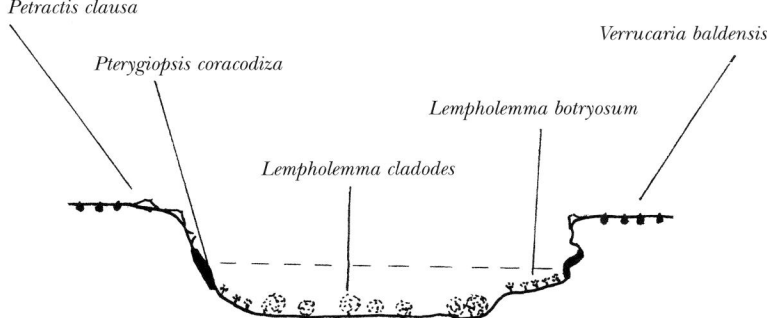

Fig. 8.7 Distribution of lichens around a solution hollow in a limestone pavement.

after rain. Technically they are known as kamenitzas. One of the first lichenologists to study them was McCarthy (1983), who reported that in the Burren, a concentric zonation of lichens developed in response to the moisture gradient. The floor of the solution hollows were thickly covered with small, black, tufts of *Lempholemma botryosum*, together with some *Collema multipartitum* and *Leptogium schraderi*. The sloping sides were occupied by *Petractis clausa*, which gave way to nondescript growth on the clint surface. The solution cups around Ingleborough are somewhat richer (Fig. 8.7). In addition to the species found in the Burren the sides of certain hollows are coloured black by a thick crust of *Pterygiopsis coracodiza*, while particularly deep ones support swards of the rare *Lempholemma cladodes*. When submerged, its balloon-like lobe ends are a conspicuous feature. It often develops into a hitherto unreported 'vagrant' (unattached) form of loose blackberry-like balls 5–8 mm in diameter; these blow about and colonise new hollows.

The luxuriant *Lempholemma botryosum* that lines solution hollows on the high-level Hutton Roof pavements was, for a time, mistaken for the rare Mediterranean species *Synalissa symphorea*. However, swards of this species have recently been discovered on pavements at Gait Barrows near the head of Morecambe Bay (Plate 9d), so it occurs in this habitat after all.

Regional variation

Regional variation in the lichen flora of the Carboniferous limestone has not yet been worked out, but certain broad trends can be recognised. The Mendip Hills and their western extension to Brean Down are distinctive for holding the only Carboniferous limestone populations of *Cladonia convoluta* and *C. symphycarpa*. The former is particularly well developed on block scree at Dolebury Warren, while the latter is a component of calcareous grassland, for example on Fries Hill, where it grows with *Placidiopsis cartilaginea*. Brean Down is famous for its higher plants and also for supporting a population of *Fulgensia fulgens* which is a member of a rich community that includes *Psora decipiens*, *Synalissa symphorea* and *Squamarina crassa*; it is associated with rocky outcrops on steep south-facing slopes of the headland. One of the few recent collections of *Lemmopsis arnoldianum* was on a small stone embedded in turf on the Down. There is also an unconfirmed record of *Squamarina lentigera* which there has never been a serious attempt to re-find.

Further west, Carboniferous limestone is well exposed in coastal cliffs and along the sides of narrow dry valleys that run down to the sea on the Gower peninsula. Though the flora is in general unremarkable there is an element of southern and western species such as *Caloplaca alociza, C. granulosa, Collema fragile, Leptogium britannicum, Moeleropsis nebulosa* and *Pannaria pezizoides*. The normally lichenicolous *Verrucaria aspiciliicola* is here locally abundant both free-living and parasitic on *Aspicilia calcarea*; it superficially resembles *Verrucaria glaucina* but has longer spores.

The lichen flora of the Derbyshire limestone is reasonably well known, but does not appear to be distinctive. All the expected species are present, but often only in small amount. The communities have been exposed to moderate levels of air pollution for many years, which appears to have caused some impoverishment and the extinction of *Caloplaca cerina* (terricolous), *Diploschistes gypsaceus* and *Poeltinula cerebrina*. An unexpected survivor is *Anaptychia ciliaris*, which is present on sheltered mossy limestone walls at three localities. *Placynthium garovaglii* has recently been discovered to be abundant in dry underhangs; *Lecania cuprea, L. rabenhorstii* and *L. sylvestris* are other recent discoveries.

The limestone flora reaches its greatest development in the Craven area of northwest Yorkshire. Here very large expanses of limestone habitat are present, often at around the 400 m level in areas where the annual rainfall exceeds 200 cm. All the expected species of *Caloplaca, Collema, Leptogium, Opegrapha, Polyblastia, Thelidium* and *Verrucaria* are present together with other less common calcicoles. Exposed limestone may support *Acarospora macrospora, Aspicilia subcircinata, Poeltinula cerebrina* and *Staurothele caesia*, while sheltered outcrops carry sheets of *Acarospora glaucocarpa, Caloplaca cirrochroa, Ionaspis epulotica, Opegrapha mougeotii* and *Placynthium subradiatum*. Ledges and moss cushions are home to uncommon species such as *Lecanora epibryon, Peltigera leucophlebia, Polyblastia wheldonii* and *Solorina spongiosa*.

The five bands of Yoredale limestone, which produce the stepped profiles of the shapely three peaks, outcrop at various heights between 400 and 610 m. They have not been fully surveyed, but the highest band on Whernside recently yielded *Placidiopsis cartilaginea*, new to Yorkshire. Further north in the Pennines, on Little Dun Fell, a high Yoredale limestone (700 m) is the sole English locality for the rare upland calcicole *Sagiolechia protuberans* and only the second locality, with Orton Scar, for *Ionaspis heteromorpha*. These, the highest outcrops of Carboniferous limestone in the country, deserve a thorough survey. They tend to be more thinly bedded and friable than the massive standard facies, so crevice species and species growing on moss are well developed. Absentees at these high levels include several *Caloplaca* and *Collema* species, *Solenopsora candicans* and *Squamarina crassa*.

Terricolous lichens are usually a minor component at most Carboniferous limestone sites and, when present, tend to be found on cliffs. An exception is the 'sugar limestone' on Cronkley Fell and Widdeybank Fell in Upper Teesdale, where a closely grazed, open, lichen-rich grassland has developed on the shallow soils. The higher plant flora contains a substantial relic component from assemblages that were widespread in lowland Europe during the Late Glacial Period 10,000 years ago; the so-called 'Teesdale rarities'. They survived here due to the special combination of geology, soil and climate. Is the lichen flora equally special? Thirty-six terricolous lichens are present, excluding six epiphytic species present on the woody stems of *Dryas octopetala* and *Helianthemun canum*, and a further ten which form a transition community to more acidic habitats. The more notable species of this tundra-like habitat are *Catapyrenium waltheri*, *Caloplaca cerina*, *Cetraria islandica*, *Lecidea hypnorum*, *Megaspora verruculosa*, *Ochrolechia frigida*, *Peltigera leucophlebia*, *Solorina saccata* and *S. spongiosa*. So, compared to the higher plants, the lichens of the sugar limestone refugium are less remarkable, though a strong northern element is present. The diversity, however, is outstanding.

The Magnesian or hidden limestone

The Magnesian limestone, of Permian age, occupies a narrow strip to the east of the Pennines stretching from near Nottingham north to the mouth of the River Tyne, a distance of some 250 km. It supports gently rolling, intensively farmed countryside studded with villages, woods and large quarries. The sparse natural outcrops are mostly overhanging crags beside rivers (Fig. 8.8), though at a few localities hard reefs in the limestone give rise to hilltop exposures. Most of the quarries are still working; abandoned ones are frequently filled with refuse. The interest of the Magnesian limestone is partly due to the chemical composition of the rock in which dolomitisation has occurred, a high proportion of the calcium carbonate ($CaCO_3$) being replaced by magnesium carbonate ($MgCO_3$), and the fact that it is a lowland limestone occurring in a dry part of the country.

The habitat has been comprehensively surveyed, 15 of the largest natural outcrops and ten disused quarries producing 106 taxa including one lichen new to science (Gilbert, 1984a). The limestone varies considerably in its physical properties and chemistry throughout the outcrop. The principal subdivision, the Lower Magnesian limestone, is typically represented by 60–70 m of

Fig. 8.8 Most natural outcrops of the Magnesian limestone are steep, overhanging cliffs which tend to be heavily shaded by trees; Cresswell Crags, Derbyshire.

relatively soft, thick-bedded, cream-yellow coloured limestone containing 60–95% dolomite. Common minerals other than dolomite and calcite are quartz, feldspars and clays. At several localities near Mansfield the rock is a sandy dolomite containing up to 25% sand, but this is unusual. The Upper Magnesian limestone subdivision is most important in County Durham where it is represented by 50 m of granular and crystalline heavily dolomitised limestone. Locally reefs are present composed of hard, unbedded pale grey limestone low in $MgCO_3$. The variability is a contrast to the uniformity of the Carboniferous limestone.

Cliffs

Most natural outcrops are steep, often dramatically overhanging cliffs beside rivers or glacial overflow channels. They are quite difficult to find so I have nicknamed it the hidden limestone. Rock faces tend to be heavily shaded by trees, rather soily and sometimes overgrown with ivy. There is often a little scree at the base. The most extensive examples studied, working north, are Pleasley Vale, Creswell Crags, Markland Grips, Brockadale by the River Went, Boston Spa by the River Wharfe, Knaresborough by the River Nidd and Castle Eden Dene, County Durham.

The huge dry overhangs which frequently make up over 80% of the cliffs are mostly devoid of lichens. Somewhere, however, the *Leproplacetum chrysodetae* association is usually present, represented by *Lecania erysibe*, *Lecanora dispersa*, *Lepraria incana*, *L. lesdainii*, *Leproplaca chrysodeta* and *L. xantholyta* (Fig. 8.9). At the margins of overhangs in better illuminated, slightly more mesic sites *Bacidia arceutina*, *Caloplaca cirrochroa* (often abundant), *Diploicia canescens*, *Dirina massiliensis* f. *sorediata*, *Opegrapha mougeotii* and *O. saxatilis* occur. Some of these species reach their most easterly sites in the British Isles on the Magnesian limestone.

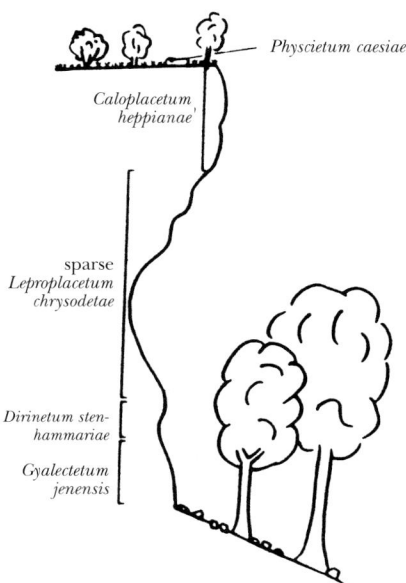

Fig. 8.9 Section through a Magnesian limestone cliff to show the distribution of the main lichen communities.

Small, often shaded, buttresses, ridges and faces on the lower part of the crags remain damp for long periods after rain; these carry variable amounts of *Acarospora heppii*, *Acrocordia conoidea*, *Bacidia arnoldiana*, *B. delicata*, *B. egenula*, *B. sabuletorum*, *Caloplaca citrina*, *Gyalecta jenensis*, *Lecanora campestris*, *Lecidella scabra*, *Rinodina gennarii*. *Thelidium decipiens*, *Verrucaria hochstetteri* and *V. viridula*. The tops of the larger crags are mostly covered in dense scrub, but locally promontories and belvederes of fast drying, well-illuminated, clean limestone occur. On these many, but by no means all, of the characteristic lichens of exposed limestone in the British Isles – the *Caloplacetum heppianae* association – can be found. The presence of moss encourages *Agonimia tristicula, Cladonia pocillum, Leptogium schraderi* and *Toninia lobulata*, but almost nowhere are niches sufficiently damp to favour members of the *Placynthietum nigri* association.

Nutrient-enriched sites on promontories and the tops of isolated boulders carry examples of the *Physcietum caesiae* association in which *Caloplaca isidiigera, Lecanora muralis, Phaeophyscia nigricans, P. orbicularis* (dominant) and many other species more normally encountered on the tops of churchyard memorials can be found.

Reef knolls and quarries

The two richest sites for lichens on the Magnesian limestone are 'The Maidens Paps' on the edge of Sunderland and, 120 km further south, Wood Lee Common near Rotherham, which yielded 53 and 52 species respectively. Both features represent the crests of hard reefs which, where they come to the surface, provide limited but rugged tor-like outcrops of pale limestone in a setting of natural calcareous grassland. The following species have the reef knolls as their only or main sites on the Magnesian limestone; *Acarospora glaucocarpa, Belonia nidarosiensis, Caloplaca aurantia, C. lactea, Leptogium turgidum, Solenopsora candicans* and *Toninia sedifolia*. Bryophilous and crevice species are also widespread. The surrounding coarse grasslands contain no sites suitable for terricolous lichens and are being invaded by gorse and brambles.

The chief interest of Magnesian limestone quarries lies in the terricolous lichens which develop on the soft, yellowish magnesium-rich subsoil of the quarry floor. Where the soil is compacted, as along paths or by vehicles, a wide range of small lichens can be found if one is willing to crawl about on hands and knees. They include *Bacidia sabuletorum, Catapyrenium squamulosum, Collema bachmanianum, C. tenax, Sarcosagium campestre, Vezdaea aestivalis* and *V. rheocarpa*, while small pebbles support *Lempholemma polyanthes*.

Comparison with the Carboniferous limestone

Certain species are highly characteristic of the Magnesian limestone. Chief among these must be *Lecidella scabra* which, particularly south of the River Tees, is a leading species at every site, though it is not normally considered to be a calcicole. Other species notably abundant are *Caloplaca citrina, C. isidiigera, Candelariella vitellina, Lecania erysibe, Lecanora campestris, Phaeophyscia orbicularis* and *Rinodina gennarii*, all of which may occur in great sheets. None are demanding calcicoles. A conspicuous, but unknown, pale to dark grey crust bearing discrete white, farinose soralia (K+ yellow; Pd+ yellow), first noticed at Cresswell Crags, has been described as a new taxon, *Lecanora campestris* subsp. *dolomitica* (Fig. 8.10). It is widespread on the Magnesian limestone of Derbyshire and South Yorkshire. Isolated records have since been made from other substrates including dolomitised Carboniferous limestone.

Calcicoles that are rare on the Magnesian limestone compared with their high frequency on Carboniferous limestone are *Catillaria chalybeia, Caloplaca holocarpa, C. lactea, Placynthium nigrum, Protoblastenia rupestris, Sarcogyne regularis* and many immersed pyrenocarps. Another group that has a high constancy on the Carboniferous limestone but is scarce on the Magnesian limestone consists

Fig. 8.10 Lecanora campestris subsp. *dolomitica*, a taxon first described from the Magnesian limestone where it is locally abundant.

of the foliose/squamulose species *Collema auriforme, C. tenax, Leptogium schraderi, Peltigera* spp., *Psora lurida, Solorina saccata* and *Toninia sedifolia*. These are primarily ledge and crevice species, a microhabitat poorly represented on the Permian limestone. The nutrient-deficient nature of most Carboniferous limestone surfaces renders them suitable for stress-tolerant calcicoles; in contrast, the mildly eutrophicated Magnesian limestone in the dusty agricultural lowlands is more suited to species of *Physcia*.

Field survey of the higher plant flora of standard and dolomitised limestones in South Wales has shown that several species, notably common rock-rose (*Helianthemum nummularium*), hoary plantain (*Plantago media*) and the grass *Koeleria macrantha*, are absent from the dolomitic but frequent in the standard limestone areas. Experiments indicated that this was caused by sensitivity to high soil magnesium. A candidate among the lichens for sensitivity to magnesium is the common but undemanding calcicole *Protoblastenia rupestris*, which is absent or very rare at most sites. Lichens favoured by, or tolerant, of high magnesium are, on field evidence, *Caloplaca isidiigera, Lecidella scabra* and *Lecanora campestris* subsp. *dolomitica*. The distinctiveness of the Magnesium limestone lichen flora is, therefore, due to the chemical composition of the rock, the morphology of the outcrops, and discernible levels of eutrophication originating from the surrounding countryside.

The scarplands

There are few natural outcrops associated with inland stretches of the Jurassic scarp, which is genuinely deficient in semi-natural saxicolous habitats. Places where the flora of the oolite and the ironstone can best be studied are all manmade features such as drystone walls, the roofs of buildings and churchyards; these are dealt with in the next Chapter. The striking orange lichen *Caloplaca aurantia*, which has a southern distribution in Britain, is particularly characteristic of Jurassic limestones (Plate 9e). Only three sites with a significant lichen flora are known to the author. There is a long-disused quarry at Wytham Hill in Berkshire, the floor of which was covered with *Psora decipiens* when I last saw it in the 1970s. Fifty kilometres further east, at Barnack Hills and Holes near Peterborough, an astonishing find was made in 1958. During a meeting of the Kettering and District Natural History Society the five members present spread out to look for terricolous lichens; on reassembling one person returned with *Solorina saccata* in his hand. Despite subsequent searches this species, which is normally found in much higher rainfall areas, has never been re-found, but the original specimen can be seen in the herbarium of The Natural History Museum. Its next nearest locality is in Derbyshire. Many people are unaware that the Jurassic limestone extends north to the Kirkbymoorside area of Yorkshire, but there, at Hutton Common, a natural hilltop exposure of oolitic limestone is a treasure trove of lichens. The rock supports species like *Caloplaca lactea*, *C. variabilis*, *Collema multipartitum*, *Petractis clausa*, *Polyblastia albida* and *Staurothele caesia*, while on ledges and in crevices *Cladonia foliacea*, *Megaspora verrucosa*, *Squamarina cartilaginea* and *Toninia sedifolia* can be found. A thorough survey of the Jurassic limestone scarp is called for.

Low level limestone in Scotland

Two complex, narrow, bands of hard, pure limestone run diagonally across Scotland in an overall north-northeast to south-southwest direction. The solid geology map (Fig. 8.1) gives a misleading impression of their area of outcrop, because in many places they are buried under glacial drift or colluvial deposits. The northernmost band, the Durness limestone of Cambrian age, can be examined on the north coast of Sutherland around the village of Durness, on the east side of Loch Eriboll, on Skye, and in the Inchnadamph area where it is particularly well exposed with white cliffs and green fields occurring intermittently for 20 km to the west of the main road north of Knockan. The Durness limestone is dolomitic, containing over 10% magnesium carbonate, and hard. Under the impact of regional metamorphism it has partially recrystallized into marble.

As they experience a different climate to the English and Welsh limestones these distant outcrops might be expected to support distinctive lichens. Though by no means fully explored, the lichen flora of these northern limestones appears to be floristically poorer than those in the south (see lists in Coppins *et al*., 1986). For example, large orange *Caloplaca* species are almost absent, as are the two grey species *C. chalybaea* and *C. variabilis* and the white *Solenopsora candicans*. Species of unusual abundance include several Collemataceae such as *Collema furfuraceum*, *C. multipartitum* and *C. polycarpon*, *Caloplaca arnoldii*, *Gyalecta jenensis*, *Lecanora crenulata*, and there is usually a good range of bryophilous species like *Agonimia tristicula* and *Toninia lobulata*. There are also rare species. For a long time Inchnadamph was the only place in Britain where one could see *Ionaspis melanocarpa*, an inconspicuous species with an immersed thallus and small sunken blackish apothecia. As with other *Ionaspis* species when scratched with a fingernail the orange of the *Trentepohlia* algal layer in the thallus is exposed. It is not uncommon on limestone beside the wild Traligill River at Inchnadamph and by the Allt nan Uamh 5 km to the south. *I. melanocarpa* has recently been found at a further locality, a high limestone pavement on Cross Fell in the northern Pennines. A careful climatic and floristic comparison of these two habitats might show they have much in common; both for instance also support *Sagiolechia protuberans* and *Staurothele bacilligera*.

The second band of limestone crossing the Highlands is a Dalradian limestone of Pre-Cambrian age. It has been given different names in the various places where it is well exposed, for example the Blair Atholl limestone, the Loch Tay limestone and the Ballachulish limestone. Many of the outcrops are very small, quarried, and chiefly of interest at the local level, for instance to the writers of county floras. Other exposures are high level, and are covered in Chapter 11, where they are shown to provide some of the most exciting botanising to be had anywhere in Britain.

At two lowland sites the Dalradian limestone is associated with localities of national significance for their lichens. The first of these is Craig Tulloch, a low-lying hill south of Blair Atholl. This site was discovered by the Rev. J. M. Crombie (1873) who recorded among other species *Catillaria picila*, *Gyalecta ulmi* and *Megaspora verrucosa*. A modern survey of the extensive north-facing cliffs (Purvis *et al.*, 1994b) revealed the largest population of *Gyalecta ulmi* in the United Kingdom. This species experienced a rapid decline as a result of the death of elms due to Dutch elm disease and is now believed to be extinct as an epiphyte. Yet it survives on six mossy basic outcrops in Scotland and two in England. *Catillaria picila* has not been rediscovered but, during a search for it, the equally rare *Polysporina cyclocarpa* was encountered together with other strict calcicoles such as *Acarospora macrocarpa* and *Polyblastia wheldonii*.

The Dalradian limestone runs into Loch Linnhe where it forms the long, narrow Island of Lismore. Due to the fondness lichenologists have for studying islands, a long list of species has been recorded from Lismore, but it is best known as the type locality of *Placynthium lismorense*, which is abundant on flat damp limestone. This close relative of *P. nigrum* can be distinguished by its more conspicuously radiating margin, an olivaceous tinge to the dark thallus, and the absence of a prothallus. It also occurs at a few places on the Continent, but is certainly one of the stars of our limestone lichen flora.

9

Village, Church and Farmland

Since Anglo-Saxon times, England has been a land of villages. These were formerly encircled by open fields, which are now enclosed by hedges. To the north and west a mixed pattern of settlement is found that includes hamlets and single farmsteads. Villages take one of three forms; the houses may be grouped around a central green or square, may be strung out along a main street, or be in a haphazard conglomeration. Whatever the layout, each has a church, with tower or spire rising to pierce the sky, often from a tuft of trees, and normally includes a fine, stone-built hall or manor house erected by the local gentry in the late sixteenth or seventeenth century. Fidelity to using local building materials is greatest in the smaller and older villages; those that have grown in recent times contain a great variety of fabrics including concrete, brick, asbestos and Marley tile (coarse-coloured cement).

The cultural landscape represented by the village and its surrounding farmland has provided lichenologists with a rich source of conveniently situated habitat for study. This is of particular importance in lowland Britain where the village may provide the only saxicolous habitats in a parish. Add to these the lignicolous habitats presented by gates, fences, seats and old barns together with the well-illuminated, dust-impregnated boles of free-standing trees and the attractions are obvious. One part of this mosaic that has proved especially popular in recent years is the churchyard. A dedicated churchyard group, led by Tom Chester, has surveyed in excess of 6,000 sites, making churchyards the best studied habitat in Britain. On the Continent they are less well surveyed lichenologically, so we are unable to judge ours in a European context.

It is fascinating to discover what a high proportion of our lichen flora has adapted to growing alongside man. Over 600 species, more than a third of the British list, have been recorded from churchyards. Since all have colonised over the last thousand years, many are catholic species with efficient methods of dispersal. Alongside these widespread species are others, formerly rare, that were apparently pre-adapted to the cultural landscape; these have prospered and now have their main centres of distribution in lowland villages. There is a group of six lichens that, in Britain, are currently known only from churchyards.

It is their wide range of saxicolous habitats that makes lowland villages distinctive, so this chapter, though mindful of the full range of opportunities for lichens, concentrates on such substrates. Figure 9.1 shows the distribution of traditional building materials in lowland Britain and adjacent areas; these control the type of lichens you might expect to find.

The village

Garden walls

A stroll round any village, armed with a hand lens and a set of chemicals, will produce a rich haul of lichens from garden walls. The walls are invariably

Fig. 9.1 The distribution of traditional building materials in England and Wales; these largely dictate the lichen flora of the older villages.

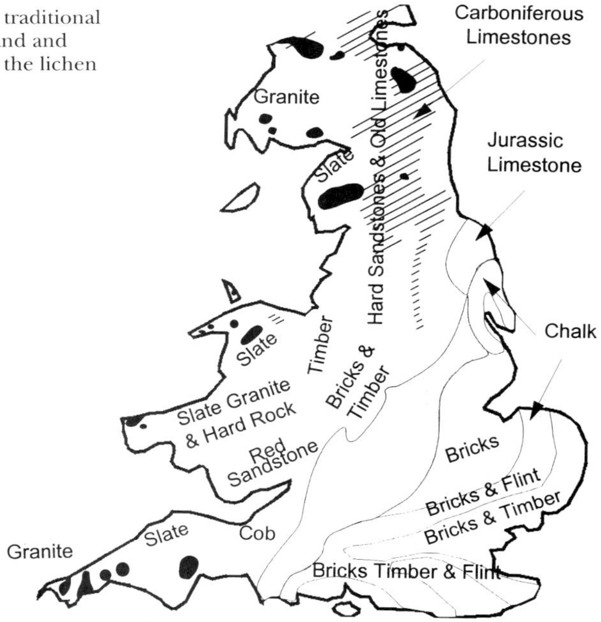

mortared, so that even sandstone walls yield a mixture of calcicole and calcifuge species. Typical of the flat top of siliceous walls are *Acarospora fuscata, Candelariella vitellina, Lecanora muralis, L. polytropa, Lecidella scabra, Parmelia glabratula, Porpidia tuberculosa, Rhizocarpon obscuratum, Scoliciosporum umbrinum* and any one of four species of *Trapelia*. If the wall is topped by metal railings, and you are lucky, *Stereocaulon nanodes* may be present on the iron-stained sandstone underneath them together with an increased density of *Lecanora polytropa*. Both can also colonise the railings directly (Plate 10a). Mortar, if of the hard cement type, will support a range of species that are also common to concrete fence posts, for example *Candelariella aurella, Caloplaca citrina, C. holocarpa, Lecanora albescens, L. dispersa, Physcia adscendens, Rinodina gennarii* and *Xanthoria parietina*. This assemblage is international, being found throughout Europe (Plate 11a).

Walls constructed prior to the twentieth century using lime mortar are well worth seeking out as the material is soft and weathers to provide a range of nooks, crannies, incipient soils and moss-covered ledges that bring in different lichens. A mossy, lime-mortared sandstone wall is just the place to look for additional species such as *Acrocordia salweyi, Agonimia tristicula, Bacidia sabuletorum, Cladonia fimbriata, Collema crispum, Diploschistes muscorum, Lepraria lesdainii* (a verdigris green felt in deep crevices), *Leproplaca chrysodeta, Leptogium gelatinosum* and *Toninia lobulata*. Such walls are concentrated in the older core of the village and support specialised higher plants which, together with the cryptogam cover, contribute to the rural ambience which is an essential ingredient of the typical English village.

The great belt of Oolitic and Liassic limestone of Jurassic age that crosses the country diagonally from Dorset to North Yorkshire (Fig. 9.1) provides some of the best building stone in the world. It is occupied by a sequence of attractive villages much of whose character is provided by the pattern of orange, yellow, white and chocolate-coloured lichen growing on the smooth, sawn masonry. The white patches are provided by *Aspicilia calcarea*, *A. contorta*, *Diploicia canescens*, *Diplotomma alboatrum*, *Lecanora campestris* and *Solenopsora candicans*, the yellow by *Candelariella medians*, the chocolate by *Verrucaria nigrescens*, and the orange by *Caloplaca aurantia*, *C. dalmatica*, *C. flavescens*, *C. saxicola*, *Xanthoria calcarea* and *X. parietina*. A range of more local species such as *Caloplaca lactea* may also be present, but garden front walls are not the place to hunt for rarities. The streets of some of our prettiest Cotswold villages, like Burford, are, due to tradition, followed by planning restrictions, built entirely of limestone. They produce rather short lichen lists, as calcifuge species are entirely missing; Burford has 58 species.

New housing at the edge of villages often has low front walls capped with precast cement blocks, with matching ones on pillars and gateposts. These are favourable sites to search for early colonisers of calcareous substrates and are likely to hold large populations of *Caloplaca decipiens*, *C. isidiigera*, *Phaeophyscia nigricans* and *Xanthoria elegans*. Between the wars, front walls in poorer districts were sometimes capped with flat blocks of cement that had been bulked up with coarse aggregate. These are now starting to crumble and are a goldmine for the village lichenologist; one such wall examined at Horton-in-Ribblesdale, Yorkshire, produced *Caloplaca flavovirescens*, *C. variabilis*, *Collema auriforme*, *C. cristatum*, *C. fuscovirens*, large black rosettes of *Placynthium nigrum* overgrown by *Toninia verrucarioides*, *Sarcogyne regularis* and much besides.

Many villages are short of strongly acidic saxicolous habitats at or near ground level (see, however, slate and pantile roofs later). In some areas this gap is filled by a special type of hard vitrified brick that is used to cap walls, particularly those associated with such institutions as schools. Frequently such walls are the only site in the village, outside the churchyard, for such common calcifuges as *Buellia aethalea*, *B. ocellata*, *Parmelia mougeotii*, *Porpidia tuberculosa* and *Rhizocarpon obscuratum*. At Harrold in Bedfordshire, this assemblage is present on a granite war memorial on the green. Another habitat that can yield different species is old river bridges for which the best stone was imported, often over a considerable distance. Their parapets become somewhat enriched by road dust and are good places to seek *Acarospora umbilicata*, *Caloplaca crenularia*, *Lecanora sulphurea*, *Lecidea fuscoatra*, *Ochrolechia parella*, *Pertusaria aspergilla* and other siliceous species favoured by slight basicity.

The manor house is usually only available for inspection by special appointment but may have an impressive entrance of massive gate piers flanked by walls of expensive ashlar. If acidic, these vertical expanses are the place to look for white, circular patches of *Haematomma ochroleucum* with its distinctive cottony prothallus, *Lecanora caesiosora* and *Psilolechia lucida*; the capitals may support *Lecanora conizaeoides*, which can be unexpectedly hard to find on village stonework. Acid walls, overhung by deciduous trees, become enriched by throughfall, pollen and honeydew. These, together with shade and a cover of moss, favour a number of large foliose lichens such as *Parmelia caperata*, *P. exasperatula* (under sycamore), *P. revoluta*, *P. subaurifera*, *P. tiliacea*, *P. sulcata* and *Physconia grisea*.

Roofs

When undertaking a village survey it is customary to look wistfully at the roofs of the buildings, long for a ladder, and pass on. Occasionally a low roof can be investigated with steps, but the majority remain unknown territory. It was the East Anglian lichenologists, driven by their region's acute shortage of acidic saxicolous habitats, who first realised the potential of roofs for producing surprises. When Shipmeadow Church, Suffolk, was being re-roofed in 1991, Chris Hitch took the opportunity to examine the old tiles and found they carried considerable quantities of an unfamiliar dark-coloured *Buellia*, which was later identified as *B. badia*, new to Britain. It has subsequently been found widely on pantile and slate roofs in the lowlands. Then, in 1992, while travelling through Debenham, Suffolk on a double-decker, Peter Earland-Bennett spotted an unusual looking yellow *Parmelia* on a slate roof. He returned with a ladder to collect it, and it was identified as *P. protomatrae*, again new to Britain. It has since been found on two further roofs in the same district. Other remarkable finds on slate roofs in Suffolk have included *Parmelia tinctina*, previously thought to be confined to three sites in the southwest, *P. conspersa*, *P. pulla* and the small *Lichenothelia convexa* which is often associated with *Rhizocarpon geographicum*.

To collect comparative data on different roofing materials Tom Chester and I investigated five small roofs in Northamptonshire. They provided a combined total of 66 species, each material having its own specialities (Table 9.1). The 30-year-old Marley tile roof was well colonised with calcicoles, but also supported species typical of only weakly basic substrata, such as *Lecidella carpathica*, *Physcia dubia* and *Tephromela atra*. The 50-year-old asbestos roof was low pitched and partly moss-covered, its flora including *Agonimia tristicula*, *Bacidia sabuletorum*, *Collema fuscovirens* and *Leptogium gelatinosum*. An 80-year-old Jurassic limestone flag roof was outstanding, supporting 20 species not found on the others, among them *Caloplaca lactea*, *Clauzadea metzleri*, *Collema auriforme*, *Thelidium papulare* and *Verrucaria glaucina*. Pantile and slate roofs both provide a fine-grained siliceous substratum, often the most acid habitat in a village; their floras have much in common. What appears from ground level to be dirty streaks on pantiles is usually a mixture of *Buellia aethalea* and *Catillaria chalybeia*.

Table 9.1 The lichen flora of five roofs in Northamptonshire compared for the number of species present.

	Marley tile	Asbestos	Jurassic limestone	Pantile	Slate
No. of species	32	28	50	7	7
No. of unique species	4	4	20	4	3

This cursory examination of house roofs in lowland England has highlighted their importance for lichens; any village survey is incomplete until a range of them has been examined. Their value in the uplands is hardly known, though a slate roof in Pitlochry, Perthshire, is notable for its lush cover of the rare *Umbilicaria hirta*. A tip worth remembering is that when roof lichens become detached, they frequently land in the gutter, get washed down the fall pipe and collect at ground level on drain grills.

Houses

As far as I am aware, only one account has been published describing the lichen ecology of a house (Raistrick & Gilbert, 1963). This involved my place of work at that time, Tarn House, Malham, one of the best-known buildings in Yorkshire. The account demonstrates how the colonisation of the building by lichens, bryophytes and algae is related to the age of the masonry and conditions of moisture, light, pH, exposure, pollution and decay. Today, many of the observations appear naïve, but some indication of the factors controlling the distribution of lichens on stucco cement can be seen in Figure 9.2. The disposition of *Caloplaca citrina* is of particular interest as it accurately marks the height of the capillary damp line, dipping in response to a cellar grid and rising over the kitchen drain. It is also conspicuous on a large rain track originating from a leaky gutter. *Protoblastenia rupestris* is concentrated on the older leached stucco, while *Lecanora dispersa* shows a preference for newer, high pH stucco and the vicinity of crumbling holes in the older surface. Lichens also picked out those parts of the building used as bird perches. One, *Psilolechia lucida*, was selective for the sandstone used in the 1852 extension.

At a finer level of detail it was observed that rainwater, entering sandstone window jambs and mullions, gravitates to the window bottom where it drains out of the mortar joint keeping a small part of the sill damper and less acid than the rest. Such sites are frequently occupied by *Candelariella vitellina* and *Lecanora dispersa*. If window frames have been treated using lead paint, they and the sills should be inspected with a lens for species of *Thelocarpon*. Studying the ecology of a building makes a first-class student project.

Fig. 9.2 The west side of Malham Tarn House showing the distribution of lichens on lime stucco.

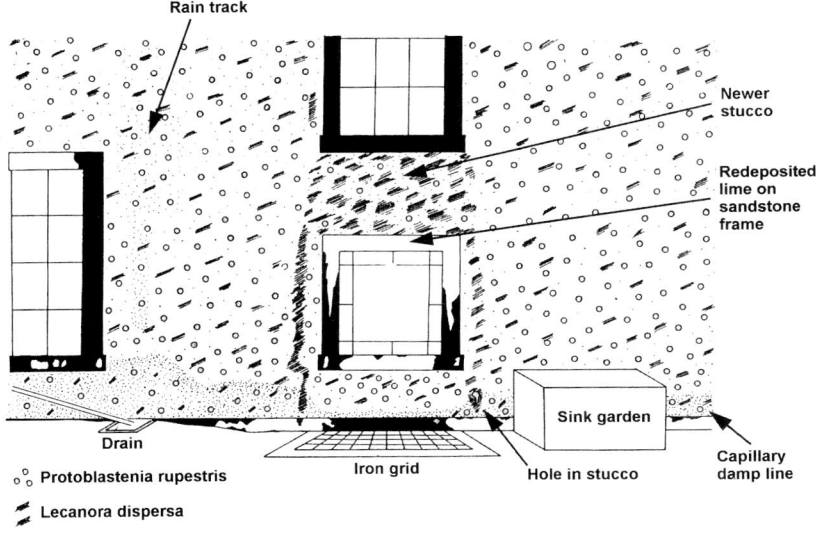

The church

Churchyards currently enjoy unrivalled popularity among British lichenologists; in several counties enthusiasts are attempting to survey every one. A national Churchyard Project has been set up by the British Lichen Society, the first phase of which is focusing on 'the lowland triangle', an area comprising 35 Watsonian vice-counties lying roughly southeast of a line from Scarborough on the Yorkshire coast to Lyme Regis in Dorset. There are few significant rock outcrops in this area, so the 10,000 churchyards represent a major saxicolous habitat. The objective of visiting at least one site in each 10 km grid square is now achieved, and with work having commenced north and west of the line, it should soon be possible to compare upland and lowland sites.

Work in Northamptonshire will be used to quantify the importance of churchyards (Chester, 1997). In this landlocked county, with no natural rock exposures and only vestiges of ancient woodland and heath, 352 lichen taxa have been recorded. Two-thirds of these are present in the 350 Anglican churchyards. Of the 217 saxicolous species in the county all but ten have been recorded from the churchyards, and many do not occur elsewhere. The significance of churchyards as a haven for saxicolous species is clear; exceptionally over a hundred may be present. Only brief reference to the epiphytes and lignicolous species of churchyards will be made here, as they are often better developed elsewhere in a parish.

Despite much talk of churchyards containing over 100 saxicolous species, Figure 9.3 shows that such counts are quite rare. The average yard contains a

Fig. 9.3 Histogram showing the number of saxicolous lichens recorded in 3,000 churchyards (data provided by members of the Churchyard Group).

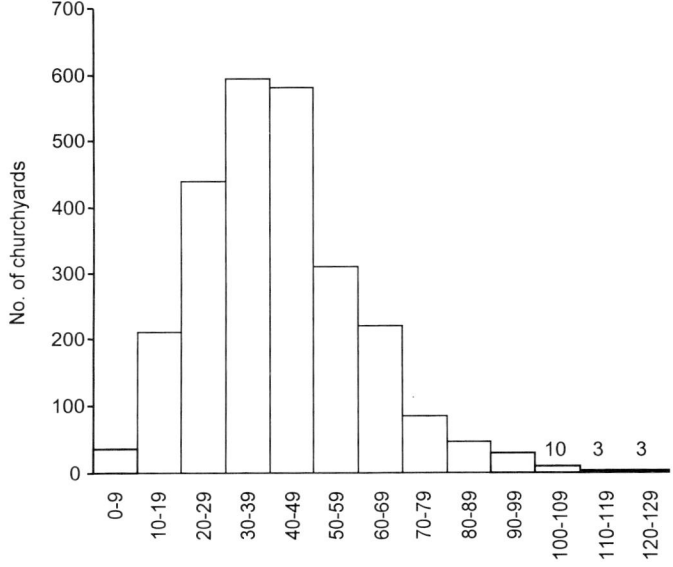

number of species somewhere in the upper thirties. I would anticipate that with resurveys this might be edged well into the forties, but the number is frequently depressed by a combination of air pollution, heavy shade and a lack of variety in the stonework. There is evidence that churchyards in the west e.g. Dorset are richer than those in the east e.g. Suffolk and East Yorkshire, with counties like Northamptonshire intermediate.

The more experienced members of the Churchyard Group use a systematic approach when surveying. They start at the southeast corner of the building, then work round in an anticlockwise direction carrying a pair of stepladders to aid the inspection of windowsills. Next, the memorials in the graveyard are investigated, paying particular attention to types of stonework not present in the building; then the boundary wall is scrutinized. Lastly, trees and other habitats such as gates, paths and loose bits of stone at the base of walls are checked. Since collecting chips of rock from 'God's Acre' is frowned on, a variety of techniques for dealing with unknown species have developed; these involve scraping, preparing slides in the field or using a strip of Sellotape to lift a film of lichen off the stone, which can then be stuck straight into a notebook.

The church building

Of the three main habitats – the church, the memorials and the boundary wall – the church is usually the most important. This is because of its age; it is a thrill to examine an unrestored Saxon or Norman tower knowing that the surfaces are close to a thousand years old. Churches are frequently composed of a considerable variety of materials as they become added to or partly rebuilt over the centuries. While the body of the church may be constructed of ragstone or flint, masons demanded the most durable stone for coins, windows, doorways and the shoulders of buttresses and it is on these that the richest lichen communities are found. Many lichens are so precise in their requirements that they can be used to distinguish between the different types of stone, not just between sandstone and limestone, but between limestones and sandstones from different quarries. For example, in Northamptonshire, *Hypocenomyce scalaris* is found on the local sand ironstone but not on other varieties. A further important factor relates to aspect and inclination of the stone. Here the traditional east-west alignment of the church helps to provide maximum contrast.

Compared with the rest of the village, several features make churches rather special habitats; these will be examined in turn. The first is the age of the stonework. A small range of species appears to be specific to ancient stonework, forming slow-developing communities that represent an advanced successional stage. The following lichens are rarely found on recent masonry *Acrocordia conoidea, Caloplaca aurantia* (in the north), *C. cirrochroa, C. dalmatica, C. lactea, C. ruderum, Gyalecta jenensis, Lecanora pruinosa, Leproplaca xantholyta, Opegrapha gyrocarpa, O. saxatilis, Ramalina canariensis* and *R. lacera*. Another dozen species are more weakly associated with medieval stonework. When ecologists evoke the concept of fidelity the challenge of finding exceptions is eagerly taken up, usually with positive results; perhaps it is best to regard these species as no more than preferential for old stone and to remember that they may be transgressive in character across Britain.

Second, a church provides considerable expanses of dry, vertical stonework, supporting distinctive communities. North walls are characterised by sterile sorediate species with a powdery or felty surface that it is almost impossible to wet. Most can be identified using combinations of form, colour and chemistry. The commonest are *Diploicia canescens, Dirina massiliensis* f. *sorediata, Haematomma ochroleucum, Lepraria* spp., *Leproloma* spp., *Leproplaca chrysodeta, L. xantholyta, Opegrapha* spp. and *Psilolechia lucida*. That they are able to use humidity as their water supply is confirmed by the large patches of *Dirina* that occur on limestone pillars inside Ely cathedral. This is a community that is better developed on churches than on most natural outcrops.

Sunny, vertical or tilted stone on the south side of the building has the most noticeable and richest lichen cover. It is best developed on limestone or mortar; acid stone can be disappointing. The fine-grained, colourful mosaic of lichens will most probably include *Caloplaca flavescens, C. saxicola, C. teicholyta, Candelariella medians, Diplotomma alboatrum, Lecanora crenulata, L. campestris, Toninia aromatica* (mortar joints), *Verrucaria nigrescens* and much else. A string course will bring in additional species such as *Verrucaria glaucina*, while the seam of moss along its back angle is a promising place to search for *Lempholemma polyanthes*. Most of these species can also be found in the village.

To many lichenologists the most fascinating habitat on the church is stone affected by metal runoff. This is seen at its simplest on vertical walls adjacent to copper lightning conductors. These are flanked by a bare, highly toxic area that grades laterally into a zone where the pale grey copper-loving species, *Psilolechia leprosa*, is dominant; its identification can be confirmed with a drop of bleach that will turn its powdery thallus beetroot-coloured. A band of yellow *Caloplaca citrina* (alkaline) or green *Psilolechia lucida* (acid) sometimes separates the toxic streak from the normal flora. As copper is widely used for protective window grills, the sills and the walls below them also support a modified, zoned flora. A typical arrangement might be a completely bare oolite sill, while the sandstone wall below carries successive zones of an unnamed dark brown *Acarospora, Psilolechia leprosa*, and a brownish form of *Candelariella vitellina*. This indicates how copper is less toxic under acid conditions. Acid window sills should be searched for the phenomenon of copper turning certain lichens green, for example *Acarospora smaragdula, Buellia aethalea, Lecanora polytropa* and *Lecidella scabra*. In this condition they look so untypical that copper-affected *A. smaragdula* has been described as a new species on three separate occasions. Within the toxic zone even lichens that show resistance are contorted and curl up so they have minimal contact with the substratum.

Iron staining can be seen on sills in association with rusting window grills and vertical iron bars, though its solubility is low under most conditions. Iron streaks are flanked by an assemblage composed of *Arthonia lapidicola, Lecanora stenotropa, Polysporina simplex* and *Scoliciosporum umbrinum*, which occur on both sandstone and mortar. At Holy Trinity Church, Melrose, iron is responsible for the narrow zone of *Lecanora subaurea* that occurs at the back of every sandstone sill.

Lead appears to be less toxic than iron or copper, though it does have a negative effect on some species. Water that has run down leaded windows has no discernible effect on the sill flora, while lead sheeting on the roof may support apparently healthy colonies of half a dozen species, with *Physcia caesia* particularly prominent. Figure 9.4 shows the pronounced effect of inset lead lettering on the lichen cover of a marble headstone. The influence of zinc is hard to isolate as leaded lights are frequently protected by zinc-galvanized iron grills

Fig. 9.4 Toxic wash from lead lettering producing a lichen-free zone on a gravestone at Avebury, Wiltshire (B.J. Coppins).

attached to copper pegs sunk in leaded sockets. This produces a general toxic wash.

Memorials

During medieval times memorials were sited inside the church, but, from the early seventeenth century, chest tombs and short stubby headstones made from local materials were placed on the south side of the church. In the eighteenth century, headstones became larger and more elaborately carved, while during the following century tastes evolved to favour crosses, angels, obelisks, draped urns and kerbed grave plots often made from imported materials such as granite or marble. Today low, simple headstones of polished, imported granite are popular, and there is a fashion for filling kerbed grave plots with glass chippings. Both resist lichen colonisation for many decades; anyone wishing to do their bit for lichen conservation should opt for a substantial, elaborately carved headstone made from local materials.

The most exciting memorials lichenologically are very old chest tombs. These have scarcity value and, although mostly found on the sunny south side of the church, their horizontal surfaces are slow to dry out. Their specialities include *Aspicilia subcircinata*, *Catapyrenium sqamulosum*, *Dermatocarpon miniatum*, *Gyalecta jenensis*, *Hymenelia prevostii*, *Leptogium plicatile*, *L. teretiusculum*, *Opegrapha mougeotii* (underhangs), *Placynthium nigrum*, *Verrucaria caerulea* and *V. pinguicula*. The ornate bale tombs of the eastern Cotswolds are so elaborate that they can take half and hour to examine; a group of five in Burford churchyard (Fig. 9.5), supported an average of 31 species each. Acid chest tombs are a good habitat for unusual *Parmelia* species. A 'tea caddy' tomb at Kirkland in Westmorland is believed to hold the British record for the number of species. This 2 m tall, highly ornate structure, carved from red Triassic sandstone in 1789, supports 47 species including *Diploschistes scruposus, Pertusaria albescens, P. pseudocorallina, Physcia pulverulenta, Ramalina canariensis, R. fastigiata* and *R. subfarinacea*. The key to its high diversity lies in the wide pH range; leached areas bear calcifuges, sheltered faces carry calcicoles, the top is mossy.

A conspicuous lichen cover develops on headstones used as bird perches which become thickly encrusted with a cap of orange *Xanthoria*, grey *Physcia* and brownish-pink *Physconia* spp. Occasionally rarer lichens such as *Anaptychia ciliaris* or *Physcia clementei* may be present. Granite headstones that become eutrophicated carry swards of *Candelariella vitellina, Physcia dubia* and *Xanthoria candelaria*.

Churchyards make a fine training ground for young ecologists due to the range of saxicolous substrates available for comparison. Though few species are completely substratum-specific, strongly acidic granite is the place to search for *Acarospora smaragdula, Buellia occelata, Lecanora polytropa, Parmelia mougeotii, Polysporina simplex, Ramalina siliquosa* (near the sea) and *Xanthoria candelaria*. Red sandstone with its high water-holding capacity and neutral pH is often dominated by large circular colonies of *Haematomma ochroleucum, Lecanora rupicola* (Plate 10b), *Ochrolechia parella* and *Tephromela atra* supported by *Pertusaria albescens, P. amara, P. aspergilla, P. pertusa* and *Phlyctis argena*. Slate is usually well colonised by *Buellia aethalea, Parmelia glabratula* and in upland

Fig. 9.5 A group of seventeenth-century bale tombs in Burford churchyard, Oxfordshire. Ornate limestone memorials such as these support a particularly rich lichen flora.

Britain *Rhizocarpon geographicum* (Plate 10c), while damp sandstone holds *Porpidia crustulata*, *P. tuberculosa* and up to four *Trapelia* spp. Ironstones are the habitat in which to seek saxicolous *Bacidia rubella* and *Lecanora pannonica*.

Churchyards provide an opportunity to study lichen colonisation using quantitative methods. Figure 9.6 shows three curves generated using a series of dated limestone memorials at Monyash, Derbyshire. It can be seen that species diversity rises sharply at first, then falls off, but is still increasing after 180 years. In contrast, the percentage cover of lichens on apical parts of the headstones reaches a maximum after only half this period levelling off at around 95% after 90 years. Though only a small number of observations were made on the growth rate of *Caloplaca flavescens*, colonies apparently show a steady radial increase over 180 years (but see pp. 38–40).

Boundary wall

The boundary wall can make the difference between an average and an outstanding churchyard. The best walls for lichens are constructed of irregularly-shaped stone that needs a lot of lime mortar to hold it together. Some of the best I have seen are around the head of Morecambe Bay, where the tradition is to build in slate held together with now crumbling lime mortar. The crevices and mossy ledges support luxuriant cushions of *Catapyrenium squamulosum*, *Collema auriforme*, *Lempholemma polyanthes*, *Psora lurida* and *Toninia sedifolia*, all recruits from the surrounding limestone hills. The acidic capstones support a different assemblage. This type of boundary wall is the exception; most are

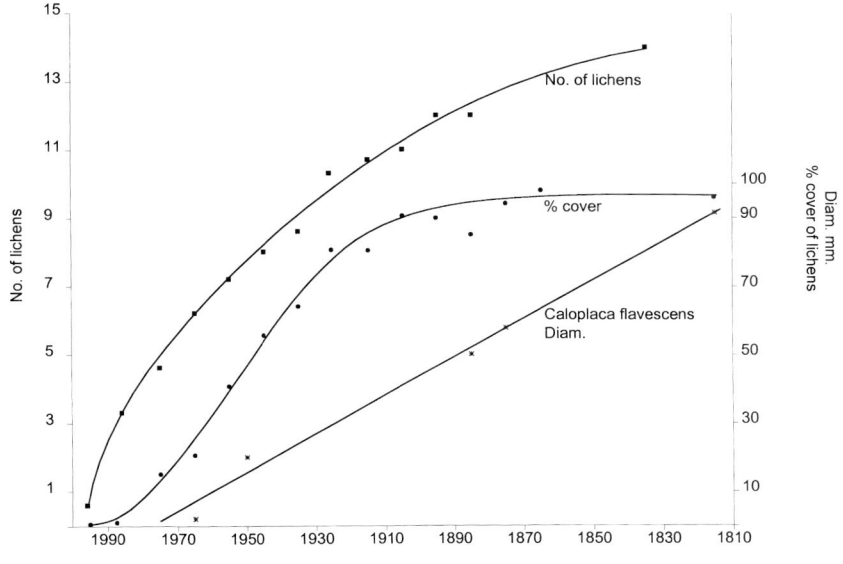

Fig. 9.6 Graph showing the increase over time, in lichen cover, diversity and thallus diameter on limestone headstones at Monyash churchyard, Derbyshire.

rather well maintained, having been repointed this century with hard cement, or are made of ashlar which requires minimal mortar to hold it in place.

Further habitats

It is always worth examining gravel paths for *Collema tenax*, and any additional toxic sites, such as brass sundials, for *Stereocaulon pileatum*, or moss around drains receiving contaminated water from the roof for the tiny pale fruits of *Vezdaea leprosa*. The wooden door of the church may support *Arthonia impolita*, close mown grassy banks can hold interesting *Cladonia* and *Peltigera* species, while the roofs of porch and lych gate may yield additional species of *Parmelia*. Fragments of stone, brick and slate lying in the shade of shrubberies, or at the base of walls, can add the ruderal species *Micarea erratica*, *Thelidium zwackhii*, *Verrucaria dolosa* and *V. simplex*. The upper part of the tower provides a habitat that differs from the rest of the building in experiencing a sharp wetting-drying cycle; it sometimes carries a spectacular growth of shrubby species including *Ramalina lacera*, *R. siliquosa* and, memorably, at several churches along the south coast and on St David's Cathedral, Pembrokeshire, a prolific growth of the dye lichen *Roccella phycopsis*.

A comparison of the village and the churchyard

With churchyards being so well worked compared with the villages in which they stand, it is difficult to assess the relative importance of the two. A general feeling has grown up that, due to their age and the variety of stone involved, the churchyard is the unrivalled saxicolous habitat in lowland Britain. To test this I surveyed 25 villages and their 25 churchyards separately, putting equal survey effort into each. To keep the results comparable, in large villages only an area within a 200 m radius of the church was surveyed. The sites covered Dorset, East Anglia, Lincolnshire, Northamptonshire, Oxfordshire and Bedfordshire in the lowlands, and Derbyshire, Yorkshire and Cumbria in upland Britain. No poor churchyards were included; the average number of saxicolous species per churchyard was 67 while the richest had 120.

Results showed that in the average settlement 64% of species were common to both habitats. The rest were almost equally divided between village (17%) and churchyard (19%). The mean number of saxicolous species per settlement was 81, so the churchyard contained, on average, 15 species not found in the village and the village contained 14 species not present in the churchyard. It had been expected that a regional pattern would be discernible with churchyards being greatly superior to villages in the lowland triangle and the position being reversed in the uplands. This was not so. It was surprising how often the numbers in each were within 5% regardless of the geographical location.

Habitats that tend to make the village special are poorly-maintained walls, mossy walls, acid brick walls, walls under trees, recent concrete, and the presence of several types of roofing material. Some of the species seen only in villages during the survey were *Candelaria concolor*, *Clauzadea metzleri*, *Diploschistes muscorum*, *Leprocaulon microscopicum*, *Normandina pulchella*, *Parmelia delisei*, *P. incurva*, *Physcia tribacea*, *Ramalina lacera* and *Thelidium papulare*. Their occurrence is mentioned to show that villages support more than everyday species. Other differences are quantitative, for example, villages contain much larger populations of *Caloplaca decipiens* and *Xanthoria elegans*. What makes the churchyard special as a habitat is the presence of ancient stonework, a range of sites influenced by toxic metals, large expanses of sheltered vertical masonry, and granite memorials. Species seen only in the churchyards included *Arthonia lapidicola, Bacidia rubella, Lecanora pruinosa, Petractis clausa, Placopsis gelida, Protoparmelia badia, Psilolechia leprosa* and *Verrucaria caerulea*. Lichens abundant in churchyards, but only occasional in the village, include *Dirina massiliensis* f. *sorediata, Haematomma ochroleucum, Polysporina simplex* and *Sarcopyrenia gibba*.

The point that needs stressing is not the difference between the saxicolous floras of village and churchyard, but their similarity. The combined list from both habitats is, on average, only 18% longer. Time and again unusual species encountered in one habitat turned up in the other, for example *Aspicilia subcircinata, Dermatocarpon miniatum, Lecanora pannonica, L. subaurea, Parmelia conspersa, Sarcogyne privigna, Stereocaulon nanodes* and *Toninia lobulata*. It was a surprise to discover how extensive and varied saxicolous habitats are in the publicly accessible parts of a typical village. Evidently, to assess fully the saxicolous flora of the cultural landscape, 'whole village' surveys are required.

Conservation

Certain churchyard lichens are national rarities and demand conservation; due to lack of survey no village species have a similar status. The following species are restricted to churchyards in the UK. *Calicium corynellum* occurs on the base of a Saxon tower in the church I was married in at Bywell, Northumberland; it has decreased in area by 95% since being discovered in 1972. The decline appears to be the result of replacing a stone slab at the base of the tower with gravel, so that water cascading from a pipe no longer splashes up the wall. A few thalli of *Acarospora chlorophana* occur on a sandstone headstone in a small churchyard in Bilsdale, North Yorkshire; it was discovered in 1991 by Don Smith who was systematically working the churchyards of that area.

Considerable excitement was generated in 1993 when Ken Palmer and Ishby Blatchley, churchyard group members, made the first record this century of *Lecanora pruinosa* from a chest tomb at Cricklade, Wiltshire; it has since been found at over 30 ecclesiastical sites. In 1997 Brian Coppins spotted an *Endocarpon* new to Britain, as an extra in a packet sent by Vince Giavarini from a Dorset churchyard. For a while *Parmelia submontana* was only known from siliceous gravestones at Sweetheart Abbey, Kirkcudbright, where it was found by Brian and Sandy Coppins in 1992; it is now known on trees at two additional sites. Very recently (1998), a spate of outstanding lichen records has been made from churchyards. These include *Thelopsis isiaca* new to Britain from Devon by Barbara Benfield, *Pertusaria lactescens*, a species considered possibly extinct, which was found by Tom Chester on the windowsill of a church near Blair Atholl, and, as is so often the case, is starting to turn up elsewhere now a few people have their eye in for it. A further recent record that reflects the intense scrutiny that churchyards are currently undergoing is the discovery of *Ramalina capitata* in Lincolnshire by Mark Seaward. This species, together with others known only from very small populations near the east coast, may be temporary colonists that have 'blown in' over the North Sea from the Continent.

Further achievements of the churchyard movement have been to alert lichenologists to the presence of major populations of species previously regarded as rare, for example *Arthonia lapidicola, Lecanactis hemisphaerica* (on plasterwork), *Lecanora conferta, L. pannonica, Lecania turicensis, Physcia clementei, Psilolechia leprosa, Rinodina calcarea, Roccella phycopsis* (on towers) and *Sarcopyrenia gibba*. Conservation literature is distributed to those responsible for church fabric, an education pack has been produced for schools, courses are arranged, and popular articles are written, to raise awareness and promote management sympathetic to lichens.

Farmland

The term 'lichen desert' was originally coined to describe those parts of a town where air pollution has depressed the lichen flora. Today it is an equally appropriate label for intensively farmed areas where hedgerow destruction and field enlargement have left few stable habitats. Such farmland is also regularly treated with agrochemicals harmful to lichens. To a few lichenologists a scenario such as this acts as a challenge and, although obviously no Mecca, a few interesting habitats have been identified.

Ditches

During 1972–3 Coppins & Lambley (1974) surveyed the largely arable parish of Mendlesham in Suffolk, as they wished to compare the present-day lichen flora with a careful study undertaken 50 years previously. They reported that only half the species from the original survey could be found and that many epiphytes had retreated from tree boles to buttresses and roots in overgrown ditches. Particularly well developed in this shady niche were *Opegrapha ochrocheila, O. sorediifera, O. vermicillifera, O. vulgata* and pinhead lichens such as *Chaenotheca brachypoda*. Additional lichens recently recorded from ditches in East Anglia include the epiphytes *Chaenotheca furfuracea, Opegrapha demutata* and *Porina leptalea*, together with *Micarea* spp. on pebbles projecting from the earth sides. This is a seriously neglected habitat; if the flora of ditches was better known, lichen lists from lowland parishes might be considerably increased.

Gates, fences and barns

Arthur Mayfield, who carried out the original investigation at Mendlesham, frequently mentions fences, gates, gateposts, barns and other worked timber from which he recorded 21 species. By the 1970s this number had been reduced to 7, partly due to habitat destruction, partly to air pollution. Despite the increasingly frequent use of iron gates, wood preservatives and barbed wire, the lignicolous habitat can still be a significant source of records in farmland. Almost restricted to worked timber is *Cyphelium inquinans* with its grey thallus and neat black fruits that leave a black smudge of spores on a fingertip. It is always a delight to find, and in the southeast its relative *C. notarisii*, which has a bright green thallus, should be looked for. Other species typical of well-weathered wooden structures on farmland are *Hypocenomyce scalaris, Hypogymnia physodes, Lecanora conizaeoides, L. saligna, L. symmicta, L. aitema, L. varia, Micarea denigrata, Parmeliopsis aleurites, P. ambigua* and *Thelomma ocellatum*. Field work has shown that structures made from hardwood, particularly oak, are considerably richer than those constructed using softwood, and that all types of wood preservative discourage the development of lichens. Close observation may reveal unexpected species such as the pale fruits of *Bacidia saxenii* clustered around rusty nails.

To give some idea of the importance of this habitat nationally, Brightman & Seaward (1977) have listed 56 species commonly found on worked timber, a further 68 that are of rare or local occurrence and 44 that are normally saxicolous but have occasionally been recorded on wood that has become impregnated with dust. Many of the species on the first two lists are epiphytes with a wide ecological amplitude but at least 20 are thought to have their original home on decorticated trees in the Caledonian pinewoods from where they have spread to this secondary habitat (Fig. 9.7).

Anybody undertaking research for a county flora will be struck by the number of nineteenth-century records from pales and old barns. The latter have almost disappeared from the countryside, so when an old timber-boarded barn was spotted on a British Lichen Society field meeting in Sussex in 1991 it

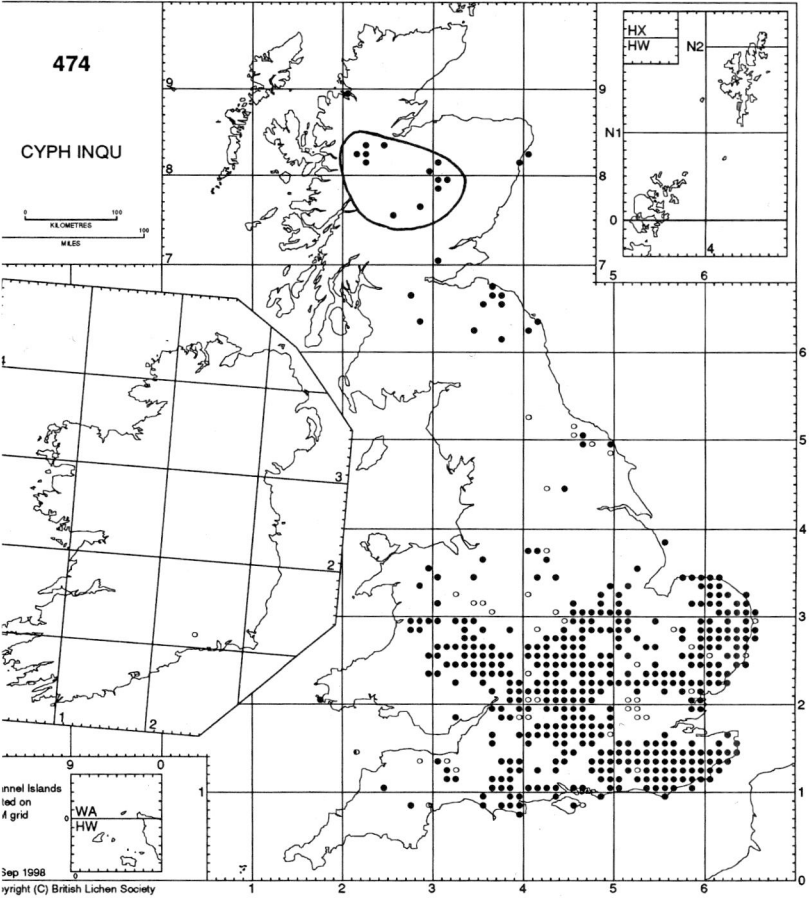

Fig. 9.7 Map of *Cyphelium inquinans*. It has two centres of distribution, an original range in the Caledonian pinewoods (circled) from where it has spread onto worked timber elsewhere (BLS Mapping Scheme).

caused much excitement (Plate 10d). The oak-clad building was found to support *Chaenotheca phaeocephala*, thought to be extinct in the UK, *Thelomma ocellatum*, new to Sussex, *Cyphelium tigillare*, regarded as extinct in West Sussex, and the very local *Candelaria concolor*.

For some time timber barns have been replaced by huge sheds clad in asbestos. To improve their appearance the Design Council (1975) advised that coloured or painted asbestos should be used for the roof so that it looked dark when seen from the middle distance. In response to this the Ministry of Agriculture, Fisheries and Food (1980) produced Advisory Leaflet No. 753, which pointed out that natural colonisation by lichens can satisfy these requirements, thereby saving farmers the considerable extra expense of purchasing coloured asbestos (Fig. 9.8). Research for the leaflet was carried out by P. G. Gibson who surveyed 40 randomly selected farm buildings assessing their colour, lichen colonisation, pH and reflectance. It was found that for the first two years the roofs remained near-white then changed progressively until after 12 years some 70% of roof was covered with dark grey lichen, mostly *Phaeophyscia orbicularis*. Reflectance was reduced in direct proportion to lichen cover and by 16 years was very close to the Design Guide recommendation.

The time taken for lichens to render an asbestos building visually acceptable in the farm landscape is very short in relation to the life span of the building. No harmful effects of lichen growth on asbestos have been observed.

Wayside trees

The well-lit boles of wayside deciduous trees support a lichen community that is dependent on the mild nutrient enrichment provided by animal excrement and dust stirred up by passing traffic. It is loosely known as the 'Xanthorion' and characterised by species of *Buellia, Caloplaca, Physcia, Physconia, Ramalina* and *Xanthoria*. Rarer members, now in decline, include *Anaptychia ciliaris, Bacidia rubella, Parmelia acetabulum* and *Candelaria concolor*. The latter species was so frequently recorded in old floras, in comparison with its distribution today, that for a time it was thought to have been confused with the much commoner *Xanthoria candelaria*. The full importance of this habitat was only realised with the publication of the lichen Red Data List (Church *et al.*, 1996); it supports five Red Data List species, *Caloplaca flavorubescens, C. luteoalba, C. virescens, Collema fragrans* and *Physcia tribacioides*, and the following four unrecorded since 1960 and therefore considered extinct: *Caloplaca haematites, Collema conglomeratum, Lecania fuscella* and *Lecanora populicola*. Some of the best trees in farmland now that elms have gone are introduced species such as sycamore and poplar. *Anaptychia ciliaris* is a particularly handsome lichen; its margins are covered with long hairs (cilia) which make it the 'wooly-bear caterpillar' of the lichen world.

Reasons for the deterioration of this habitat include the spread of industrial air pollution, the death of elms, tree removal associated with the enlargement of fields, road widening and the use of agrochemicals, in particular fertilisers (Coppins, 1996). It is the latter that is the most worrying as it is almost universal, affecting both arable and pastoral areas. For some time lichenologists have been aware that isolated trees in farmland supported an unexpectedly poor lichen flora and that the surviving species were becoming smothered by the green alga *Desmococcus*. The scale of the decline took people by surprise and has been compared to the disappearance of once common farmland birds. It

Ministry of Agriculture, Fisheries and Food   Leaflet 753
Published 1980

Lichen on Farm Roofs

A group of farm buildings shortly after erection.

The same buildings nine years later, showing colour damages due to lichen growth.

Fig. 9.8 A group of farm buildings photographed shortly after erection and nine years later. The pronounced colour change of the roofs is due to lichen growth and helps the buildings blend into the landscape. From Ministry of Agriculture, Fisheries and Food, 1980.

was believed to be a direct effect of contamination by fertilisers and confirmation of this has come from the small rural parish of Plymtree in Devon where Benfield (1994) kept ten hedgerow oaks under observation throughout a period when liquid slurry from dairy farming started to be sprayed on the fields. The result was threefold; a rapid increase in *Desmococcus* so that it thickly covered the bark and any lichens present; a decrease or extinction of species typical of slightly acid bark such as *Hypogymnia physodes*, *Parmelia sulcata* and *Pertusaria amara*; and an increase of 'weedy' nitrophilous species of *Physcia* and *Xanthoria*. Elsewhere, an increase in the density of the circular white colonies of *Diploica canescens* has been noted as indicating contamination with artificial fertiliser.

Now there is an additional concern on the horizon. Animal units and dense stocking rates create localised areas of high ammonia pollution in farmland. These are often associated with exactly the symptoms described above. Work in the Netherlands suggests that this is a consequence of elevated pH caused by the alkaline ammonia gas rather than a nutrient effect (de Bakker, 1989).

The two landscapes of lowland England

We are familiar with the division of England into a highland and a lowland zone, but tend to forget that the latter can be further divided. On the one hand there is the 'Ancient Countryside' of, for instance, Essex and Herefordshire, described by Rackham (1986) as 'the England of hamlets, medieval farms in hollows of the hills, lonely moats and great barns, pollards and ancient trees, holloways, footpaths, fords, heathland and thick crooked hedges full of oak and ash, an intricate land of mystery and surprise'. On the other there is the Cambridgeshire type of 'Planned Countryside', the England of 'big villages, thin hawthorn hedges, windswept brick farms, ivied trees in corners of fields, little bracken or broom, a predictable landscape of straight lines'. One is the product of at least a thousand years of continuity, the other of the eighteenth and nineteenth century Enclosure Acts.

The reputation of lichens as environmental indicators could be extended by determining those species typical of Ancient and Planned Countryside. I would predict that rather local epiphytes such as *Enterographa crassa*, *Graphis scripta*, *Pertusaria flavida*, *P. hemisphaerica* and *Thelotrema lepadinum*, together with the terricolous species *Cladonia coccifera*, *C. portentosa*, *C. uncialis* and several *Peltigera* spp. will be preferential for 'Ancient Countryside'. Detailed mapping at the tetrad (2 x 2 km) scale would show this relationship most clearly.

Local character is an increasingly prized commodity; so a demonstration of the way lichens contribute to this by selectively colonising building stone would also help to raise appreciation of them. The geological background is available in books such as *Oxford Stone* (Arkell, 1947), *The Pattern of English Building* (Clifton-Taylor, 1987) and *English Brickwork* (Brunskill & Clifton-Taylor, 1978). These provide copious information on the diverse substrata; lichenologists now need to tabulate the communities typical of each.

10

Work, Wealth and Wheels

The activities of man, involving work, the accumulation of wealth and the provision of transport facilities, have had a major effect on the environment, creating many new habitats. This chapter explores those lichens that have become closely associated with urban areas.

Towns, characterised by high levels of disturbance, would not normally be considered suitable for lichens, which are noted for their slow growth rate. However, a surprisingly large number of species are early colonisers and can be accommodated in the strategy group, stress-tolerant ruderals, while a few possess the character of ruderals, a term used to describe plants that are adapted to living in disturbed conditions. Urban areas are ideal for studying those pioneer species typical of early successional stages. It was the chance discovery of the ruderal lichens *Cladonia humilis*, *Peltigera didactyla* and *Sarcosagium campestre* on a recently cleared inner city site in Sheffield that first stimulated my interest in this habitat.

It might be thought that air pollution would depress the urban lichen flora to the point of insignificance, but sulphur dioxide does not influence terricolous species to anything like the extent it affects epiphytes, and the alkaline nature of many city habitats also provides protection. No-one was examining these sites during the period 1945–65, when air pollution was at its height, so we may never know whether the highly mobile, opportunistic urban lichen flora now present has persisted from the industrial period or has recently invaded.

Urban areas are characterised by large numbers of introduced higher plant species, most of which have arrived accidentally through trade and commerce. Their typical pattern of spread is a slow and uncertain start that eventually speeds up. It is remarkable that no unequivocal example of an introduced lichen spreading in Britain has been reported. The most likely candidate is *Lecanora conizaeoides* (see p. 56), other contenders being the epiphytic brown *Parmelia* species *P. elegantula*, first seen at Knole Park, Kent in 1965, *P. exasperatula*, first noted in 1888 and *P. laciniatula*, first found in 1933 at Dorking, Surrey, although their spread could equally be associated with aspects of air pollution. Very recently a fourth species *P. submontana*, has been found in three sites in Scotland, growing on non-native trees, gravestones and wooden garden tubs. These raise the number of possible introductions to five, or 0.3 per cent of our lichen flora. In contrast, urban areas in Britain contain at least 20 bryophytes that are accepted as recent immigrants, a number of them from the Southern Hemisphere. The UK, with 500 years of worldwide trade, extensive disturbed habitats, and the best documented lichen flora in the world, is well fitted for the detection of immigrant lichens so successful imports must be rare. Certain introductions have possibly gone undetected, in which case studies on the genetic variability of populations should settle the question, for example with regard to the brown *Parmelia* species.

Work

The urban common

Ecologically, one of the most significant recent events in towns has been the expansion of wasteland. This habitat principally occupies sites in the inner cities where heavy industry, and its associated substandard high-density housing, was demolished during the decade 1975–85. Plans exist assigning most of these sites to a new use but, following demolition and grading out, the majority are left dormant for a while during which they slowly green over as vegetation spontaneously colonises the brick rubble; people wear paths across them, bonfires are lit, fly-tipping occurs and they gradually become unofficial public open space (Fig. 10.1). These urban commons range in size from 0.1 to 10 ha. A wave of new building and landscape work has greatly reduced the extent of this habitat, but it is still present in most cities, providing a first-class opportunity to study early successional stages and to become acquainted with lichens that are rare in the countryside.

A nationwide survey of urban commons during their heyday revealed a lichen flora of some 100 species; there was an average of ten lichens on sites three to four years old rising to 29 after 13 to 16 years (Gilbert, 1989; 1990). The flora consists of the following communities.

Terricolous communities

It is unusual to find many terricolous species until a site has lain dormant for around eight years. By this time, higher plants will have colonised the more

Fig. 10.1 Demolished industrial premises where brick and concrete rubble, wood, rubbish and spreads of shallow gritty soil lying over vast concrete floors provide a superb habitat for pioneer lichens.

favourable areas, restricting terricolous lichens to spots where stress factors, such as nutrient deficiency, drought, compaction or toxicity, limit the vegetation to swards of low-growing mosses.

The pioneer lichen of such spots is the cup-lichen *Cladonia humilis*, which locally forms extensive patches of pale, but bright, fruit bodies. Occasionally it is joined by *C. fimbriata* or *C. rangiformis*. Where there is light trampling, such as beside paths, *Collema tenax* and *C. limosum* should be searched for. The latter species, seen in five towns, is uncommon in Britain, and was formerly regarded as a lichen of chalk grassland. Other strict calcicoles seen on these thin, mortar-rich soils of *c.* pH 8, include *Leptogium turgidum* and *Toninia sedifolia*, both at Swansea Docks, and the insignificant-looking *Verrucaria bryoctona*, which may well have its headquarters in urban areas.

Old bonfire sites and places where cinders have been incorporated into the soil hold a distinctive community composed of the prolifically fruiting ephemeral species *Sarcosagium campestre*, *Steinia geophana*, *Vezdaea leprosa* and *V. retigera*. Mostly they occur among or overgrow the mosses *Barbula convoluta*, *Ceratodon purpureus* or *Funaria hygrometrica* during the brief period before higher plants take over. Occasionally they occur on soil away from ash, the *Vezdaea* species often adjacent to pieces of metal lying on the ground. Additional species, more or less restricted to these nutritionally unbalanced, ash-enriched soils, are the small, dark brown, lichenised agaric, *Myxomphalina maura*, which is usually associated with a jelly-like green algal sward, and *Peltigera didactyla*, the only *Peltigera* common on these sites.

Areas of dumped furnace slag or cinders are a feature of certain sites; after 8 to 10 years surface leaching has reduced the pH of the top centimetre of the substratum from *c.* 7.5 to *c.* 5.5 with the result that extensive swards of *Cladonia* may develop. Bryophytes pioneer the succession, building up a surface layer of organic matter, which is colonised by *C, chlorophaea*, *C. coniocraea*, *C. subulata*, *C. furcata* and the red-fruited species, *C. coccifera* and *C. floerkeana*. Two unusual taxa that may be present are *Stereocaulon vesuvianum* var. *symphycheileoides*, which has sorediate margins to the basal squamules, and the rare *Thelocarpon lichenicola*. Its minute lime-green dots were recorded on soil in Hull.

Gravel-sized fragments of rock, mortar and clinker, lying on the soil surface or embedded in moss carpets, are often colonised by a single crustose lichen. It is traditional to consider these pebble communities as part of the terricolous flora. Calcareous pebbles commonly support *Caloplaca citrina*, *Lecanora dispersa*, *Sarcogyne regularis* or *Verrucaria muralis*, while the more acid fragments hold *Micarea lithinella*, *Thelidium minutulum*, *Trapelia coarctata*, *T. involuta* or *T. obtegens*. All are favoured by the moist sheltered conditions. A splinter of sandstone from a four-year-old site in Sheffield produced the first British record of *Lecidea polycarpella*; otherwise known only from loose stones by tracks in Austria and Germany, it is clearly an early colonist.

To comprehend fully the terricolous lichens of urban wasteland it is necessary to survey sites in bright weather with the sun high in the sky to minimise shadows, and to spend most of the time on hands and knees. One's first acquaintance with the tiny fruits of ephemeral species in such an inhospitable habitat is like discovering hidden treasure. After 15 years most sites pass into a grassland-scrub stage, and lichen habitats become buried under vegetation and litter; their heyday is now over, as other groups rise to prominence.

Fig. 10.2 A brick rubble site four years after demolition. The mortar-encrusted bricks already carry a 5% cover of lichens.

Lignicolous communities

Lignicolous habitats in the form of worked timber are widespread at most sites. They range from large items such as pallets, beams and cable drums, through planks, chairs and sheets of plywood, down to small blocks and fragments. Being at ground level, wind-blown dust and rain-splash quickly impregnate their surface with mineral particles. As a result, over half the 28 species recorded are better known as saxicolous lichens with a wide ecological amplitude, for example, *Caloplaca holocarpa*, *Candelariella aurella*, *Lecanora dispersa* and *Lecanora muralis*. A typical lignicolous flora develops only where such contamination is reduced.

One of the most successful colonists of this habitat is *Micarea denigrata*, often present as a form with few apothecia but numerous white-tipped pycnidia. Associated species are *Lecanora saligna*, *Placynthiella icmalea*, *P. uliginosa*, *Trapeliopsis flexuosa* and *T. granulosa*. A notable feature of this flora is the regular presence of *Thelocarpon laureri*. It can be widespread, but does not favour damp or rotting wood, the vicinity of nails (a niche favoured by *Scoliciosporum umbrinum*), charred material or shaded sites. A baulk of rotting wood in Birmingham supported a large patch of *Stereocaulon vesuvianum* var. *symphycheileoides*, while a similar baulk in Bristol held *Thelocarpon intermediellum*. The scarcity of foliose lichens is probably a result of air pollution, but this cannot explain the rarity of *Lecanora conizaeoides*, which remains an enigma.

Saxicolous communities

The major saxicolous habitat of demolition sites is mortar-encrusted brick. Bricks provide a nutritionally unusual substratum, being rich in phosphorus, potassium, calcium and magnesium, which are abundant in the clay from which they are manufactured (Fig. 10.2). Engineering bricks have a dense, strong, semi-vitreous body with a hard, smooth exterior. They support a flora of calcifuge species such as *Porpidia tuberculosa* and *Rhizocarpon obscuratum*. The community that develops on the softer, more water-retentive common brick, the 'commons' of the trade, is very different. These have a surface pH of 7 even after weathering. The colonisation of brickbats starts very early; after only four years they may have a 5% cover of *Trapelia coarctata* and *Verrucaria muralis*, together with *Lecanora dispersa*, *Thelocarpon laureri* and *Trapelia obtegens*. As rosebay willowherb and other species spread over the site, and the microclimate at ground level becomes increasingly damp and sheltered, a range of pyrenocarps with minute fruiting bodies invade such as *Pyrenocollema monensis*, *Thelidium minutulum*, *T. zwackhii*, *Thelocarpon intermediellum* and *Verrucaria simplex*. A further group of lichens colonising damp, mortar-encrusted brick have pale fruits that may be ephemeral; they include the strongly urban species *Bacidia caligans*, *B. chloroticula* and *B. saxenii*.

Lumps of concrete provide a markedly calcareous habitat; the pH of a freshly broken surface is 12. Five years exposure to the elements lowers this to around 8, at which stage the pioneer alliance *Lecanoretum dispersae* starts to colonise (Plate 11a). Prominences that are used as bird perches hold nitrophilous species including, on occasion, *Acarospora heppii*, observed in Hull. A striking community is sometimes found where iron girders or reinforcing rods project from horizontal concrete. An iron-stained area with a lowered pH develops at these points, which favours *Lecanora stenotropa*, *Scoliciosporum umbrinum* and *Thelocarpon laureri*.

Coarse limestone chippings frequently find their way onto urban commons and are always worth examining. They are a good habitat for a range of *Verrucaria* spp., *Protoblastenia rupestris* and *Sarcogyne regularis*. A hand lens trained on a chipping picked up in Carlisle revealed a scatter of yellowish-green discs; the specimen was later identified in Edinburgh as the first British record of *Acarospora heppii* f. *luteopruinosa*.

Sandstone is mostly present as a component of dumped material, often along with subsoil. Early colonisers are *Lecanora dispersa*, *Trapelia* spp. and *Verrucaria muralis*. After a while, more acidophilous lichens such as *Lecanora stenotropa*, *Micarea erratica*, *Porpidia crustulata*, *P. tuberculosa* and *Rhizocarpon obscuratum* may appear, but remain subordinate. Flushing by dust and rainsplash create mesotrophic conditions, so that strongly acid surfaces are rare on urban sites. Pieces of clinker, which can have a surface pH below 5, are favoured by *Micarea lithinella*.

Other habitats

Plastic, cardboard, rubber tyres and cloth have rather neutral properties so their lichen flora is largely determined by the environment in which they come to lie. Species with a wide substratum amplitude are common on these man-made materials, though it was observed that dust contamination accelerates colonisation. This can be seen on old car tyres, where the initial establishment of *Candelariella aurella*, *Lecanora muralis* and *Scoliciosporum umbrinum* is around dust-filled depressions.

Metal objects abound on demolition sites. Once they have become well rusted, horizontal surfaces are the place to look for the thin, pale green thalli of *Bacidia saxenii*, which has been found on oil drums, tin cans and cast iron. The closely related *B. chloroticula* has a habitat range that embraces plastic and leather, while *B. caligans*, usually found on moss, was collected off a rotting blanket.

When well compacted, cardboard provides conditions not dissimilar to soil. Boxes rot too quickly, but may acquire a fleeting cover of *Steinia geophana*. The best type of cardboard for lichens is provided by the thick sides of cable drums, which remain discrete for long enough to be colonised by *Cladonia* species. The first British record of *Micarea excipulata* came from a cable drum on the site of a demolished steelworks in Sheffield.

Broken pottery fragments are rather acidic (*Porpidia crustulata*), hardboard resembles lignum (*Micarea denigrata*), cotton waste is similar to soil (*Cladonia humilis*), tarmac is mildly basic (*Buellia punctata*, *Lecanora muralis*), pieces of coke are always worth examining (*Bacidia saxenii*), while glass remains uncolonised.

General attributes

When lists from different towns were compared, it was seen that variation within towns is greater than variation between them. It had been hoped that it might be possible to identify regional patterns of urban lichens, as has been done for higher plants (Gilbert, 1989), but the only suggestion of a climatic effect was the greater frequency of *Collema* and *Leptogium* spp. in the west of Britain. Surprisingly, towns surrounded by Carboniferous limestone, like Bristol, had a very similar urban lichen flora to the small Pennine town of Glossop, which is largely built of and surrounded by Millstone Grit. The common factor is the universal presence of mortar.

The majority of British lichens can be classified as stress tolerators, characterised by a slow growth rate, longevity, and devoting only a small proportion of their annual production to reproduction (Grime, 1979). Many of the lichens encountered on urban wasteland, however, fit more closely into the category stress-tolerant ruderals. Such plants are typical of habitats where there are moderate intensities of stress and disturbance. *Bacidia chloroticula*, *B. saxenii*, *Peltigera didactyla*, *Steinia geophana*, *Verrucaria bryoctona* and *Vezdaea* spp. are candidates for the ruderal strategy, being ephemeral colonists with a rapid growth rate and a short life cycle that involves fruiting early, then dying, having diverted a high proportion of their annual production to reproduction. Recent work has shown that the life span of *Peltigera didactyla*, from soredium to death of the fertile thallus, can be less than 12 months. It would be rewarding to carry out further investigations into the fecundity and longevity of pioneer species.

Wealth

For thousands of years man has been winning wealth from the rocks in the form of metals. This activity has left a profusion of scars on the landscape, which are regarded as horrific or romantic, depending on one's point of view (Plate 11b). Whatever their visual merits, such sites are now compulsory stops for lichenologists. It was only in the 1970s, in the UK, that their potential for lichens was realised (James *et al*, 1977; Gilbert, 1980a, b). Since then, in a flurry of catching up, most metal mining areas have been surveyed during British Lichen Society field meetings, through contract work, or as a result of personal initiatives. Steve Chambers was financed for two years to survey the lichens of mine heaps in mid-Wales. All this endeavour has confirmed that these environments support a rich and specialised lichen flora, having added at least ten species to the British list, while others, new to science, are waiting to be described. The surveys also revealed that the sites are almost universally threatened by reclamation schemes, reworking using modern mining technology, or recreational activities such as off-road driving. In 1988 it was estimated that 68,000 ha of land in England and Wales were contaminated by metalliferous spoil.

A recent publication by the Department of the Environment (1994) listed the following numbers of specialist metallophytes (plants showing a close association with metalliferous habitats) in Britain: vascular plants 15; bryophytes 13; lichens 32; so lichens are twice as well represented as any other group. The nature of the metal enrichment is all-important in determining which species are present; iron-rich, copper-rich and lead/zinc-rich environments each have their characteristic assemblage (Table 10.1). Alongside the metallophytes many lichens occur that are generally widespread and in no way restricted to such sites, for example, common species of *Cladonia, Lecanora, Lecidea, Peltigera, Porpidia* and *Trapelia*. They are sometimes known by the ugly and unnecessary term pseudometallophytes.

Table 10.1 A selection of specialist metallophyte lichens occurring in the British Isles (after Purvis, 1996)

Iron sulphide-rich rocks	Copper-rich rocks	Lead/zinc-rich rocks
Acarospora sinopica	*Lecidea inops*	*Bacidia saxenii*
Arthonia lapidicola	*Psilolechia leprosa*	*B. viridescens*
Lecanora epanora	*Stereocaulon leucophaeopsis*	*Cladonia cariosa*
L. handelii	*S. symphycheilum*	*Gyalidea roseola*
L. subaurea		*G. subscutellaris*
Lecidea silacea		*Peltigera neckeri*
Miriquidica atrofulva		*P. venosa*
Rhizocarpon furfurosum		*Placynthiella hyporhoda*
Stereocaulon leucophaeopsis		*Sarcosagium campestre*
		Steinia geophana
		Stereocaulon condensatum
		S. dactylophyllum
		S. delisei
		S. glareosum
		S. leucophaeopsis
		S. nanodes
		S. pileatum
		Vezdaea (all six British species)

Lead and zinc mines

Lead and zinc mines occur in many parts of upland Britain. Ecologically they can be grouped into two types according to the nature of the rock in which the mineralisation occurred. In northeast Wales, the North Pennines, Derbyshire and the Mendips the mines are associated with an alkaline, limestone environment, whilst in Scotland (Strontian, Tyndrum, Leadhills), Cumbria, central Wales and the Isle of Man mineralisation has taken place in acid rocks. The

Ordovician areas of Cornwall exhibit both acid and alkaline environments due to secondary deposition of calcite.

Abandoned mines are complex places frequently comprising ruined buildings amid spoil heaps, collapsed shafts, tunnels, dressing floors, leats and settling lagoons. Some understanding of working methods helps to focus a survey. The highest contamination is usually where ore processing or smelting took place; here deposits containing up to 50% total metal can be found, whereas dumps of rock created during the development of the mine may have metal levels little above background.

The alkaline mines of the Northern Pennines and Derbyshire were among the first to be studied in detail (Gilbert, 1980a, b; Coppins & Gilbert, 1981). At that time, many of the spoil heaps were being reworked for fluorspar and barytes, so the sites were a mixture of old and recently disturbed areas, some grazed, others ungrazed. Older, lightly grazed workings were covered with *Cladonia rangiformis*, which occasionally was so abundant as to give the ground a frosted appearance. Due to the large amounts of waste limestone in the spoil heaps they carried an open fescue-thyme turf in which the calcicoles *Bacidia muscorum, B. sabuletorum, Cladonia pocillum, Collema tenax, Diploschistes muscorum, Leptogium teretiusculum, Peltigera rufescens, Polyblastia gelatinosa* and *Vezdaea aestivalis* occurred. Rarer species included *Pannaria pezizoides, Peltigera leucophlebia* and *P. neckeri*. Litter accumulation resulting from retarded decomposition is a feature of all heavy metal sites. This microhabitat brings in a mildly calcifuge element including *Baeomyces rufus, Cladonia cervicornis* subsp. *verticillata, C. scabriuscula, C. subulata, C. uncialis* and *Coelocaulon aculeatum*, which are found on the undecayed bases of grass tussocks and on peaty overhangs at the edge of terracettes.

Many older heaps become surface leached, so that reworking, which stirs them up, restores toxicity. It was a major revelation to find that restored sites regularly support a group of inconspicuous metallophytes with ephemeral fruit bodies. In spring and autumn, the mineral soil, bryophytes and plant debris are covered with tiny fruits belonging to *Bacidia saxenii, B. viridescens, Sarcosagium campestre, Steinia geophana, Verrucaria bryoctona, Vezdaea acicularis, V. leprosa, V, retigera* and *V. rheocarpa*. This assemblage also occurs along older veins where there has been disturbance from vehicles or animals. The seasonality of this community would be a rewarding field for research.

Mine workings in the Pennines have provided an open, competition-free refuge for relic populations of lichens well south of their normal range. For example, several mines on the Northumberland-Cumberland border support fine colonies of the Red Data List species *Peltigera venosa*, and lead/zinc rakes in Derbyshire are home to *Cetraria islandica*.

Mining over the last 2,000 years has considerably extended opportunities for metallophyte lichens. A further habitat for these species has come to light which owes less to man's activities. These are patches of heavy-metal shingle beside rivers downstream of ore fields. The best known are by the Rivers Tyne and South Tyne in Northumberland (Fig. 10.3). They carry an open vegetation, which includes the higher plant metallophytes Young's helleborine (*Epipactis youngiana*), spring sandwort (*Minuartia verna*) and alpine pennycress (*Thlaspi caerulescens*), together with a wide range of the species in Table 10.1 including *Peltigera venosa*. Other examples, by the Rivers Rheidol and Ystwyth in mid-Wales, support populations of *Baeomyces placophyllus, Cladonia cariosa, C.*

Fig. 10.3 Heavy metal shingle by the river South Tyne at Lambley provides a rare natural habitat for metallophytes.

uncialis subsp. *uncialis, Stereocaulon condensatum* and *S. glaerosum.* These sites may have remained open throughout the postglacial forest period, but the extent to which the heavy metal burden has been increased by later mining activity is not known.

The acid mine spoils of Wales, Cumbria and Scotland are taxonomically challenging sites yielding long lists of saxicolous species. Due to the unusual nature of the substratum the lichens frequently appear poorly developed or otherwise untypical; this is no place for beginners. In addition to lead/zinc specialists, species of iron-rich rock are also abundant (Table 10.1), together with many lichens that have no special affinity with heavy metals. When working the spoil heaps, or along walls constructed from low grade vein material, the genera most frequently encountered will be *Acarospora, Lecanora, Placopsis, Porpidia, Rhizocarpon, Stereocaulon* and *Trapelia*. My own favourites, all of which can be identified in the field, are *Acarospora smaragdula, Baeomyces roseus, Placopsis lambii, Stereocaulon delisei, S. leucophaeopsis* and *Trapelia mooreana*. In 1962, Peter James and Dougal Swinscow made the first and only UK record of the beautiful *Gyalidea roseola*. It was growing at the entrance to an adit at Feish Dhomhnuill Mine in Strontian; all attempts to re-find it have failed.

Disused settling lagoons are present at the larger sites, recognisable as a series of terraces covered with highly toxic silt. These usually resist colonisation by plants, but locally spreads of moss or grass may be present. Any terricolous lichens tend to be associated with these or with gravelly areas. A search of this habitat is likely to reveal *Baeomyces roseus*, which only regularly fruits at mine sites (Plate 11c), *B. rufus* (Plate 11d), often with its citrine-green parasite *Epilichen scabrosus*, *B. placophyllus* and, possibly, *Cladonia cariosa* and *C.*

fragilissima. Extending the search to hands and knees should add several minute species if what appears to be 'algal scum' is viewed through a lens, for example *Belonia incarnata*, *Bryophagus gloeocapsa*, *Epigloea soleiformis* and *Thrombium epigaeum*.

At acidic mines it is mandatory to examine the site of buildings and other mortared structures. Alan Fryday and Steve Chambers were the first to observe how the enhanced pH in such sites brings in a wide range of additional terricolous species such as *Collema tenax*, *Leptogium gelatinosum*, *Peltigera didactyla*, *Polyblastia agraria*, *P. wheldonii*, *Sarcosagium campestre*, *Thelocarpon epibolum* and *Vezdaea* spp. A similar increase in pH is also typical of ancient ash dumps associated with the smelting process.

The value of metal mine spoil as a persistent weedkiller has been recognised for a long time, so metalliferous communities can turn up in surprising places. In mid-Wales the Cambrian Railway Company used spoil from the Van Lead Mines to surface the track up to 50 km away (Woods, 1988). Elsewhere, foresters and farmers have used it as a top dressing on unsurfaced roads; the first report of *Vezdaea acicularis* came from one such track in Camarthanshire.

Copper mines

Copper mines are not common, but can be found in Cornwall, North Wales and the Lake District. Their lichen flora is highly distinctive. There can be few greater thrills for a lichenologist than being the first to explore a major complex of workings. Such an experience fell to William Purvis and Peter James

Fig. 10.4 View of the disused Coniston Copper Mines. The hillsides are scared with old workings; ore-dressing took place in the foreground (O.W. Purvis).

when, in spring 1984, they spent two days surveying the Coniston Copper Mines in the Lake District (Fig. 10.4), (Purvis & James, 1985). At this site, natural surface exposures of the ore-rich veins (chalcopyrite) are poor, so artificial habitats like mine dumps, ore-dressing floor walls and large, sunny boulders were the main ones available for study. At the end of their visit they had found five species new to Britain, including *Lecidea inops*, a specific indicator of copper mineralisation. Later an entirely new lichen community – the *Lecideion inopsis* alliance – was described to accommodate lichens on copper-rich rocks; so far it is known only from Central Scandinavia and two sites in the Lake District (Purvis & Halls, 1996). Other British species characteristic of copper mineralisation are *Psilolechia leprosa*, which occurs in sheltered high pH sites such as mortared walls, chimneys and inside flues, and *Stereocaulon leucophaeopsis* and *S. symphycheilum.* found on exposed acid boulders. At Coniston these species occur alongside, but separate from, members of the *Acarosporion sinopicae* alliance as the copper ore is flanked by associated material rich in iron sulphides.

A number of the lichens found on copper-rich rocks are an anomalous green colour. Several were originally described as new to science, but it is now realised that they are just copper ecotypes of well-known species. British species exhibiting this phenomenon include *Acarospora smaragdula* (Plate 11e) and *Buellia aethelea* in which the coloration is due to copper complexing with norstictic acid, and *Lecanora polytropa* in which the origin of the green inclusions is unknown. Green ecotypes are only sporadically present; factors determining their occurrence would be a fruitful field of research for amateurs, with Cornwall the best place to start. In addition to copper mines, green copper ecotypes have been reported from an epidiorite cliff in the north of Scotland, beside the copper lightening conductor on St David's Cathedral, Pembrokeshire, from church window sills, on fence posts treated with copper preservative, and colonising vine supports in Austria that had been sprayed with Bordeaux Mixture (copper sulphate).

Iron-rich rocks

In a typical mine, most veins are occupied by secondary materials in which strings of the metal ore occur. When excavating the ore the secondary material is also removed and accumulates as conspicuous waste heaps close to the mine entrance. One of the commonest types of this secondary material is rock rich in ferrous sulphide which, on weathering, creates a low pH environment. Waste heaps and walls constructed from lumps of sulphide-rich rock can have a pH as low as 2, due to the release of sulphuric acid.

Rocks rich in ferrous sulphide support a highly characteristic lichen community typified by rust-coloured species such as *Acarospora sinopica* and *Rhizocarpon oederi*. Two variants of the community are involved, which can be used to illustrate the continental approach to vegetation classification in which units are given Latin names and arranged in a hierarchy. Table 10.2 shows how the entire lichen flora of iron-sulphide rocks belongs to an alliance named the *Acarosporion sinopicae*, typified by five Character Species, each with a strong preference for this vegetation unit. The alliance is further divided into two associations exploiting exposed and sheltered niches respectively, each with their own Character Species. At most large, acid mines, such as those at Coniston and Parys Mountain, both associations can be found.

Table 10.2 An example of the continental approach to classifying lichen communities. The *Acarosporion sinopicae* alliance, typical of iron sulphide-rich rocks, and the two associations into which it can be divided (after Purvis & Halls, 1996).

***Acarosporion sinopicae* alliance**
Character species of the alliance
Acarospora sinopica, Lecanora polytropa, Rhizocarpon furfurosum
R. oederi, Scoliciosporum umbrinum

***Lecanoretum epanorae* association**	***Acarospotetum sinopicae* association**
Vertical, dry, sheltered overhangs	Exposed, sunny, horizontal surfaces
Lecanora epanora	*Buellia aethalea*
L. handelii	*Lecanora soralifera*
Lecidea endomelaena	*L. subaurea*
Lepraria spp.	*Lecidea lapicida*
	L. silacea
	Miriquidica atrofulva
	Stereocaulon nanodes
	Tremolechia atrata

Acid, sulphide-rich sandstones also occur away from mine sites. In the Pennines, sandstones interbedded with shales are sometimes of this type. Where they have been used to construct drystone walls, sheltered faces may be covered with *Lecanora epanora* and *L. handelii*, while nearby cliffs support *Lecidea endomelaena* and *Rhizocarpon furfurosum*. Terricolous lichen communities associated with ferrous sulphide mineralisation are generally species-poor and composed of common and widespread heathland species; a change from the richness of this habitat at lead/zinc mines.

Pylon lines

The National Grid estimates that there are 25,000 pylons in England and Wales with another 5,000 in Scotland. Their importance to lichenologists lies in the toxic shadow produced by water dripping off the metal. Under acid conditions, grass growth is inhibited below the dripline, often being replaced by a moss-dominated community in which lichens figure (Fig. 10.5). Toxicity is greatest under new pylons which are made from galvanised (zinc-coated) steel. After ten years the towers are coated with a protective resin-based paint so metal additives to the soil decline. Under older, heavily painted pylons, the chief source of toxicity is rainwater dripping from the barbed wire defences incorporated to deter climbers.

Pylons crossing arable land, improved pasture or similar fertile sites have few if any lichens on the ground underneath them, which is densely colonised by rank grasses. On less fertile land, open conditions predominate under the pylons and soil, moss and stones frequently carry a selection of early colonising lichens such as *Bacidia arnoldiana, B. chloroticula, B. saxenii, Lecanora dispersa* and *Trapelia coarctata*, any of which may be dominant. Regularly present at the best lowland sites are the metal-tolerant species *Sarcosagium campestre, Steinia geophana, Thelocarpon laureri* and *Vezdaea leprosa*. Pylons in the uplands can also provide surprises. Beside the A939, east of Tomintoul, a transmission line crosses the Lecht Pass at 620 m. Here the ground under several towers is white with the rarely recorded *Stereocaulon glareosum* while toxic stripes, bare of higher plants but supporting *B. saxenii* and *V. leprosa* extend down the hillside from each leg.

Taking their lead from Britain, lichenologists in Canada have started to examine the ground under pylons. They found the same range of species, which may in consequence be an international assemblage with a circumpolar distribution. What pleased them in particular was that two were new to North America.

Quarries and other workings

Disused quarries and other workings are less exciting than metal mines. However, working on the hypothesis that every site holds some lichen interest, a large number have been examined over the years. Their floors are reliable sites for tracking down the commoner species of *Cladonia* and *Peltigera* which, if the site is rabbit-grazed, frequently dominate the ground. In my experience *Cladonia cariosa* and *C. fragilissima* have their UK headquarters in quarries, ranging from the Millstone Grit in Yorkshire to ball-clay pits in Dorset. Chalk pits and flint mines in Breckland were once legendary as the home of the Breckland rarities, and are still the main sites in southern England for *Psora decipiens*. The few pit heaps that remain in our coalfields bring red-fruited species of *Cladonia* deep into the lowlands, and it is frequently forgotten that Wimbledon Common, with its spreads of *Coelocaulon aculeatum*, is a series of disused gravel pits.

Wheels

Roads

Studying busy roadsides is unattractive and not often attempted. Though vehicle fumes are not toxic to lichens, some have difficulty coping with the dust and the 'traffic film' that builds up. The first evidence that trunk roads can support interesting species came from Germany where it was reported that

Fig. 10.5 The toxic shadow produced by water dripping off pylon towers. Metallophyte lichens thrive on the bare ground created.

Vezdaea leprosa, well known for its tolerance to heavy metals, regularly occurred in the drip zone below metal crash barriers, where it is part of a community that also includes *V. aestivalis* and *V. acicularis* (Ernst, 1995). Having read this I immediately went out and found *V. leprosa* on soil and dead vegetation under the third crash barrier I examined beside the A1(M). During the search, dead, grit-covered leaves bearing the white pycnidia of *Bacidia arnoldiana* were also encountered.

Another feature of road verge ecology is the number of maritime vascular plants invading them in response to winter applications of salt (sodium chloride). These species, which can be spotted from moving vehicles, grow in the strips of open ground beside the carriageways referred to by engineers as salt burn. To date, no maritime lichens have been reported from inland salt burn, possibly because special searches have not been made. The presence of the normally maritime *Solenopsora vulturiensis* on the salt-splashed wall of a bridge in Northumberland may be the first example of such an occurrence. To minimise the risk of chemical corrosion, urea is used for de-icing purposes on the Forth and Severn road bridges; its use runs to many tonnes per year. It would be an attractive project to assess its influence on the lichens of walkways and other structures.

Numerous papers have pointed out the increased levels of heavy metals, particularly lead, that occur in the vicinity of roadsides. Lead from petrol additives can accumulate till concentrations in soil and dust reach several thousand parts per million. These levels are not necessarily toxic to lichens, but can shift the balance of competition in favour of lead-tolerant species. The rise in car use since the Second World War has been matched by the advance of certain *Stereocaulon* species from the uplands, where they are often present on lead mine spoil, into towns where they are now widespread; a phenomenon that would have amazed Victorian lichenologists. The species involved are *Stereocaulon nanodes*, *S. pileatum*, *S. vesuvianum* and *S. vesuvianum* var. *symphycheileoides*, all of which are now prevalent on walls in built-up areas, including London. It should be stressed that the link with car usage is circumstantial and appealing rather than conclusive.

Along suburban roads a subtle interaction can be observed between dogs and pavement trees. The area against which they urinate is known to botanists as the 'canine zone'; it carries a different epiphytic flora to the rest of the trunk (Fig. 10.6). In towns this zone appears dark in contrast to the remaining trunk, which is covered by the pale green lichen *Lecanora conizaeoides* or the bright green alga *Desmococcus*. When well developed, the epiphytic flora of the canine zone is dominated by the green alga *Prasiola crispa*, present in its juvenile filamentous *Hormidium* stage. It forms a velvety felt on the bark, that is almost black when wet, glossy emerald-green when dry. Marginal to the strongly eutrophicated zone, up to a dozen lichens, including *Candelariella reflexa*, *Physcia tenella*, *Phaeophyscia orbicularis* and a form of *Lecanora dispersa* with a conspicuous white thallus, can be found.

Railways

The permanent way is the name given by railwaymen to the ballasted road bed on which the track is laid. It is composed of a 30–50 cm thick basal layer of cinder ballast (the cess), on which strips of much coarser ballast (the track bed) are put down to take the sleepers and rails. Once genuine ships ballast was

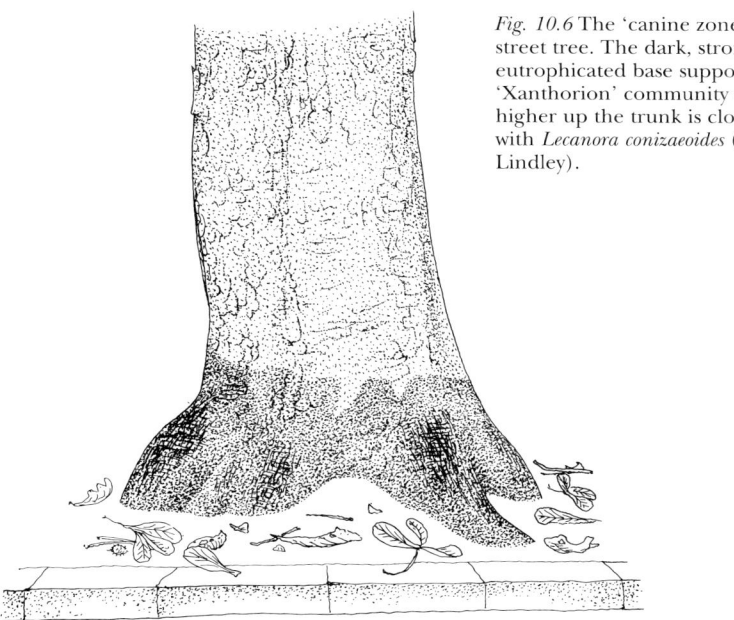

Fig. 10.6 The 'canine zone' on a street tree. The dark, strongly eutrophicated base supports a 'Xanthorion' community while higher up the trunk is clothed with *Lecanora conizaeoides* (M.J Lindley).

used, but now the rail companies have their own quarries. Every 10 years or so the track is taken up and the ballast mechanically cleaned to prevent a build-up of humus. An inspection of seldom used and disused tracks has revealed a small but characteristic lichen flora associated with the permanent way.

The corroded iron of railway chairs, used to secure the rails, sometimes becomes colonised by *Stereocaulon vesuvianum*, which may spread on to the sleepers (Fig. 10.7). Being at ground level, the wooden sleepers become ingrained with mineral matter and support a flora more typical of a saxicolous than a lignicolous habitat, for example *Candelareiella vitellina*, *Lecanora muralis*, *L. polytropa*, *Lecidella scabra* and *Scoliciosporum umbrinum*. As the wood rots this community is replaced by one containing species of *Cladonia* and *Trapeliopsis*. Railway ballast is too disturbed to support species of much interest, but where it is of limestone early colonisers such as *Acarospora heppii* and *Thelidium minulutum* may be present.

The cess quickly becomes colonised by a *Cladonia*-rich community in which two species are particularly characteristic; the cup-lichen *C. humilis*, with its patches of bright, pale green cups, and *C. rei*. The latter, regarded as rather rare in Britain, is frequently overlooked owing to confusion with the morphologically similar *C. subulata*. Recent work in Germany has shown that the two species are ecologically distinct. *C. rei* is most frequently encountered in disturbed secondary sites, such as railway cess, growing on bare mineral soil with a relatively high pH, a low humus content, and accompanied by species such as *Peltigera didactyla*, *C. humilis* and the moss *Ceratodon purpureus*. By contrast, *C. subulata* is usually found on raw acid humus in stable heathland communities associated with red-fruited *Cladonia* species.

Fig. 10.7 Stereocaulon vesuvianum *growing on the corroded iron of a railway chair and spreading onto the sleeper along a disused line in North Norfolk.*

Another railway habitat that deserves special mention is bridges. These provide a stonework habitat in areas of lowland Britain where it may otherwise be scarce. Though many are constructed predominantly of heavily glazed, acidic, blue engineering brick, the parapets and jokers (massive capstones) are often made of stone that has been transported a considerable distance. The habitat encompasses a variety of conditions, ranging from exposed to sheltered, acid to alkaline, damp to dry, fine to coarse textured, eutrophic to oligotrophic, etc. (Fig. 10.8). Up to 70 lichens can be found on a single bridge, which compares well with an entire churchyard. The following description is of four railway bridges near Adlestrop in Gloucestershire where Edward Thomas' famous train 'drew up'.

Parapets of blue engineering brick hold a distinctive acidophilous community that includes *Buellia aethalea, B. occelata, Catillaria chalybeia, Lecanora polytropa, Micarea erratica, Porpidia tuberculosa* and *P. soredizodes*. Sandstone jokers bring in additional species of dust-contaminated stonework such as *Caloplaca crenularia, Lecanora fuscoatra, Ochrolechia androgyna, Lecanora sulphurea, Pertusaria amara, Rinodina teichophila, Tephromela atra* and *Trapelia* spp. Parapets and jokers carved from well-illuminated oolitic limestone carry the expected range of *Caloplaca aurantia, C. teicholyta, C. variabilis, Solenopsora candicans* and *Verrucaria glaucina* with *Agonimia tristicula* on moss. Vertical, damp oolite on the wings, often shaded by vegetation, supports *Bacidia sabuletorum, Collema auriforme, Dermatocarpon miniatum, Leproplaca chrysodeta, Opegrapha saxatilis, Placynthium nigrum, Physciopsis adglutinata* and *Physconia grisea*. Due to the use of mortar, wings constructed of brick have a rather similar flora with additions including *Collema fuscovirens, Leptogium turgidum, Porina chlorotica* and *Toninia aromatica*. Studies in other parts of Britain have shown *Xanthoria elegans* to be frequent on railway bridges, with *Stereocaulon evolutum* and *Parmelia verruculifera* both occasional. For lichenologists in lowland Britain the above assemblages include exciting species. I suspect that railway botany has many more surprises in store.

Disused airfields

Ordnance Survey maps show that the countryside of Britain is littered with disused airfields, many dating from the Second World War. Their runways provide an extensive, secondary saxicolous habitat. Though subsidiary activities take place on many, it is always possible to find little-used areas of tarmac or concrete that are slowly being taken over by nature. An examination of a dozen examples, from Yorkshire to Suffolk, showed that the majority produce a list of under 30 lichens, most of them common also on asbestos roofs, so they do not add much to regional biodiversity. Typically, arable crops lap the edges of the runways, which are eutrophicated to the extent that the most abundant species are *Candelariella aurella*, *Caloplaca citrina*, *Lecanora dispersa*, *L. muralis* and *Physcia caesia*. Slightly unexpected was the high frequency shown by *Caloplaca isidiigera*, *Lecanora stenotropa* (the narrow-spored form of *L. polytropa*), *Sarcogyne regularis*, the presence of *Leptogium turgidum*, and, in Suffolk, fruiting *Collema tenax* var. *ceranoides*. The most exciting species is *Caloplaca crenulatella*, previously known only from a single nineteenth-century record in Cumbria. This lichen, reminiscent of a diminutive *Caloplaca flavovirescens*, was found on all the concrete runways examined, so probably has its British headquarters on airfields.

A quarter of the airfields examined were buffered from the surrounding farmland by woodland or heath, which was starting to take over. These were more interesting lichenologically. In places spreads of *Cladonia* including *C. cervicornis*, *C. furcata*, *C. rangiformis* and *C. squamosa* were invading the tarmac together with *Peltigera didactyla*, *P. lactucifolia*, *P. neckeri*, mosses and white stonecrop (*Sedum album*), which provided a humus base. Elsewhere, mildly

Fig. 10.8 Railway bridges are complex structures; the best support as many lichens as an entire churchyard.

Fig. 10.9 Extensive patches of *Stereocaulon pileatum* whiten the tarmac runway of a disused airfield at Skipwith, Yorkshire.

calcifuge species such as *Baeomyces rufus*, *Lecanora polytropa*, *Lecidella scabra* and *Trapelia placodioides* were present. At Skipwith in Yorkshire the tarmac runway is locally whitened by *Stereocaulon pileatum*, and the infrequently recorded *Leptogium biatorinum* is also present (Fig. 10.9).

Abandoned wartime airfields are renowned for ghostly appearances. Skipwith airfield, however, has been the setting of an unexplained disappearance. In 1980 the main runway was covered with an equal mixture of the normal yellow form of *Candelariella vitellina* f. *vitellina* and the citrine-green form f. *flavovirella*. There are voucher specimens in the herbarium of The Natural History Museum, London. Both were still present a year or two later, but by 1997 the citrine-green variety had completely disappeared, only the yellow form remaining.

Airfields in the north and west are rather different. They are more regularly tarmacked and support a fascinating range of *Parmelia* species such as *P. caperata*, *P. conspersa*, *P. loxodes*, *P. mougeotii*, *P. perlata*, *P. revoluta*, *P. tiliacea* and *P. verruculifera*, together with *Caloplaca arenaria*, *Candelariella coralliza*, *Lecanora rupicola* and *Placopsis gelida*. There are clear similarities with communities present on basalt. Several sites are now known with over 50 species, a number that could be raised with further work as the flora of individual runways can be astonishingly different. The richest airfield so far encountered is Smeatharpe in Devon which has yielded the rare *Phaeophyscia sciastra*, in fruit and *Staurothele frustulenta* new to Britain. The grit-covered margins of taxiways in the west of Britain support a lichen flora reminiscent of sand dunes with *Leptogium corniculatum*, *Peltigera necheri*, *Usnea articulata*, and *Cladonia rangiformis* as a vagant forming loose balls the size of a tangerine.

Disused airfields provide pleasant, if undemanding, botanising, with those influenced by adjacent farmland being less diverse than ones in a heathland setting. Examples in the north and west tend to be the richest. Together with the other environments covered in this chapter, they prove the point that any habitat can produce interesting species if it is lovingly scoured by lichenologists willing to return again and again. It is never worth getting dismayed by sites that fail to live up to expectations. Increasingly one sets out to discover what is there and to appreciate ecological relationships rather than to hunt for rarities, though occasionally I find the following lines of poetry pulsing through my head:

'Oh brother, not every sea has pearls,
Not every branch will flower,
Nor will the nightingale sing there on.'

11

Discovering the Montane Lichen Flora

Some 2.5% of Britain lies above 610 m with 0.2% located above 915 m. The higher zones, where heather (*Calluna vulgaris*) grows prostrate and plant communities are dominated by lichen, moss and sedge heath, are among the scarcest and most natural habitats in Britain. At one time our montane lichen flora was thought to be an impoverished version of that present in Scandinavia, but it is now recognised as distinctive in its own right. Due to their oceanic situation the Western Highlands experience a combination of low snow cover, high humidity and heavy rainfall that is unique. The lichen flora consists of some 400 species, many with specialised requirements; a number of them are believed to be endemic. Nearly a third (52) of the British Red List lichens are montane species found chiefly above 800 m in Scotland. A division of the montane flora can be made on climatic grounds, with certain species restricted to the drier, cooler Eastern Highlands and others to the strongly oceanic Western hills. Scotland is the gem, it holds all the principal sites, although high ground in North Wales, the Lake District, the west of Ireland and the northern Pennines also have a fascination.

The discovery of the montane lichen flora has progressed gradually and is still incomplete. During the late eighteenth and early nineteenth centuries such leading field botanists as Lightfoot, Dickson, Borrer and Hooker collected lichens while undertaking general investigations of the flora of the Highlands, but no-one made a particular study of them. The second half of the nineteenth century saw the emergence of lichen specialists who visited all the main montane areas, paying less attention, however, to the Lake District and Ireland. The railway network made previously remote areas accessible, with the result that from around 1860 until 1890 the Highlands were repeatedly visited by some of the most active and discerning lichenologists of the day, including Holl, Admiral Jones, Carroll, Crombie, Lindsay, Stirton and Ferguson. Between them these workers added numerous species to the British flora and described many new taxa. Though the information remained largely scattered in the form of herbarium specimens or short notes in journals, a milestone had been passed; for the first time there was a substantial framework of knowledge into which new discoveries could be fitted. After 30 years this surge of activity drew to a close and it was to be many years before their achievement was built upon, for lichenology was about to enter the 'lean years', when for half a century little field work was undertaken and almost none of it in the mountains.

With the revival of interest in lichens during the early 1960s, the Scottish mountains were visited by Ursula Duncan, Peter James, Dougal Swinscow, Pauline Topham and David Walkinshaw. Much of their work remained undocumented or appeared only as unpublished reports. An exception was Peter James's account of the lichens of Ben Lawers (James, 1965), which set, in its ecological detail, a standard for the future. It reports many species not re-found since their discovery nearly a century earlier together with a large number of new records, including seven species new to Britain. As part of the Breadalbane Survey organised by the Royal Botanic Garden, Edinburgh, Brian Coppins visited a number of less well-known high level mica-schist sites, discovering additional species (Coppins, 1978; 1979).

Between 1981 and 1994, with a group of friends, I made a special study of our montane lichen vegetation. Each summer a small party set out to explore a new area working from high camps or bothy accommodation. Before the ecology of the montane lichen communities is described or an attempt made to convey something of the excitement of these expeditions, the issue of the Crombie rarities needs airing.

The Crombie rarities

During the above period of rediscovery it gradually became apparent that a high proportion of the Rev. J. M. Crombie's most spectacular discoveries, although represented by herbarium specimens in The Natural History Museum (Fig. 11.1), could not be re-found. This became known as 'The problem of the Crombie rarities'. His records of *Baeomyces carneus*, *Hypogymnia vittata*, *Parmelia glabra*, *Ramalina dilacerata* and *Umbilicaria rigida* were omitted from the 1980 checklist (Hawksworth *et al.*, 1980) on the grounds that they were considered to have been incorrectly labelled and to originate outside the British Isles. By the time the 1993 checklist (Purvis *et al.*, 1994a) was published, the number of dubious species for which Crombie was the sole collector had risen to 31 (montane 20, epiphytic 7, coastal 4). In addition, his records of *Cetraria cucullata* and *Cladonia alpestris* were coming under suspicion. With the help of Peter James I recently examined all the doubtful material at The Natural History Museum so that the reasons for misgivings could be summarised. They are, for the first time, set out below.

First there is circumstantial evidence. Two-thirds of the species involved are relatively conspicuous, easily identifiable macrolichens, many of which have been repeatedly searched for at the localities given by Crombie, but without success. In addition, Scotland is outside the expected range of species such as *Bryoria nitidula*, *Candelariella arctica*, *Cladonia alpestris* (records from the west Highlands), *Ophioparma lapponica*, *Ramalina thrausta* and *Usnea cavernosa*, while the luxuriance of certain specimens, for example *Cetraria cucullata*, *Cladonia amaurocraea* and *Umbilicaria arctica* is surprising in species for which Scotland would be the very edge of their range.

The detailed examination of herbarium specimens sometimes provided direct evidence of locality mislabelling. Collections of *Ramalina dilacerata* labelled 'branches of stunted larches, Craig Cluny, Braemar, Sept. 1872' also support three other non-British species of lichen (Krog & James, 1977); and it was noticed that a specimen of *Ochrolechia upsaliensis* labelled 'Morrone, Braemar, n.d.' has the non-British *Rinodina turfacea* as an associate. The specimen of *Physconia muscigena*, reputedly collected from Craig Tulloch (420 m), has the unlikely associates *Minuartia sedoides*, a plant of the high tops, and a scrap of a *Ditrichum* that does not match any British species. There are other inconsistencies; for example, the only two British collections of *Bellemerea cinereorufescens*, from Ben Lawers and Blair Atholl, appear to be part of the same gathering; and Crombie's first British record of *Pertusaria bryontha*, labelled from the Cairngorms, is a chemotype that has otherwise only been reported from northern Scandinavia (Purvis *et al.*, 1992). There is a correctly

Fig. 11.1 Herbarium specimens of two Crombie rarities that are under suspicion of having originated outside the British Isles (The Natural History Museum, London).

identified Crombie specimen of *Baeomyces carneus* from 'Rannoch, Perthshire, Sept. 1888', in the Natural History Museum, though confusingly he later wrote (Crombie, 1894) that it was not a British species. A few of the Crombie rarities, for example *Acarospora peliscypha*, *Cetraria odentella* and possibly *Hypogymnia vittata* (poor material) are now discounted as misidentifications, but the scale of locality mislabelling suggests a deliberate attempt to mislead; this is more than just muddle, it is glory-seeking. Despite these lapses, I prefer to remember Crombie for his outstanding and enlightened contribution to lichenology.

The Ben Alder treasure trove

While carrying out survey work in the Highlands, Derek Ratcliffe discovered a previously unknown limestone buttress at 1,030 m in the Ben Alder range that was outstanding for its vascular plants. A few years later, in 1964, David Walkinshaw made the first lichen collections from the site. They were impressive, including *Fulgensia bracteata*, *Halecania alpivaga* and *Polyblastia helvetica*, all new to Britain; in 1980 Pauline Topham added *Pertusaria bryontha*. This site, consisting of a single rib of pale 'sugar limestone', 80 m high and 30 m wide, flanked by a sill of epidiorite (Fig. 11.2), was the objective of our first expedition. The party of three, myself, Brian Fox and William Purvis, who was still an undergraduate, walked in on three days from Culra Bothy.

Though approached in dense mist, the site was immediately recognisable by the abundance of *Fulgensia bracteata*, which colours the lower parts of the buttress yellow (Plate 12a). In places, the soft easily eroded surface of the limestone is tinged cream by an extensive crust of *Ionaspis epulotica*, which, together with *Gyalecta jenensis*, *Ionaspis heteromorpha*, *Polyblastia cupularis*, *P. theleodes*, *Sagiolechia protuberans* and *Toninia aromatica*, dominates much of the rock. Elsewhere, deeply weathered surfaces support *Belonia russula*, *Collema callopismum* var. *rhyparodes*, *C. glebulentum*, neat rosettes of *Lempholemma radiatum* and *Placynthium pluriseptatum*. While the crumbling, coarsely granular surface of the limestone is a poor substrate for most saxicolous lichens, the loose material that accumulates on ledges is an outstanding terricolous habitat. Here lichens cover the surface of slow-growing mosses, the dead bases of higher plants and the general organic debris accumulated in fissures. The species involved include such rarities as *Catapyrenium cinereum*, *C. waltheri*, *Cladonia symphycarpa*, *Collema ceraniscum*, *Peltigera venosa*, *Polyblastia sendtneri*, *P. terrestris* and *Thelopsis melathelia*. This is the only site in Britain where six taxa within *Solorina* are present including all three varieties of *S. bispora*.

Another special alpine habitat well represented on the crag is mats of slow-growing prostrate woody plants such as dwarf willow (*Salix herbacea*), net-leaved willow (*S. reticulata*) and mountain avens (*Dryas octopetala*). These become half buried in the soil so provide a niche that is part terricolous and part epiphytic. The older parts of these plants support abundantly fertile colonies of *Biatora tetramera*, the golden-fruited *Brigantiaea fuscolutea*, *Lecanora epibryon* (Plate 12b), *Lecidea berengeriana*, *Megaspora verrucosa*, *Pertusaria glomerata* and *P. oculata*. What is notable is the sheer abundance of species represented at other localities in the Highlands by only a few thalli.

The epidiorite is less well exposed, forming short ribs and towers jutting out from the steep grassy slope. Soil pH here ranges from 4.4 to 6.6, compared with 6.6 to 7.7 on the sugar limestone. The commonest species are weakly basiphilous and include *Koerberiella wimmeriana* and the rare *Pannaria hookeri* which, in the Highlands at least, is selective for epidiorite.

With all identifications made, this small outcrop was found to support eight calcicolous montane lichens and one lichenicolous fungus unknown elsewhere in Britain. As knowledge of the Scottish mountains has increased, this number has dropped, but for a site of less that 1 ha its flora is exceptional. The ecological factors combining to single it out are altitude, exposure and the high pH of the limestone soil, which exceeds that present in the mica-schist mountains of Breadalbane. The juxtaposition of pure limestone and chemically complex epidiorite increases habitat diversity. Possibly the most important local factor is the spreads of loose calcite 'sand', which cover ledges and surrounds the outcrop, promoting intermediate levels of disturbance. The 'intermediate disturbance hypothesis' predicts that alpine plant species diversity will be maximised at intermediate levels of vegetation disturbance (Fox, 1981). Though we did not realise it at the time, never again were we to encounter such richness. Elsewhere rarities had to be diligently sought; on Ben Alder they parade in profusion.

The Cairngorm plateau

The Cairngorm Mountains are the largest area of high ground in Britain. Four summits rise above 1,220 m and their forms are massive so that 30 km^2 lie above 1,100 m. Situated on the east side of the Highlands, they occupy a relatively continental position and, consequently, low winter temperatures are a notable feature and throughout summer the plateau is dotted with late snow patches (Fig. 11.3). Underlain by granite, they provide an extremely acidic habitat. This area was the objective of the second and third expeditions undertaken in 1982 and 1983. We had little idea of what to expect (Gilbert & Fox, 1985). My diary for the first day records 'Within five minutes of setting off from the top of the White Lady chair lift we were in a wonderland of lichens; each step brought new and intriguing species; as Brian Fox observed, "the only way to make progress was to close your eyes". I was on an adrenaline high and completely disorientated; there was so much new information coming in that it was many hours before we started to make sense of the communities and could anticipate what lay at our feet.'

Fig. 11.2 The richest site for montane lichens in Britain is the pale outcrop of sugar limestone in the centre of the photograph; Ben Alder.

Summit heath

Much of the summit plateau is covered with a rush-heath comprising various combinations of the three-leaved rush (*Juncus trifidus*), woolly hair-moss (*Racomitrium lanuginosum*), bare granite gravel and low boulders. The commonest macrolichens are *Cetraria islandica*, various *Cladonia* spp. and *Coelocaulon aculeatum*, while crustose lichens like *Ochrolechia frigida* are frequent but rarely conspicuous. Slightly damp hollows and runnels are the

Fig. 11.3 General view of the Cairngorm Plateau in summer showing a landscape dotted with late snow-beds and with a granite boulder field in the foreground (D. Gowans).

characteristic habitat of *Cetrariella delisei*, which is like *Cetraria islandica* but with richly divided apices. The Cairngorm rarity, *C. ericetorum*, with its inrolled, almost tubular thallus is present among boulders, particularly on the northern flank of Ben Macdhui. Where wind speeds are too high for woolly hair-moss to form a carpet, circular tussocks of three-leaved rush with decaying centres occur; these must be hundreds of years old. Their moribund shoot bases support small crustaceous lichens such as *Lecidea berengeriana*, *Micarea leprosula*, *Ochrolechia tartarea* and *Pycnothelia papillaria*, while the surrounding granite gravel is semi-stabilised by colonies of *Catillaria contristans*, *Frutidella caesioatra*, *Lecidea limosa*, *Lecidoma demissum* and *Micarea turfosa*. The distribution of the lichens is controlled by the life cycle of the rush. Other montane species present in the summit heath are *Alectoria nigricans*, *Cladonia macrophylla*, *Pertusaria dactylina*, *P. geminipara* and *Porina mammillosa*, these occurring where the rather simple niche structure is enhanced by the presence of hummocks, hollows or terracettes.

Saxicolous habitats on the plateau are small isolated tors and vast numbers of low (to 75 cm) rounded granite boulders. The latter are curiously bare of lichens, those present tending to be poorly developed, closely adpressed to the rock and often damaged. This is interpreted as being the result of abrasion from windblown gravel and ice spicules which, during blizzards, are a potent and unusual ecological factor. McVean & Ratcliffe (1962) have reported how wooden posts and the painted metal surfaces of ski-tow pylons on the plateau quickly become scoured by these agencies. The summit of Cairn Gorm holds the national wind speed record with a gust of 173 mph recorded in March 1986. Assemblages that can survive these conditions have been termed anemophylic (wind-loving), and on Cairn Gorm include *Fuscidea gothoburgensis*,

Miriquidica nigroleprosa, *Orphniospora moriopsis* and *Rhizocarpon geographicum*. Fryday (1997) has observed how the lichens on these low boulders are arranged in colour bands similar to those on the shore, a dark basal zone passing into a yellow mid-zone while the tops of the boulders are grey. No explanation for this chromatic sequence has been forthcoming.

Larger boulders, and those in slightly sheltered positions, which avoid the blast of ice spicules, are somewhat richer, being colonised by typical montane calcifuge species such as *Allantoparmelia alpicola*, *Cetraria commixta*, *Cornicularia normoerica*, *Pseudephebe pubescens*, *Rhizocarpon alpicola*, *Umbilicaria hypoborea*, *U. proboscidea* and *U. torrefacta*. As usual, the largest boulders carry the richest lichen cover. The Cairngorm speciality, *Hypogymnia intestiniformis*, appears to have experienced a decline: though well known to the Victorian lichenologists, it was last collected near the summit of Ben Avon by Peter James in 1964.

Lichen-rich dwarf Calluna heath

This community is restricted to exposed crests, ridges and cols between 700 m and 950 m altitude. It is composed of dwarf shrubs that take on a wave form, creeping downwind and dying away behind as they slowly advance over the stony surface. A number of rare mountain lichens reach their maximum development in this community, which grows on sites blown clear of snow in winter (Fig. 11.4). The sulphur-yellow species *Alectoria sarmentosa* and *Cetraria nivalis* are regularly present, caught up among the vegetation, while on the northern spurs of Cairn Gorm the Red Data List species *A. ochroleuca* has its headquarters in this community. Each sequential niche in the growth and decay of the dwarf shrubs favours a different lichen assemblage.

Fig. 11.4 Wave Callunetum at 720 m on Carn Fiaclach; the col is blown clear of snow in winter.

Fig. 11.5 Britain's most permanent snow-bed in Garbh Corrie, Braeriach. Photo taken 24 August 1983.

Old dead and detached heather stems are well colonised by *Lecanora symmicta*, *Ochrolechia frigida* and *Pertusaria xanthostoma*. Among stems that are still attached to the soil, fruticose species such as *Alectoria nigricans*, *A. sarmentosa*, *Cladonia subcervicornis* and *Coelocaulon aculeatum* are prominent, while *Cetraria islandica* and *C. nivalis* occur firmly packed among the living shoots of the ericaceous plants. The 'bare' gravelly surface between bushes, which may occupy 50% of the ground, is the favoured habitat of *Cladonia zopfii* and *Thamnolia vermicularis*.

Opinion is divided on the authenticity of the Cairngorm records of *Cetraria cucullata*, a conspicuous sulphur-yellow species with a purple base. It was reported several times in the eighteenth and nineteenth centuries, but the only specimens with locality data are Crombie's. If present it is probably in the wave Callunetum of the eastern Cairngorms rather than the declared localities, which are the summits of Cairn Gorm and Cairn Toul.

Snow-beds

Until the 1982–3 expeditions the so-called perpetual snow-beds of the Cairngorms had never been seriously studied by lichenologists. Consequently, particular attention was paid to this habitat, and from it several lichens new to Britain, various undescribed taxa, and a plethora of rare species were recorded. The best known permanent snowfield, known locally as 'The Old Snowman', lies at 1,100 m in Garbh Choire on the flank of Braeriach (Fig. 11.5). It is an awe-inspiring place, so remote that the only practical way to investigate it is to camp. This deep, cliff-girt corrie is ideally suited for collecting and holding snow, being surrounded by huge gathering grounds at over 1,200 m, sheltered from most winds by its easterly aspect, and lying in perpetual shadow except briefly in the morning and around midday in summer. Once considered permanent, even an incipient glacier, the climatic amelioration since around 1910 has been reflected in its total disappearance in 1933, 1959 and 1996. Westerly winter winds, which promote snow accumulation, are more important than total snowfall or summer temperature in controlling persistence.

As the corrie is lined with block scree, saxicolous and bryophilous lichens are well developed, the latter colonising the small cushions of *Andreaea* and *Grimmia* which grow thickly on the boulders. Specialities of these boulders are *Frutidella caesioatra*, the handsome yellow *Lecanora leptacina*, *Lecidea griseoatra*, *Lecidella bullata*, *Rhizocarpon jemtlandicum* and *Toninia squalescens*. An insignificant byrophilous lichen with small reddish fruits turned out to belong to an undescribed genus now known to be widespread around Scottish snow-beds. It is to be called *Amelia* to celebrate Alan Fryday's baby daughter of the same name. These snow-bed specialists are accompanied by lichens that, while tolerant of the conditions, have a wide ecological amplitude, for example *Cladonia bellidiflora*, *Lepraria neglecta*, *Stereocaulon evolutum* and *S. vesuvianum*. The back wall of the corrie is a wet granite cliff where, as the snow melts down, a series of vegetation-covered ledges is exposed from which *Bryophagus gloeocapsa*, *Cladonia phyllophora*, *Micarea turfosa*, *Pertusaria oculata*, *Solorina crocea* and *Sporastatia polyspora* were collected, the later representing another genus new to Britain.

The two days spent exploring the Garbh Choire included an investigation of meltwater streams which drain through the block scree to issue as springs. Submerged boulders in the cold water were bare, apart from occasional dark brown, circular colonies of a crustaceous lichen with conspicuous pycnidia; it could not be determined at the time, but was eventually matched with the continental *Gyalidea diaphana*. A meltwater issue in the adjacent Coire an Lochain Uaine contained an additional species, *Staurothele areolata*, again new to Britain.

The other well-known persistent snow-bed in the Cairngorms is Ciste Mhearad (Margaret's Coffin), which survives about eight years in ten (Plate 12c). Being underlain by gravelly soil, the zonation of the vegetation in relation to length of snow cover is well displayed. In the heart of the snow-bed, under a steep north-facing wall, a group of three large, rounded boulders is completely devoid of lichens; they would make a useful biological monitoring site to detect signs of global warming. At this locality, snow patch vegetation can be divided into five communities, each with a distinctive combination of lichens and higher plants (Fig. 11.6).

1. Plateau zone. Lichens such as *Alectoria nigricans*, *Allantoparmelia alpicola*, *Cladonia uncialis*, *Cornicularia normoerica* and *Thamnolia vermicularis* are typical of the three-leaved rush community and are unable to survive more than about 18 weeks burial.
2. Zone of moderate snow-lie. Burial of around 30 weeks favours the development of an alpine *Nardus* grassland that is home to *Cetrariella delisei*, *Cetraria islandica* and the very rare *Cladonia maxima* and *C. stricta*. The latter occur deep in the sward and were only discovered during transect work.
3. The next zone, dominated by stiff sedge (*Carex bigelowii*), is where *Stereocaulon saxatile* reaches its maximum development.
4. Inner snow-bed zone. Here true snow-bed specialists dominate among a

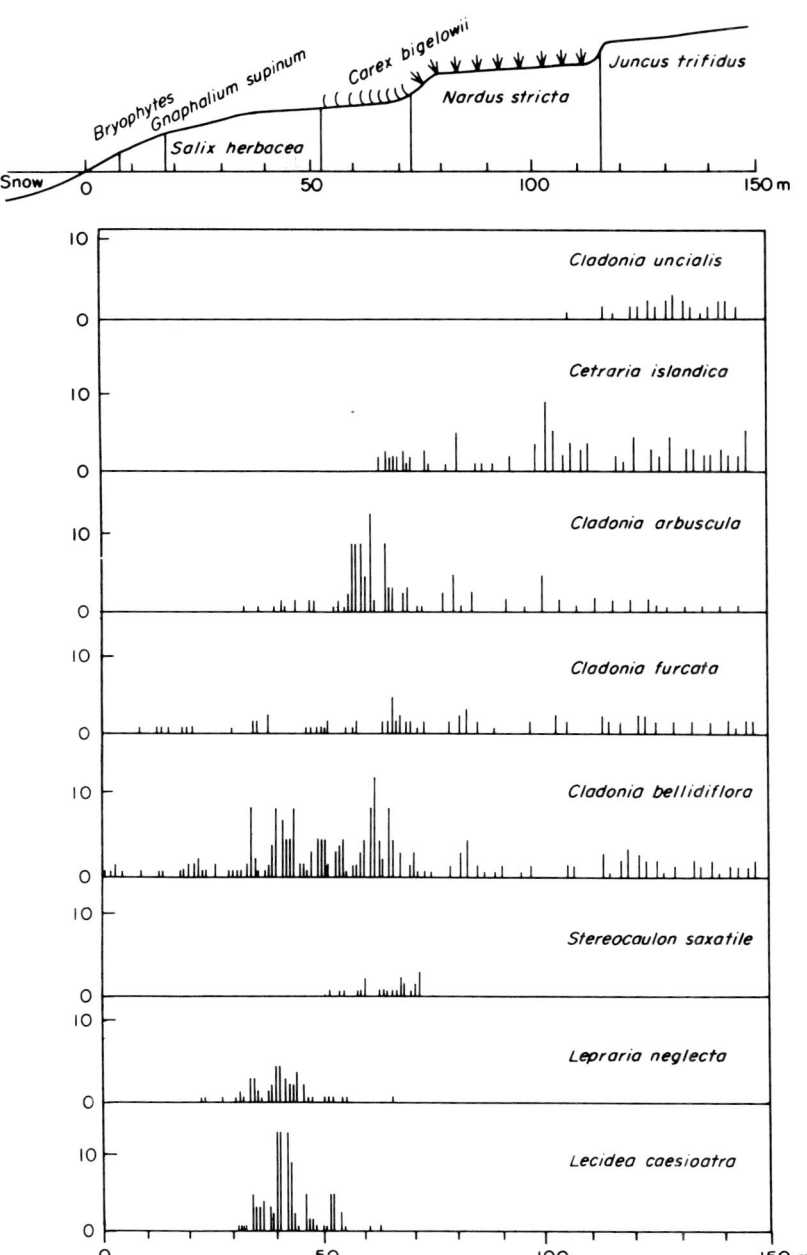

Fig. 11.6 Transect showing the zonation of terricolous lichens around Ciste Mhearad (Margaret's Coffin) snow-bed on Cairn Gorm. The profile indicates dominant higher plants. (Note recordings continuous 1–75 m, alternate 77–150 m).

vegetation in which dwarf willow (*Salix herbacea*) and dwarf cudweed (*Gnaphalium supinum*) are common. The most prominent terricolous lichens are *Catillaria contristans, Frutidella caesioatra, Lecidea limosa, Lecidoma demissum, Lepraria neglecta, Micarea viridiatra, Protothelenella sphinctrinoidella* and *Toninia squalescens*. These are found where the soil has a 'rubbery skin' formed of small liverworts. Saxicolous species include *Lecanora leptacina* and *Lecidella bullata*.

5. Central zone. This is dominated by the bryophytes *Kiaeria starkei* and *Polytrichum sexangulare*, which can survive several years of burial. A number of the terricolous zone 4 species also occur on bare soil on the plateau; a feature common to both habitats is reduced competition from higher plants.

In 1983 several colonies of *Bellemerea alpina* were found growing on quartz in the central area of the Ciste Mheared snow-bed. Though this species had been recorded last century from Ben Lawers, this was the first modern record. By 1994 no trace of it could be found. It is speculated that the factor responsible for its disappearance is either an intensification of acid precipitation concentrated in snow melt, or the result of increasing recreation in the form of snow-boarding, snow-holing and trampling, to which the site is subjected. As the Scottish snow-beds represent the last relics of Late Glacial vegetation remaining in Britain, and support a lichen vegetation apparently unique in Europe, they deserve a high conservation priority.

The Far North

North of the main mass of the Highlands three residual hills rise from a flat boggy landscape; they are Ben Loyal, Ben Hope and Foinaven. Being the most northerly hills in Britain they experience a distinctive climate, but what makes them of particular interest to ecologists is the rocks from which each is formed (Fig. 11.7). Ben Loyal is composed of syenite, a rather acid igneous rock; high ground on Foinaven is carved from hard, white Cambrian quartzite, while the Moine granulites of Ben Hope contain a thick bed of epidiorite, which weathers to give base-rich soils and is the second richest site (after Ben Alder) for alpine vascular plants in the northern Highlands. Apart from a brief visit to Ben Hope by James and Rose in 1972 these hills are little known lichenologically, so they were targeted by the 1984 expedition (Gilbert & Fox, 1986). Again, high-level camps proved their worth: if the weather was fine, a 12 hour day could be spent on site; if unsettled, full advantage could be taken of any breaks.

Fig. 11.7 Section, 40 km long, illustrating the structure of a group of geologically distinctive mountains in the far north of Scotland.

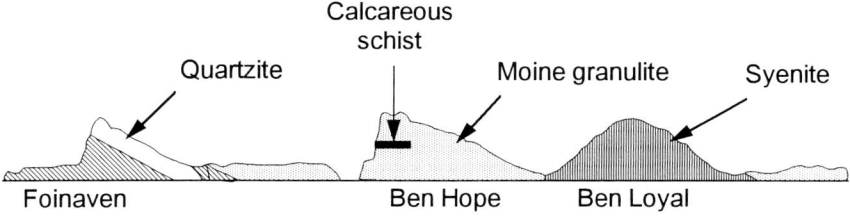

The grassy summit ridge of Ben Loyal rises to 764 m and is capped by enormous tors covered by lichens of a luxuriance rarely encountered on rock in Britain. The dense, multi-layered, shaggy coat is composed of adpressed species such as *Allantoparmelia alpicola* and *Fuscidea cyathoides*, overgrown with abundant *Hypogymnia physodes, Parmelia omphalodes, Platismatia glauca* and *Sphaerophorus globosus*. Lodged in this sward are *Cladonia coccifera, Coelocaulon aculeatum, Ochrolechia frigida, O. tartarea* and *Pseudevernia furfuracea*. The most spectacular feature, however, is the fine development of shrubby filamentous species, such as *Alectoria nigricans, A. sarmentosa, Bryoria bicolor, B. chalybeiformis, B. fuscescens, B. lanestris* and *Pseudephebe pubescens*. The environmental conditions under which this prodigious lichen growth occurs – maritime air close to saturation point constantly pouring over the ridge, frequent low cloud, and vertical faces which encourage up-currents – suggest the community has affinities with fog-induced and cloud-zone assemblages. In particular, the high cover of fruticose species parallels the structure of fog-zone communities described from cliffs on the west coast of America.

On sheltered areas of tor, where rain lash has leached the rock less deeply, the slightly basic nature of the syenite expresses itself through the presence of *Immersaria athroocarpa, Lithographa tesserata* and, locally, the handsome *Lecanora chlorophaeoides*, unknown elsewhere in Britain. Other features of note on Ben Loyal are the lichen zonation around semipermanent, water-filled rock basins which includes *Rhizocarpon geminatum* and windswept cols with a rich terricolous flora that includes *Pertusaria dactylina, P. oculata, Solorina crocea* (Plate 13a) and *Thamnolia vermicularis*.

The lichenological interest of Ben Hope is largely restricted to the gigantic escarpment that forms the west side of the mountain. Three kilometres long, and in places 600 m high, it is one of the major precipices of Scotland. Halfway up this escarpment a band of calcareous hornblende-schist, up to 30 m thick, outcrops, but is poorly exposed, much of it being covered by acidic scree. Four days were spent surveying this complex site. The largest exposures of basic rock are associated with the walls of gullies, but the richest sites are the scarce, well-illuminated, fast-drying buttresses (Fig. 11.8). Once located, it was found most profitable to examine basic exposures minutely rather than to work in a more extensive fashion. The best 'lead-in' species (a 'lead-in' species tells you you are getting warm) for the hornblende-schist are *Pannaria pezizoides* and *Peltigera* spp. of which nine were present including *P. venosa*. Basic soil and bryophyte cushions are home to *Catapyreniun cinereum, Gyalecta foveolaris, Placidiopsis cartilaginea, Psoroma hypnorum, Sagiolechia rhexoblephara, Solorina bispora* and *Gyalideopsis scotica* for which this cliff is the type locality.

Free-draining buttresses added the alpine rarities *Gyalidea fritzei, Lecanora atrosulphurea, Polyblastia inumbrata, P. melaspora, P. verrucosa, Staurothele succedens* and *Psora decipiens*, while damp schist supported *Belonia russula, Lempholemma radiatum, Pannaria hookeri, Placynthium asperellum, P. pannariellum* and *Euopsis pulvinata*. A severe restraint on the alpine lichen flora of the site is the low upper limit (600 m) of the calcareous strata, which results in most ledges being occupied by vigorously-growing higher plants. By contrast, the richest localities on Ben Lawers and Ben Alder are at altitudes of 1,030–1,210 m where there is far less competition. As a result, most of the nationally rare lichen species on Ben Hope are present at very low frequencies and take a great deal of finding. There is no comparison with the lichen-dominated ledges present on Ben Alder.

Fig. 11.8
Cliffs of calcareous schist on Ben Hope are a major site for alpine lichens.

The layman associates the name Foinaven with a famous racehorse, but he was named after an equally magnificent peak that dominates the far northwest. Its long quartzite ridges, flanked by dazzling white scree, are of international importance for nature conservation. Its flora is described in the section on quartzite in Chapter 6 (pp. 105–106).

A distinctive feature of these three prominent hills is the influence of salt spray carried inland on Atlantic gales. Though it is 30 km from the coast, Ferreira (1958) interpreted unexpectedly high sodium levels in soil on Ben Hope as resulting from this cause, which also explains the presence there of species with maritime tendencies, such as *Ramalina siliquosa, R. subfarinacea, Rinodina confragosa* and *Trapeliopsis wallrothii*.

Glory days on Ben Lawers

Ben Lawers (1,214 m) and its immediate hills are the foremost area in the country for calcicolous arctic-alpine plants. This is due to the presence of locally extensive deposits of calcareous mica-schist. Several types of schist are present, but the richest botanically is the chlorite-muscovite schist (Ben Lawers Schist), which locally contains 10–20% (-40%) by volume of calcite. This weathers to produce crevice soils with a pH of 6.4–7.2. Though the lichen riches of Ben Lawers were discovered by the Victorians, direct knowledge of them died out until 1963 when the summer field meeting of the British Lichen Society, led by James and Duncan, spent three days in the area rediscovering many species not seen for nearly a century, and making a number of new records. The field meeting report, one of the most detailed ever produced up to that time (James, 1965), contained so many unfamiliar names that few lichenologists ventured near the mountain for the next 20 years. Even with Ben Alder and Ben Hope behind us it was with trepidation that we organised Ben Lawers expeditions in 1985 and 1986 (Gilbert *et al.*, 1988). The inclusion of Brian Coppins in the party was essential both for his experience in the field and his taxonomic skills.

The Ben Lawers schist stretches in a band from Creag na Caillich in the west, via Cam Chreag, Meall nan Tarmachan, Ben Ghlas and Ben Lawers, to the Lochan nan Cat corrie in the east, a distance of 11 km. As the geology is complex it is worth copying it onto an Ordnance Survey map and carrying this in the field. The survey started in dense cloud and drizzle from a camp under the west cliffs, it ended with a 5 am start from tents by Lochan nan Cat to take advantage of a bright morning.

At the foot of the west cliffs a jumble of large boulders below the main face was soon producing species we had never seen before. These included *Decampia hookeri*, *Pannaria praetermissa* and a *Peltigera* with a turquoise tinged thallus that was later identified as the first definite British record of *P. ponojensis*. Moving to the base of the main cliff (Fig. 11.9) the number of unfamiliar species increased to bewildering proportions. Wet mossy faces locally supported a community made up of two of Ben Lawers greatest rarities: *Acarospora rhizobola*, composed of large, chestnut-coloured squamules, (bright green when wet), with white edges anchored by conspicuous long white rhizinae (there is nothing else remotely like it in the British flora); and *Psora rubiformis* with distinctive, elongated, grey, white-margined squamules bearing dark button-like fruits. They were first discovered by Admiral Jones in 1864. The Victorians had some trouble separating this *Psora* from the darker *P. globifera* which occurs in the same community; it was originally found by Crombie (1868) then not seen again until a very small population was rediscovered by Alan Fryday in 1989. These are three of Britain's rarest lichens.

Caloplaca spp. are scarce in the mountains, so any found deserve close examination. Two overgrow bryophytes on the west cliffs; they are a beautiful

Fig. 11.9
The summit of Ben Lawers viewed from Ben Ghlas. The southwest cliffs, below which lie a jumble of boulders, are in the middle distance (D. Mardon).

Fig. 11.10 Tor-like outcrops of calcareous schist on the summit of Ben Lawers are exceptionally rich in Red Data List lichens. Good weather and ample time are needed to do them justice.

muscicolous variety of *C. cerina* with fruits covered in golden pruina and *C. ammiospila*, a pale grey species with rust-coloured apothecia, which is another Lawers speciality. Underhangs produced the arctic species, *C. approximata*, new to Britain. Ben Lawers is famous as the only British locality for *Caloplaca nivalis*, which overgrows and is apparently parasitic on the moss *Racomitrium heterostichum*. It was described as 'not rare' in the mid-nineteenth century and there are many herbarium specimens to substantiate this. In the modern period *C. nivalis* has proved difficult to find, with only two small populations having come to light; a special search in 1994 was unsuccessful. It was last seen during our second expedition high on the east ridge; reasons for its decline are not known, though over-collecting in the past has been suggested.

While the west cliffs provide a good introduction to the lichens of Ben Lawers, the summit area is arguably the richest locality in the Highlands. To do it justice, good weather is essential. Under normal conditions with anorak hood up, gloves on, glasses and hand lens steamed up, wet packets disintegrating in one's bag, and the wind whipping specimens out of collecting tins before the lid can be replaced, it can appear disappointing. The constellation of features that makes it an exciting habitat are tor-like outcrops of highly calcareous schist (Fig. 11.10), a shallow gully 200 m long, and a depression known as 'The Crater' that holds late snow. The rock weathers to form soil-filled crevices, in the seams of which can be found the ivory-white fruits of *Biatora carneopallida*, the large reddish apothecia of *Biatorella hemisphaerica*, cushions of *Collema ceraniscum*, *Gyalecta foveolaris*, *Gyalideopsis scotica*, *Micarea crassipes*, *Schadonia fecunda* and *Solorina bispora*. These outcrops illustrate one of the tantalising features of the Ben Lawers schists; highly calcareous seams and lenses that weather to give soils of pH >7, are scarce, occurring almost at random among much larger areas of less interesting habitat. The schists erode to produce such an intricate topography that once a calcareous face has been located it can take several hours to examine properly.

'The Crater' adds a snow patch element to the flora with *Amelia andreaeicola, Lecanora leptacina, Lecidella bullata, Miriquidica griseoatra* and *Protothelenella corrosa* together with more widespread small terricolous species such as *Belonia incarnata* and *Catillaria contristans*. The only other late snow patch in the area occupies a small hollow on the north ridge of An Stuc, it supports a population of the rare arctic species *Stereocaulon spathuliferum*.

The corrie containing Lochan nan Cat incorporates more rock outcrop and displays a greater diversity of habitat than any other site in the Ben Lawers range. Interesting lichens can be found throughout the corrie, but Cat Gully is particularly rich with *Bacidia herbarum, Biatora vernalis, Bryonora curvescens, Gyalecta geoica* and *Mniacea nivea*. On rock faces in general the weak calcicoles, *Buellia leptocline, Caloplaca concilians, Catillaria scotinoides* and *Pertusaria amarescens* can be found. Large boulders in stabilised scree on slopes below the cliffs provide niches that experience a sharper wetting-drying cycle than those on the cliff face. They support the largest British population of *Acarospora badiofusca*, which, until rediscovered, was considered a doubtful record. Other specialities of these boulders include *Belonia russula, Lecidea paupercula, Peltigera elisabethae*, with a dark underside, and *Bryoria bicolor*.

While the lochan and its inflow streams contain many unusual species (Gilbert *et al.*, 1988), of equal interest are some small wet outcrops in grassland to one side of the corrie which were investigated one evening after a day confined to the tents. These have ledges carrying acid vegetation, but the presence of rose-root (*Sedum rosea*) suggests that the water draining over them is base-rich. On any other mountain in Scotland their flora would be considered outstanding, but on Lawers it represents a fairly standard wet rock assemblage of *Euopsis pulvinata, Koerberiella wimmeriana, Leptogium tenuissimum, Miriquidica complanata, Peltigera leucophlebia, Placynthium flabellosum, Pyrenopsis grumulifera* and *P. subareolata*.

Two kilometres west of Ben Lawers lies Ben Ghlas, which tends to get ignored in favour of its more famous companion, though it is a site for *Pertusaria glomerata* (Plate 13b). Further west the range is split by a pass that carries a minor road linking Loch Tay to Glen Lyon. Above this road the 1.5 km long east-facing cliff, Creag an Lochan, is the lowest station for many rarities. Due to its sheltered location this is a good wet weather site. Its specialities are the abundance of the bryophilous *Strigula stigmatella* var. *alpestris* and the terricolous *Polyblastia wheldonii; Cladonia cyathomorpha* is well represented on wetter faces while a feature of drier ledges is *C. symphycarpa*. Among the rarer lichens present are *Brigantiaea fuscolutea, Lempholemma radiata, Pannaria hookeri, Peltigera venosa* and *Thelidium fumidum*. One can also enjoy the many alpine higher plants.

Some of the most pleasant botanising in the area is to be had on Meall nan Tarmachan where a long south-facing line of cliffs at 950 m provides an uncomplicated habitat. Wet rock faces are dominated by *Koerberiella wimmeriana* with the lurid fruits of *Protoblastenia siebenhaariana* lining damp crevices. Dryer faces are the main habitat of the rare and distinctive *Lecanora frustulosa, Pannaria hookeri* and *Porpidia superba* that lives up to its name, having a thick lumpy white thallus contrasting with the large, dark brown discs and black margins of the abundant apothecia. All three can be found growing together. The front edge of mossy ledges, where there is a thin accumulation of mineral and organic debris, support *Polyblastia terrestris*, *Psora decipiens* and *Thelopsis melathelia*. Smaller outcrops in this area are the type locality for the Scottish endemics, *Catillaria gilbertii*, *Halecania bryophila*. *H. micacea* and *H. rhypodiza* (Fryday & Coppins, 1997).

Our expeditions confirmed and enhanced the international reputation of the area by adding 118 lichens to the reserve list, 13 of them being new to Britain. Since then, Alan Fryday, who lived in Killin from 1989–92, has added more and extended the range of others down the mountain through intensive work at intermediate altitudes. Amounting to 500 species, lichens are the outstanding botanical feature of high ground in the Ben Lawers range, both numerically and in terms of their rarity. Twenty-nine are on the British Red List and a dozen are found nowhere else. This remarkable concentration results from the coincidence of strongly calcareous bands within the mica-schist, an unusually severe climate that reduces competition from higher plants, and the rugged topography which provides extensive outcrop up to the summit which, at 1,214 m, is the ninth highest in Great Britain.

Caenlochan

Lying at the head of Glen Isla, Caenlochan contains the greatest number of alpine vascular plants of any single corrie in Britain. Nineteenth-century lichenologists paid little attention to the area, so the first serious lichen exploration was as late as 1964 when Duncan and James recorded many notable species. Following recent surveys of other calcareous sites in the Highlands, a clearer

Fig. 11.11 Brian Coppins prepares for a day in the field; Caenlochan.

Fig. 11.12
Boulders of Lawers Calcareous Schist lying in grassland are a major lichen habitat at Caenlochan.

appreciation was wanted of the relative value of the Caenlochan lichen flora, which was investigated from a high-level camp by Brian Coppins (Fig. 11.11), myself and various visitors over a week in 1989 (Gilbert & Coppins, 1992). On the east side of the glen an outcrop of the Ben Lawers Calcareous Schist forms a belt of small grassy cliffs rising to an altitude of 850 m, while on the west side, a band of hornblende schist outcrops on the great head-wall. The remaining rocks are strongly acidic.

On the first day, setting off for the east crags, we were immediately delayed by large boulders of Lawers Schist lying in the grassland (Fig. 11.12). Weathering had etched their surfaces, providing a profusion of microhabitats. The two dominant lichens were *Acarospora glaucocarpa*, filling seams in the rock and spreading along small horizontal ledges, and the pale pink sheets of *Ionaspis epulotica* occupying recesses. Sheltered alkaline furrows supported many common calcicoles and also rarer species such as *Caloplaca approximata*, *Eiglera flavida*, *Polysporina cyclospora*, *Pannaria praetermissa* and *Staurothele bacilligera*, the latter previously known only from Carboniferous limestone in the north of England. The mossy, table-like tops of the boulders added *Acarospora badiofusca*, *Collema parvum*, new to Britain, *Lecidella wulfenii* and *Megaspora verrucosa*. Well pleased by this start we moved up to the cliffs from where the boulders had originated; these took two days to explore and are the major lichen site in the corrie.

We each have our own way of working. Brian Coppins likes to stop every so often for a drink of tea from his thermos, have a bite of cake and to enjoy a quiet pipe. He brought ten boxes of matches to Caenlochan to make sure it stayed alight. Sometimes he imagines the old collectors like Lindsay, Holl and Admiral Jones standing at his shoulder giving advice 'No over there, a little to the left sonny'. I am more obsessive; if the weather is fine I rarely stop, eating Mars Bars to maintain energy.

Many of the Ben Lawers rarities are present including *Catillaria scotinodes*, *Halecania rhypodiza*, *Porina mammillosa*, *Thelopsis melathelia* and *Toninia fusispora*. Seams of buff-coloured marble yielded *Sagiolechia protuberans*, while wetter outcrops held sheets of *Belonia russula*, *Collema callopismum*, *Lempholemma cladodes*, *L. radiata* and *Porocyphus coccodes*. Ledges and moss cushions were disappointing compared with those on Ben Lawers and Ben Alder as the relatively low altitude (800 m) meant that vigorous vascular plants occupied most available space and there were few tassels of vegetation which is an important habitat at high levels.

The large hornblende schist cliff of Craig Doubs, on the opposite side of the glen, illustrates a frustration that lichenologists frequently suffer. Though it is the major site in the corrie for arctic-alpine higher plants, it is far less good for lichens. The rock is too wet, too hard, and insufficiently alkaline at the surface for most calcicolous lichens, though vascular plants find seepage lines sufficiently base-rich for this not to be a limiting factor. Lichens that have their main population in Caenlochan at the foot of this cliff are *Peltigera elisabethae*, *Protoblastenia siebenhaariana*, *Psora decipiens* and *Solorina bispora*. Despite thorough exploration, the upper levels proved to be less interesting as, with height, the cliff becomes progressively more acid due to a decrease in flushing. The Glasallt Burn, which runs through the glen, frequently caused us to linger by producing species such as *Peltigera venosa*, *Pertusaria amarescens*, an abundance of *Phaeophyscia endococcina* and *Rinodina parasitica*, new to Britain.

Caenlochan is the third most important upland calcareous site for lichens in Britain, after Ben Lawers and Ben Alder. The interest is dispersed in many small pockets, which makes a comprehensive survey difficult to achieve. The relatively low altitude of the calcareous outcrops (610–850 m) is a severe restraint on the alpine lichen flora, which amounts to 322 species compared with 500 on Ben Lawers.

Western Scotland

The mountains along the western seaboard of Scotland have a different climate and a somewhat different flora to those in the east. We made a start on them by investigating the Ben Nevis range with two expeditions in 1990 (Gilbert *et al.*, 1992), while Alan Fryday has subsequently worked Glen Coe, Beinn Eighe and the Trottenish Ridge on Skye. The summit of Ben Nevis was explored with the aid of a helicopter; a contrast to the transport difficulties the earlier lichenologists must have experienced.

The summit of Ben Nevis (1,344 m) is a vast boulder field of hard, fine-grained igneous rock that can turn a chisel edge inside five minutes. The boulders mainly support a crustose community composed of common upland acidic species, some of which are encouraged by human eutrophication (excrement). As a result of the constantly wet, humid and misty conditions, all the rocks carry a number of lichens that are associated with damp habitats, for example *Amygdalaria consentiens*, *A. pelobotryon*, *Fuscidea gothoburgensis*, *Ionaspis odora* and *Verrucaria margacea*. The altitude brings in rarer species including *Lecidea paupercula*, *Sporastatia polyspora* and large colonies of the greenish *Tephromela armeniaca*.

A species that is commoner in the Ben Nevis range than anywhere else is *Catolechia wahlenbergii* (Plate 13c). This elegant lichen forms vivid yellow-green rosettes on the mossy rocks. When Vince Giavarini first encountered it he called us over to look at 'this yellow *Rhizocarpon* growing on moss', a mistake easily made as both taxa contain rhizocarpic acid. He had just discovered the largest population in Britain. The northern edge of the summit is a series of sunless boulder-strewn ledges, occupied by cornice snow-beds till midsummer. On these mossy rocks most of the expected snow-bed species are present together with *Ionaspis cyanocarpa, Micarea granulosa* and a pyrenocarp with a chocolate brown thallus, this was later identified as *Staurothele arctica*, otherwise known only from Scandinavia. So the highest point in Britain has a suitably distinguished lichen flora.

A feature of the lichen flora of these western hills, which sets it apart, involves the high plateaux. In the Cairngorms, terricolous lichens are prominent and varied with *Cladonia arbuscula, C. rangiferina* and *C. uncialis* dominant, diversity being provided by species such as *Alectoria ochroleuca, A. sarmentosa, Cetraria ericetorum, Cetrariella delisei, C. nivalis, Cladonia maxima* and *Thamnolia vermicularis* (Fig. 11.13). In the west these terricolous species are either missing or subordinate to carpets of the woolly hair-moss. This change is a consequence of the unsuitability, for many terricolous lichens, of the permanently waterlogged soils in the west. The specialist lichens of these hyperoceanic mountains are mostly saxicolous such as *Catolechia wahlenbergii, Claurouxia chalybeioides, Coccotrema citrinescens, Micarea paratropa, Pilophorus strumaticus, Trapelia mooreana, Toninia thiopsora* and *Vestergrenopsis elaeina*. This general shift from terricolous to saxicolous as the preferred habitat in the west involves both percentage cover and species diversity, while certain species, such as *Pertusaria oculata*, that are strictly terricolous in the east, become saxicolous or bryophilous in oceanic areas.

There are no more than half a dozen significant late snow-beds in the west of Scotland, but they contain a number of distinctive lichens such as *Micarea paratropa, Stereocaulon plicatile, S. tornense* and *Toninia squalescens*. Certain terricolous lichens identified as snow-bed specialists in the Cairngorms are more widespread in oceanic areas where they occupy generally damp habitats that

Fig. 11.13 Thamnolia vermicularis, a pure white lichen of summit heath, can look like the tentacles of snake-lock sea anemones (I.C. Munro).

are free from competition. Conversely, the distinctive saxicolous flora of snow-beds is restricted to that habitat throughout the Highlands.

Fryday (1997) has pointed out that the terricolous lichen vegetation of the Cairngorms is a fragmented, species-poor outlier of that present in Scandinavia and, consequently, of only national interest. Conversely, the predominantly saxicolous lichen vegetation of the hyperoceanic mountains in western Britain is of international importance, having a number of rare and apparently endemic species; it is possibly unmatched elsewhere. Figure 5.5 (p. 77) demonstrates the uniqueness of its climate on a European scale.

A little further north the strongly oceanic Torridon Hills support small areas of calcareous heath, a habitat that is very rare in Scotland. These were first discovered by McVean & Ratcliffe (1962) who reported *Nephroma arcticum*. Thirty years elapsed before lichenologists sought these out and extended the list of upland calcicoles they contain by the addition of *Buellia papillata*, *Rinodina mniarea* and *Schadonia fecunda*, which occur in the closely grazed summit turf. The only other area of upland limestone heath in the Highlands is on the Blair Atholl limestone south of Beinn a' Ghlo in Perthshire; at one point it supports an intimate mixture of calcicole and calcifuge lichens including a flourishing population of *Cetraria nivalis*, the first to be found outside acidic lichen-moss heaths in Britain (Purvis *et al.*, 1994b).

West of Ireland

Brandon (954 m) in west Kerry is the single most important site for alpine vascular plants in Ireland. Consequently, a week was booked at O'Connor's guest house, which has been the traditional base for botanists since Robert Lloyd Praeger stayed there in 1912–13. It is a homely place of peat fires, soda bread, fresh salmon, and the bar routinely serves hot whisky to saturated fishermen and lichenologists on their return from the field.

The montane element in the lichen flora was found to be strictly limited, populations being small and isolated, but what was notable was the unusual abundance of several species (Gilbert & Fryday, 1996). Though we did not fully realise it at the time, these comprise the hyperoceanic element and form rare, possibly endemic communities.

The perpetually damp, acid rock on Brandon has a surface that is sub-optimal for lichens, being 60% bare or covered with an algal crust. The most frequent lichens included *Cladonia subcervicornis*, *Claurouxia chalybeioides*, *Coccotrema citrinescens*, *Ephebe lanata*, *Fuscidea lygaea*, *Lecidea pycnocarpa* var. *sorediata*, *Micarea lignaria* var. *endoleuca*, *Pertusaria pseudocorallina*, *Porpidia tuberculosa*, *Stereocaulon vesuvianum* (incl. var. *symphycheileoides*) and *Toninia thiopsora*. The presence on general rock surfaces of species that elsewhere in Britain are usually found in seepage tracks or by streams is a testament to the overwhelmingly wet nature of the climate. The presence of varieties, a rank rarely used by lichenologists, also suggests that the habitat is extreme. Additional species present that have a strongly oceanic distribution are *Gyalidea hyalinescens* and *Gyalideopsis scotica*, both commoner here than in Scotland. Saxicolous *Thelotrema lepadinum* is also a feature of these strongly oceanic areas.

It is perhaps unsurprising that the most remarkable lichen habitat in these moisture-laden hills is the margins of upland tarns; a habitat that must have been widely and continuously present since the late glacial period. The presence of mildly basic rock in the catchment of the deep watered tarns, high on

Brandon, has been important in counteracting leaching and maintaining a pH in the basins close to that of the period when the landscape was covered with fresh moraines. Under these conditions a rarely encountered, probably relic community has persisted on marginal boulders influenced by wave splash. It involves 50 to 60 species including *Aspicilia recedens*, *Collema dichotomum*, *Lecanora achariana*, *Nephroma tangeriense*, *Pseudocyphellaria intricata* and *Rinodina fimbriata*. Similar soft-water tarns are known in the Lake District, Snowdonia and northwest Scotland; to discover one is the experience of a lifetime. They are fully described in Chapter 12.

The acid, heavily grazed ridges of Brandon produced a short list of interesting lichens including *Aspicilia epiglypta*, *Catillaria contristans*, *Lecidea paupercula*, *Miriquidica complanata* and *Tephromela aglaea*. The mountain's most celebrated lichen, *Solorina crocea*, was not re-found, nor was any habitat suitable for it encountered; unfortunately, neither the herbarium label in The Natural History Museum nor the announcement of its discovery (Carroll, 1865) gives any indication of its location on the peak.

The survey of the montane lichen flora in the west of Ireland was extended to Muckanaught (Connemara), Ben Bulben (Sligo) and Slieve League (Donegal). This confirmed its limited nature. Though it is possible to find rare montane lichens in Ireland they are scattered and do not form proper communities. The mica-schist on Muckanaught is well known for its arctic-alpine high plants, but the extremely wet and dirty rock is unsuitable for lichens except locally under overhangs. Ben Bulben is notable for a large population of *Lempholemma cladodes*, *Peltigera leucophlebia* and *Staurothele succedens*. Slieve League, which has a narrow band of impure gritty limestone irregularly exposed at heights of between 300 m and 500 m supports *Gyalidea hyalinescens*, *Solorina saccata* and *Toninia fusispora*. A more thorough survey of the cliff-girt corrie surrounding Lough Agh would undoubtedly produce additional species.

The relative paucity of montane lichens in Ireland can be explained by the absence of genuinely alpine conditions. Very few hills exceed 915 m, and these are situated in the southwest, which experiences the mildest climate. As a consequence, habitat stresses associated with frozen ground, a short growing season and long snow-lie are not experienced, and the mountaintop vegetation grows luxuriantly even at sites exposed to strong winds. This results in few niches suitable for terricolous lichens. The development of the saxicolous lichen flora is impeded by the local geology, which is often coarse sandstone. Finally, the investigation of the flora is frequently hampered by atrocious weather.

The Lake District and Pennines

High ground in the Lake District was neglected by the early lichenologists so knowledge of its lichen flora has been slow to accumulate. In the early 1980s a systematic exploration commenced over several Whitsun weekends continuing sporadically until 1991 when it culminated in a week-long survey of two contrasting areas; base-rich sites in the Helvellyn range and acid outcrops in Langdale (Gilbert & Giavarini, 1993).

Many of the Lake District hills have flat rounded summits at 700 m to 800 m that support montane heath. Though their lichen flora is depressed by heavy sheep grazing, *Cetraria islandica*, *Cladonia bellidiflora*, *Micarea leprosula* and *Ochrolechia frigida* are regularly present, together with small amounts of the more demanding species *Alectoria nigricans, Baeomyces placophyllus, Lecidoma demissum* and *Thamnolia vermicularis*. The last species prefers the Skiddaw Slates to the Borrowdale Volcanics, the slates providing nine of its ten records; *Cladonia rangiferina* is also virtually confined to this formation (Corner, 1992). A number of montane lichens reach their maximum development on large erratic boulders that litter the floors of corries or lie at the foot of steep slopes. The richest examples are exposed to the full force of the wind so experience strong wetting-drying cycles; they also lack a peaty cap of vegetation. In the Lake District the specialities of this habitat are *Bryoria bicolor, Cetraria commixta, C. hepatizon, Cornicularia normoerica, Lecidea fuliginosa, Pseudephebe pubescens, Umbilicaria proboscidea* and *U. torrefacta*. The sides of these boulders may support *Lithographa tesserata*, and those used as bird perches hold additional species.

Due to the wetness of the climate the lichen flora of acid cliffs is mostly unremarkable, though good stands of the western *Cladonia cyathomorpha* with its large rounded squamules, the undersides carrying a pronounced pattern of raised pale pink to brown veins, may be encountered. Areas of late snow-lie on northeast-facing slopes are picked out by stands of the grass *Nardus stricta*. These are characterised by lichens that require moist conditions, such as *Lecidea phaeops*, but no snow-bed specialists have been found.

The principal refugia for calcicolous, arctic-alpine higher plants in the Lake District are in the Helvellyn-Fairfield range where carbonates are associated with certain lava flows, tuffs, shatter belts and fault planes (Fig. 11.14). Base-rich sites are few and far between, but an exploration of numerous dripping cliffs and gullies eventually revealed a matching group of lichens which, like the higher plants, are often present as very small populations hanging on in

Fig. 11.14 The cliffs above Brown Cove on Helvellyn are a refugium for calcicolous montane lichens; most are present as very small populations.

sub-optimal conditions. Rock intermittently flushed with base-rich water yielded *Belonia russula, Collema glebulentum, Gyalidea lecideopsis, Koerberiella wimmeriana, Peltigera leucophlebia, Polychidium muscicola, Porpidia superba* and *Pyrenopsis grumulifera.* Some of the dryer niches on the cliffs are underhangs, which occasionally provide a habitat for *Gyalideopsis scotica, Lecidea hypnorum* and *Strigula stigmatella* var. *alpestris*. The huge gullies that split the crags above the Wastdale Screes were also investigated but proved disappointing.

In general, acid rocks have not received the same attention as base-rich ones because good material is often difficult to collect. Their flora includes the 'difficult' genera *Lecidea, Porpidia* and *Rhizocarpon*, and few interesting higher plants are encountered while inspecting them. To redress the balance a special study was made of a complex of acid igneous rocks in Langdale. These proved to support a far wider range of species than had been anticipated. Freely drained areas of rock carried well-demarcated mosaics of *Rhizocarpon* spp. that included the small grey *R. simillimum*, new to Britain, while sheltered recesses yielded *Enterographa crassa* and *Opegrapha lithygra*. Short, vertical, periodically irrigated faces support a characteristic assemblage of *Claurouxia chalybeioides, Euopsis pulvinata* and *Pyrenopsis subareolata,* which suggests a local mildly basic influence. Hard tuffs carried a specialised 'Fuscideetum' community while ledges and crevices favour additional species. Once the acid rock flora has been learnt it is nearly as diverse, and contains as many interesting species, as that present at more basic sites.

Botanically, the Northern Pennines is a poor relative of the Lake District, but supports a few species that are missing from that area. *Frutidella caesioatra* and *Umbilicaria hyperborea* occur on the summit of Cross Fell (835 m) at a site where snow lies until early summer, while a high Yoredale limestone on the adjacent Knock Fell (760 m) recently provided records of *Ionaspis heteromorpha* and *Sagiolechia protuberans.* There is a nineteenth-century record of *Brigantiaea fuscolutea* from Upper Teesdale, a modern one of *Hypogymnia bitteriana* from Dead Stones, Co. Durham, and *Allantoparmelia alpicola,* present on several Pennine summits, has so far not been recorded from the Lake District. Finally, there is

Fig. 11.15 The back wall of Cwm Glas showing dark bedded pyroclastic rock outcropping at the base of a cliff that is otherwise composed of acid rhyolite (A.M. Fryday).

an enigmatic record of *Cetraria nivalis* from Craven. In the late 1950s several experienced botanists staying at Malham Tarn Field Centre reported this large, sulphur-yellow lichen from the summit of Fountains Fell, noting that it gradually declined with increasing visitor pressure. No specimen was collected. Having questioned the people involved, who knew the lichen well from seeing it in Iceland, I believe that a small, temporary population was present.

Snowdonia

The Lichen Flora of Gwynedd (Pentecost, 1987) provides an account of the montane species present on acid rock and soil in Snowdonia. These amount to around 20 lichens, most of which also occur on high ground in the northern Pennines. Basic sites in Snowdonia were not expertly surveyed until the 1990s when Fryday (1996) paid three visits to seek out major outcrops of calcareous rock, high level tarns and area of late snow-lie. Sites were located using geological maps, references to vascular plants in the county flora, and by consulting local botanists. The main concentration of base-rich rocks was on the south side of Llanberis Pass in Cwm Uchaf, Cwm Glas and Cwm Glas-bach (Fig. 11.15). This work added a valuable 31 taxa to the Welsh checklist, many of them rare montane lichens otherwise only known from the Scottish Highlands. They included many species that no-one had suspected of occurring so far south; for example, *Catillaria scotinodes*, *Lempholemma radiatum*, *Peltigera elisabethae*, *Placynthium pluriseptatum*, *Polyblastia terrestris*, *Protoblastenia siebenhaariana* and *Stereocaulon tornense*. These lichens are associated with basaltic tuffs and particularly with narrow bands of orange-brown metamorphosed limestone. Most are very rare, growing in only one or two places; as in the Lake District they are probably remnants of once richer communities that have been gradually attenuated by leaching and climatic amelioration.

The longest lying snow patch south of the Highlands occupies a gully on Y Ffoes Ddyfn on Carnedd Llewelyn; in most years it persists into July. The lichen

Fig. 11.16 The number of lichen taxa belonging to each montane zone for a selection of upland areas in Great Britain and Ireland. Data from Fryday, 1996.

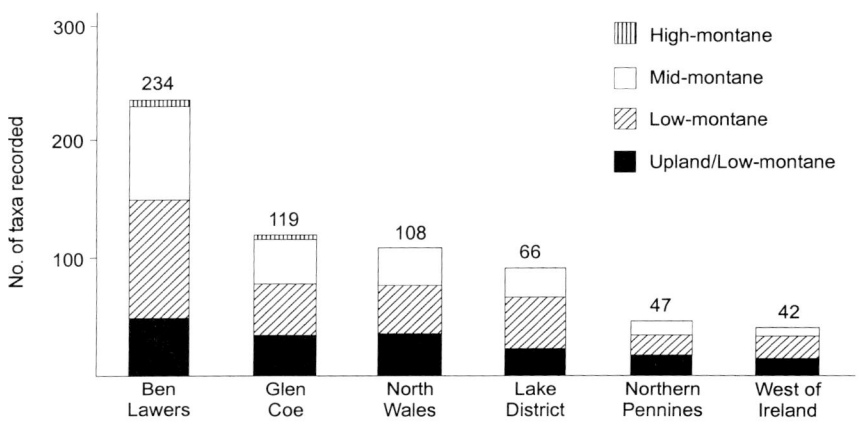

flora here shows little influence of snow-lie, though a kilometre away a set of gullies under the summit plateau has a more varied lichen assemblage of *Frutidella caesioatra, Protothelenella corrosa* and other species typical of damp rocks in north-facing corries and circumstantially associated with late snow patches. The hunt for true snow-beds outside Scotland has gone on for ten years and proved negative with regard to the most characteristic lichens, though a few of the typical bryophytes are present.

Comparing different areas

Until recently the total number of montane species recorded from an area was used as an index of the relative importance of its lichen flora. For example, North Wales with 87 was superior to the Lake District with 55, Western Ireland with 44 and the Northern Pennines with 41 (Gilbert & Giavarini, 1993; Gilbert & Fryday, 1996). Fryday (1996, 1997) has refined this by dividing the montane zone into the following sub-zones, which correspond with those used in continental Europe and recalculating the numbers.

Upland/montane Lichens with their centre of distribution in the uplands, also frequently found above the tree line (111 species).

Low montane Lichens with the centre of their distribution above the tree line but below the upper limit of ericaceous shrubs (159 species).

Mid-montane Species with the centre of their distribution above the level of ericaceous shrubs in moss/lichen heaths and stone deserts (102 species).

High montane Lichens associated with areas of prolonged snow-lie (10 species).

The total resource consists of some 382 lichens, though this is a provisional figure only, as it is a disappointing major expedition that does not add a new species to the British list and there is a backlog of undescribed taxa. When the number of species belonging to each zone is plotted, it can be seen that a high score in the mid-montane and high montane zones is characteristic of the best areas (Fig. 11.16). The total number of upland montane species, however, remains a good rough guide to the importance of an area. Whichever way the figures are treated it is clear that the richest individual sites in Scotland exceed in interest the entire montane floras of England, Ireland and Wales combined. Fryday's contribution has been valuable in providing a national perspective; meanwhile, much survey work remains to be undertaken, investigations continue, and many more surprises are anticipated.

The pleasures of pursuing montane lichens are recommended to all. They include the sheer exhilaration of being in lonely places surrounded by natural landscape. For much of the time one is detached from the surroundings, concentration is absolute, the only important things in the world are you and the lichens. In this heightened state of awareness, if a few minutes are taken off to relax, colours appear brighter, lark song more dulcet, the air purer, the view flawless, the sense of wellbeing utter. One would rather be here in the mountains than anywhere else in the world.

12

Rivers and Lakes

Fresh water encompasses lakes and watercourses ranging from tiny mountain rills, through those that Isaac Walton called 'that blue brook where leaps the speckled trout', to wide lowland rivers 'moating with emerald old cathedral towers'. It is difficult to imagine a more delightful habitat to explore on a hot summer day, wading around in gym shoes, accompanied by an abundance of wildlife, and working in an environment where the distribution of lichen species is elegantly and obviously related to site conditions. Fresh water provides beautiful examples of ecology in action. It is not without its frustrations, however, collecting off rounded boulders results in bruised knuckles, collections reduced to powder, and choice specimens ending up at the bottom of deep pools. A higher than average number of lichens needs to be collected as the regularly submerged communities are dominated by lookalike crusts and dark cyanophilic species.

All the richest freshwater habitats are in upland Britain where steep gradients result in mile after mile of stream bed dominated by outcrop or lined with boulders. The outstanding sites are frequently rapids and gorges associated with waterfalls (Fig. 12.1). In the lowlands, where deposition predominates, river channels tend to be silty and the lichen interest is confined to occasional stones projecting from vertical banks of soil, or spreads of pebbles where there is a slightly faster flow around a bridge or bend. An exception is the headwaters of spring-fed chalk streams flowing over a clean flinty bed; in these, every pebble may be completely covered with aquatic lichens.

Several factors can restrict the richness of the aquatic lichen flora. These are eutrophication, which in the uplands is caused by agricultural practices and in the lowlands by sewage, siltation and low flow resulting from impoundment or abstraction. In addition, many lakes have had their water level raised to improve the fishing or to convert them into reservoirs, so their shorelines are no longer natural. As a result of these factors, in many districts it is rare to find a pristine freshwater habitat.

The zonation across streams and rivers

There is a strong vertical zonation of lichens across a stream related to length of submergence. Four overlapping bands can be recognised, each characterised by a different assemblage. This situation is paralleled on the seashore, but here the bands are not distinctively coloured. They run as follows: 1) the submerged zone (aquatic lichens); 2) the fluvial mesic zone (amphibious lichens); 3) the fluvial xeric zone (terrestrial lichens tolerant of submergence); and 4) the fluvial terrestrial zone (terrestrial lichens). Zone boundaries are not always clear cut. Species may ascend to the next band where the rock is slow to dry out due to shade, seepage or a lush cover of bryophytes or conversely, on freely draining ridges, they may descend into the zone below. In gorges where flood levels cover an exceptional range the pattern can become very confused.

Fig. 12.1 The richest sites for river lichens are rejuvenated stretches where the channel is flanked by shelving rock; Aysgarth, River Ure.

Zonation is most conveniently studied in rivers where the channel is flanked by outcrop or on the side of large, mid-stream boulders.

Submerged zone (aquatic lichens)

The normally submerged zone is home to a number of highly specialised lichens, the majority of which occupy its upper levels and are exposed for study only during periods of drought. It is rare to find aquatic lichens more than 30 cm below summer water level, though *Verrucaria rheitrophila* and *Collema dichotomum* have been seen down to 50 cm. The latter, known as the river jelly lichen, resembles a small brown seaweed and can sometimes be detected under the water by feeling for its gelatinous thallus, rather like tickling for trout. It is a Red List species and sponsorship has been sought from various confectioners whom it was hoped might be interested in promoting its conservation, but so far without success. A River Jelly Lichen Steering Group has been set up to safeguard its interests (Plate 14a).

In acid streams the commonest lichens of this zone are the leathery, multi-lobed thalli of *Dermatocarpon luridum*, that are brown when dry and distinctly green when wet; *Hymenelia lacustris* forming cream-coloured mosaics on rock with abundant small, sunken, pink to orange coloured fruits; and a range of dark-coloured pyrenocarps such as *Staurothele fissa*, *Verrucaria funckii*, *V. hydrela* and *V. pachyderma*. In the uplands these species are joined by the rarer *Placynthium flabellosum* and *Porocyphus kenmorensis*, which form small rosettes on boulders in swift flowing streams. About 15 species have been assigned to the submerged zone in acid streams, though it is unusual for more than seven or eight to be present at any one site. Their cover may reach 60–80%. Clarifying the distribution of aquatic *Verrucaria* species relative to each other is a problem for the future.

Submerged lichen communities are not well developed in limestone streams due to active deposition of tufa, which is too soft a substratum for most species. Only variable amounts of dark *Verrucaria elaeomelaena* and greenish *V. rheitrophila* can be expected. The latter occurs down to 60 cm below normal water level whereas *V. elaeomelaena* is not often found below 20 cm depth. As the water level drops in summer it often reveals a crust of cyanobacteria covering the limestone. This excludes lichens which, in consequence, may achieve a cover of only 1%.

Chalk streams are habitats of great beauty. The majority arise from springs at the base of chalk escarpments, then flow through water meadows. The banks tend to be earthen and well colonised with water plants, so the sole habitat available to lichens is the gravelly bottom composed of flints and other stones that form shoals between beds of water crowfoot (*Ranunculus penicillatus*), water starworts (*Callitriche* spp.) and watercress (*Rorippa nasturtium-aquaticum*). At the finest sites, such as the headwaters of the River Test above Whitchurch, nearly every stone in the riverbed is covered with patches of *V. rheitrophila*. It is present in two forms, the typical morph with a greenish thallus and sterile black spots predominates, but in addition, there is a form with a black thallus and an uneven surface. It may represent thalli that have been grazed or damaged in some way. The submerged flints also support a conspicuous crustaceous red alga, *Hildenbrandia rivularis*, and a crustaceous green alga, *Protoderma viride*. Where the riverbed is silty, due, for example, to slack flow associated with the presence of weirs, or disturbance by cattle trampling, this colourful community of crustaceous algae and lichens is not present; instead the flints are dirty and covered with filamentous algae.

Amphibious zone (fluvial mesic)

This is the zone where the richest freshwater lichen flora is found. In acid streams around 25 lichen species are involved of which 8 to 10 can be found along any 100 m reach. The zone is regularly submerged so is only available to species requiring a constantly damp surface. Typically present are *Bacidia inundata, Porina chlorotica, Pterygiopsis* spp., *Rhizocarpon lavatum* and *Verrucaria aethiobola*. At the richer sites they are accompanied by a selection from *Collema flaccidum, Dermatocarpon leptophylloides, D. meiophyllizum, Ephebe lanata, Polyblastia cruenta, Porpidia hydrophila* and *Thelidium pluvium*. These species are themselves to some extent zoned. For example, *D. leptophylloides* with its interlocked button-like thalli and mid-brown undersurface, is mostly found at a lower level than the related *D. meiophyllizum*, which is monophyllous and has a dark brown underside. The lichens of the amphibious zone are not very well known, but include *Endocarpon adscendens*, a Red Data List species with a thallus composed of small, overlapping scales, beige when dry, green when wet, the tips of which turn up slightly at the ends. It is known from the Rivers Usk, Dart and Wharfe where it grows on and among mosses.

In limestone streams the amphibious zone supports around 20 species of lichen together with bryophytes and bare rock. Most are strict calcicoles and entirely different from those found by acid watercourses. Leading species include the limestone specialists, *Eiglera flavida, Leptogium plicatile, Staurothele guestphalica, Thelidium decipiens, T. fontigenum* and *T. zwackhii*, together with the generalists, *Pterygiopsis coracodiza, P. lacustris, Verrucaria aethiobola* and *V. praetermissa*. Regularly present are a number of taxonomically difficult lichens with cyanobacterial photobionts. Most of these have a thin reddish-brown thallus and perithecia-like apothecia.

The foremost limestone rivers in Britain are the head waters of the Aire, Wharfe, Ure, Swale, Tees and Ribble, which drain the Great Scar limestone of the Yorkshire Dales. A day spent collecting on any of these will provide a first class introduction to the lichens of limestone watercourses. Rivers associated with the Durness limestone in Sutherland, such as the Traligill at Inchnadamph, are notable for supporting rarities like *Gyalidea lecideopsis*, *Ionaspis melanocarpa* and *Staurothele bacilligera*.

Terrestrial lichens tolerant of submergence (fluvial xeric)

This belt, transitional between the aquatic and terrestrial zones, contains a mixture of weakly aquatic and terrestrial species that favour damp rock. It is swept by floods on average once a month. In a survey of acid streams (Gilbert & Giavarini, 1997) 50 species were recorded from this zone, which includes several elements. It is marked by the first appearance of normally terrestrial lichens with a wide ecological amplitude such as *Baeomyces rufus*, *Catillaria chalybeia*, *Scoliciosporum umbrinum* and *Trapelia coarctata*. There is also a large component that requires damp rock or seepage, these include *Amygdalaria pelobotryon*, *Massalongia carnosa*, *Porocyphus coccodes* and *Thelidium papulare*. A third element, composed of lichens favoured by the mild eutrophication that is associated with stream corridors, includes *Aspicilia caesiocinerea*, *Caloplaca isidiigera*, *Phaeophyscia orbicularis* and *Physcia caesia*. In addition, the sheltered faces of gorges at this level support a wide range of *Porina* spp. The same range of elements is present beside limestone streams (Gilbert, 1996) where riparian influence is indicated by the continued presence of *Leptogium plicatile*, *Thelidium decipiens* and *Verrucaria aethiobola*.

Terrestrial zone (fluvial terrestrial)

This zone is at the level where outcrops are kept more or less free of higher plants by occasional winter floods. The lichen communities here are richer than those found on outcrops away from the river due to the lack of competition from higher plants, and the slight flushing caused by the deposition of silt and enhanced faunal activity. The proportion of typical terrestrial lichens is high and too numerous for individual mention. Due to the habitat conditions the community, even on acid rock such as granite or Millstone Grit, has a general resemblance to those of slightly basic substrata such as basalt. It includes mild nitrophiles like *Lecanora muralis*, *Parmelia verruculifera* and *Physconia grisea*. The undersides of boulders support distinctive shade communities containing species of *Enterographa*, *Lecania*, *Micarea* and *Opegrapha*. Specialist fluvial species are absent, but this zone can produce surprises like the large colony of the Red Data List species, *Poeltinula cerebrina*, on limestone by the River Ure at Aysgarth Falls, and the second British record of *Rinodina fimbriata* from slate by the River Dart. Where *Cladonia* spp. like *C. chlorophaea*, *C. furcata*, *C. pyxidata* and *C. subcervicornis* appear in the lichen vegetation, prevailing conditions are no longer fluvial.

General observations

The margins of a watercourse provide a first-class example of an ecotone. An ecotone is defined as a narrow and fairly sharply defined transition zone

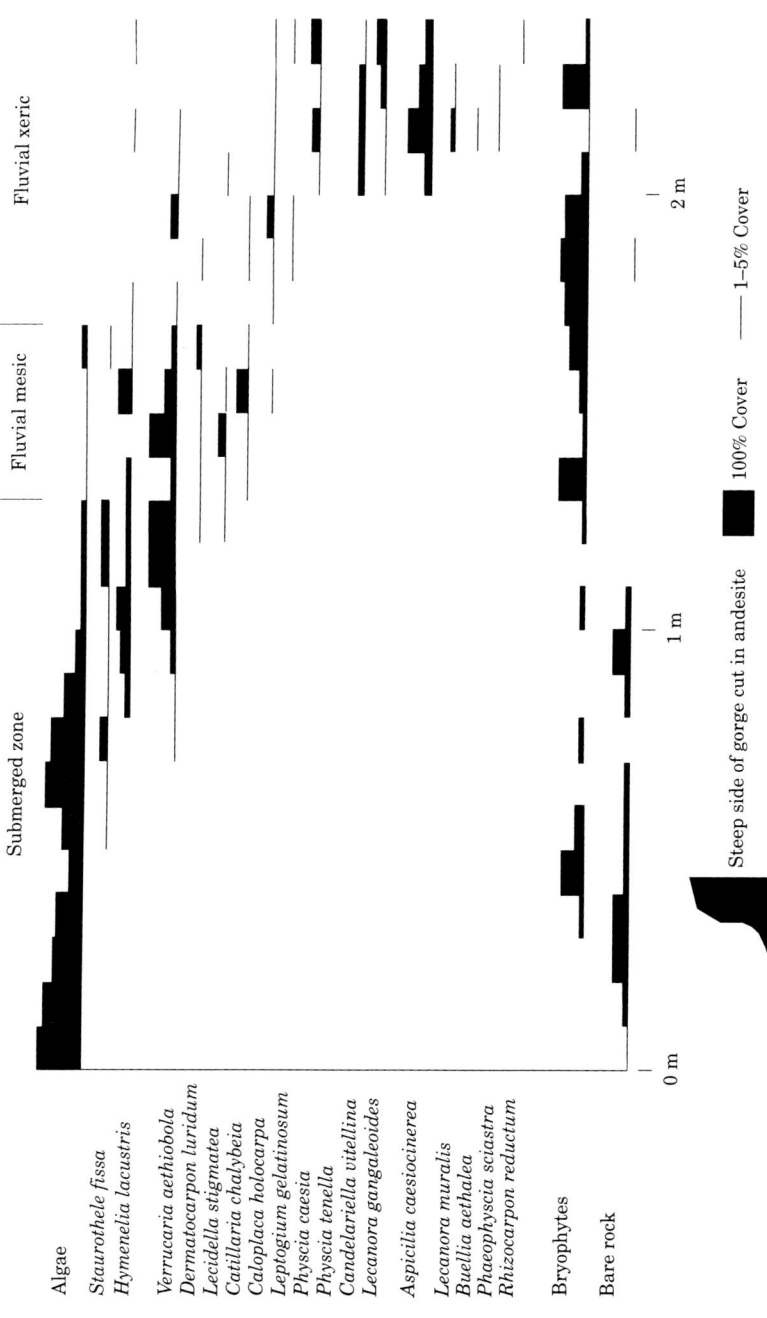

Fig. 12.2 Transect up the side of a gorge on the River Coquet to illustrate lichen zonation in relation to length of submergence; Cheviot Hills.

between two communities with characteristics of its own and of both types of adjacent vegetation. Ecotones are typically species-rich because of the wide range of habitat conditions present; in this instance the transition is from wet to dry, but ecotones also occur where habitat conditions change from shade to full illumination, or from acidic through neutral to alkaline. Surveying 100–300 m lengths of over 30 rivers showed that at least 50 lichens can be expected per site. The richest sites in terms of species diversity are quite simply those with the greatest area of exposed rock. These tend to be rejuvenated stretches in the middle reaches where there are waterfalls or rapids, and gently shelving beds of rock flank the watercourse, for example Aysgarth Falls on the River Ure; Ghaistrill's Strid on the River Wharfe; High Force on the River Tees; New Bridge on the River Dart; the River Greta at Keswick and The South Tyne at Warden Rocks (Plate 14b). These are mostly well-known beauty spots, so have to be surveyed under the curious eye of holiday makers who regularly mistake lichenologists for fossil hunters.

The reality of the zonation described above can be demonstrated by recording transects at sites where the margins of a stream are reasonably uniform. Figure 12.2 is a transect up the side of a gorge cut into andesite at Linbriggs in the Cheviot Hills. The submerged zone is dominated by an algal crust that contains few lichens. The narrow amphibious zone is also species-poor and somewhat untypical as *Verrucaria aethiobola* extends lower, and *Dermatocarpon luridum* higher than normal; the latter apparently because it is scoured off at lower levels. The fluvial xeric zone is well developed and includes the expected complement of species favoured by slight nutrient enrichment.

Variation from source to estuary

Rivers can be classified into distinctive segments down their long profile according to gradient, discharge rate, dominant type of sediment input and transport processes. These factors have a considerable influence on the lichens which, in addition to being classified according to tolerance of submergence, can be categorised into headwater, torrent, colluvial and alluvial zone species. There is still much work to be undertaken in this area, so the following observations should be regarded as preliminary.

The 'headwater' species of acid streams are found in mossy rills and pebbly flushes. *Hymenelia lacustris* is dominant, while wet pebbles may be entirely covered with a single species such as *Polyblastia cruenta*, *Porina guntheri* or *Thelidium pluvium*. In Scotland, the headwaters of montane, mica-schist streams in the Breadalbane Mountains and Clova support the *Ionaspidetum suaveolentis* association that contains many rare species, it is described in Gilbert *et al.* (1988) and Gilbert & Coppins (1992).

The next section downstream is the 'torrent zone', where lichen cover falls to less than one percent due to instability and scouring of the bedrock by the traction load which acts like a gigantic flexible file. Rock surfaces appear almost polished, the few lichens present being worn and abraded or restricted to sheltered niches and declivities. Even bryophytes are almost absent. The attenuated flora is composed mostly of common and widespread freshwater lichens, though in the Lake District the rare *Rhizocarpon amphibium* was found in the 'torrent zone' of Greenup Gill, Stonethwaite.

The next unit is the 'colluvial zone' where flow rates are gentler and there is a framework of large stable boulders in the stream bed (Fig. 12.3). In such

Fig. 12.3 Typical colluvial site on the River Greta at Keswick. Large stable boulders in the river bed are well covered with lichens.

reaches, lichen diversity recovers and is marked by the first appearance of *Dermatocarpon luridum, Porpidia hydrophila, Verrucaria praetermissa* and much else. Such reaches are the most rewarding to survey and should be targeted if time is limited.

In the 'middle zone' silting becomes heavy, so lichens are restricted to prominent upstream-facing edges of rock that are kept clean by the force of the current. The abundance of *Hymenelia lacustris* falls off dramatically. The richest sites in this section are where rejuvenation has created gorges, rapids and waterfalls. The pH of the water in acid rivers gradually increases downstream and by now usually exceeds 6, allowing cyanophilic lichens such as *Collema* and *Leptogium* to thrive.

In the 'lowland zone' the banks and bed are typically cut into soft materials that are unsuitable for lichens. The only potential habitats for fluvial lichens in the riverbed are occasional shoals of pebbles or perhaps the odd brick, together with semi-stable stones projecting from earth banks. These support a severely attenuated flora of infrequent *Verrucaria* spp. or *Porina chlorotica*. The high alluvium load of lowland rivers is strongly detrimental to lichens. However, a pebble projecting from a river bank in the New Forest provided an undescribed species of *Polyblastia*, so the habitat may be worth more attention.

Spring-fed chalk and limestone streams have a different structure. Water table fluctuations regularly cause the higher springs to fail, so the upper reaches routinely dry out; such lengths are known as winterbournes and during summer take on the appearance of a rutted track crossing the fields. The dominant lichen of these stretches is *Verrucaria aethiobola*, which covers most stones in the seasonally dry riverbed and is so vigorous that it has been seen colonising broken pottery and rusty metal. *Thelidium decipiens, T. fontigenum, T. minutulum* and *T. zwackhii* may also be present together with *Verrucaria elaeomelaena* present in damp niches such as the underside of stones. Travelling downstream to where the watercourse is permanent, the bed becomes silty fairly quickly due to the inwash of material off fields and trampling of the margins by stock. The lichen cover falls off in step with the increased silting, disappearing last from large items like bricks which, due to the unfavourable conditions, carry a variable and ephemeral cover of young colonies.

Permanent springs issuing from the chalk are the optimal habitat for *Verrucaria elaeomelaena* and *V. rheitrophila*, which form mosaics on stones associated with the clear upwelling water; they grow with equal vigour on limestone, flint, chert, sandstone, brick or fragments of old tufa. In limestone country the springs are unusually cold, running at around 8°C throughout the year. This has enabled relic populations of aquatic organisms such as diatoms and flatworms to survive well south of their normal range. No lichens with northern affinities have so far been observed in such springs.

Water chemistry or substratum?

The age-old question of the relative importance of water chemistry and substratum seems to have been answered as far as freshwater lichens are concerned. Evidence has repeatedly been found that substratum is the more important. Take the example of Littledale Beck in West Yorkshire, which rises from calcareous springs high on Whernside then flows for many kilometres over Carboniferous limestone. It supports classic communities of freshwater limestone lichens, yet sandstone boulders (originating from glacial drift) in the stream bed carry a typical acidic flora of *Hymenelia lacustris, Rhizocarpon lavatum* and *Verrucaria funkii* despite being continually bathed in strongly alkaline water (Fig. 12.4). The same community is also present on lenses of chert that are interbedded with the limestone. A further example can be seen on Widdybank Fell in Upper Teesdale, where calcareous springs emerge from the base of the sugar limestone before flowing directly onto the siliceous Whin Sill, which forms the bed of rivulets such as Red Sike. Despite the strongly buffered nature of the water the stream bed carries typically calcifuge lichens.

Water chemistry, however, appears to influence the occurrence of jelly lichens (Collemataceae) in acid watercourses. The pH of acid streams increases gradually as they progress to the lowlands, but species of *Collema* and *Leptogium* are not found until the pH is well buffered at or above 6 because their cyanobacterial symbiont, *Nostoc*, requires a neutral or alkaline pH if it is to function satisfactorily.

The effect of shade on rivers

All those who have worked on freshwater lichens have considered that light intensity is important in determining their distribution. In my experience, lichens that reach their greatest abundance along tree-lined watercourses include *Bacidia arnoldiana, B. carneoglauca, B. fuscoviridis, B. inundata, Baeomyces rufus, Enterographa hutchinsiae, Lecania* spp., *Porina lectissima, Trapelia placodioides, Verrucaria aquatilis, V. hydrela* and *V. praetermissa*. In shaded, humid sites they extend onto damp rock some distance above their usual upper level. This elevation of the zones in streams flowing through woodland can be considerable and confusing. Along shady brooks I have seen the normally submerged *Hymenelia lacustris* extending to 1.5 m above summer water level and wondered what it could be, it was so out of context.

Fig. 12.4 Little Dale Beck, Yorkshire, where a limestone flora occurs on the bedding planes while sandstone boulders in the stream support acidic assemblages despite the alkaline water.

An additional effect of shade is that it favours a dense growth of bryophytes which compete with lichens to such effect that only the largest, like *Peltigera* spp., may be present, and then only temporarily. In 1976 there was considerable excitement when *P. lepidophora* was found on mossy ledges by a shaded section of the River Ericht in East Perthshire. This was the first British record and the species is now listed on Schedule 8 of *The Wildlife and Countryside Act, 1981* where it has been given the English name of Ear-lobed dog-lichen. An attempt to re-find it in 1994 was unsuccessful, which demonstrates the rapid turnover of thalli in such situations. It was thought that it might well reappear, which it did in 1998. Within streams flowing through woodland it has been observed that algae have a greatly reduced cover, due partly to the shade, and partly to the algicidal effect of twigs and dead leaves in the riverbed.

Water quality and rivers

English rivers are classified into six chemically-based classes according to the degree to which they are polluted. They range from Class A rivers containing water of the highest quality to Class F, which are grossly polluted. Provided that suitable substrata are available, most Class A streams have an average of 20 lichen species per site with a cover of 15–30%. Where streams drop to Class B and C their lichen flora deteriorates rapidly, becoming characterised by a very low cover (< 0.1%), which is restricted to the largest, steepest and most prominent boulders as these experience maximum scour by the current. Other surfaces become encrusted with a fertile mixture of soft silt and algae that supports rapidly growing bryophytes and seedling higher plants. The submerged and amphibious lichen zones are entirely absent, having been replaced by swards of filamentous algae. The occasional thalli in the higher zones are algal-covered, fragmenting and often sterile. Such severely attenuated assemblages comprises isolated thalli of four to six species among which *Bacidia inundata* and *Trapelia coarctata* are commonest, plus a selection from *Bacidia caligans*, *Hymenelia lacustris*, *Porpidia crustulata*, *P. tuberculosa*, *Porina chlorotica*, *Rhizocarpon lavatum*, *Trapelia placodioides* and *Scoliciosporum umbrinum*.

Class D and E stretches have an even lower lichen cover and a narrower range of species that includes ephemerals such as *Bacidia chloroticula* and *Pyrenocollema monense*. All surfaces are caked with algae, mud and bryophytes. Walking in the stream bed stirs up clouds of suspended solids that are slow to settle, there are unsavoury smells, and pebbles are covered with a white deposit.

Chemical data shows total oxidised nitrogen levels increasing five to tenfold as water quality drops from Class A to Class C and D, the same is true for ammoniacal nitrogen. Phosphorus also increases many times, but less regularly. So, in acid streams, eutrophication appears to be the main cause of lichen decline, with field evidence suggesting that suspended solids are also detrimental. The deposition of fertile silt is particularly harmful. In streams flowing over chalk and limestone, lichen deterioration is also most strongly correlated with silting. Episodic pollution, involving nutrients, fungicides and algicides introduced as a result of accidents at fish farms, is currently unevaluated. In addition to water quality it should be recalled that water abstraction resulting in low flow, tufa deposition, and disturbance caused by farm stock or people paddling also reduce lichen diversity.

To help the recovery of aquatic vegetation in a number of key rivers a programme of phosphate removal has been agreed, while elsewhere, the Ministry of Agriculture is investigating the effectiveness of buffer strips and the re-creation of water-fringe habitats to help control agricultural runoff. Despite these measures, the heyday of chalk streams is thought to be over.

Lakes and tarns

The lichen flora of lake margins is currently under investigation, so a complete interpretation of their richness is not possible. It is difficult to understand why they have remained relatively neglected as few habitats are more enjoyable to work. Lakes are best tackled during hot weather, wearing gym shoes, knee pads and shorts or bathing trunks so there are no constraints on wading out to emergent boulders, kneeling in the shallows or on the retrieval of 'lost' chips from deep water. It usually takes a day to survey a lake or tarn properly as a shoreline may be interspersed with outcrop, covered with massive boulders, have stony or gravelly beaches or be fringed with wave-lapped turf; levels of eutrophication and grazing also affect the lichen flora. The habitat has a charm equalling that of seaside rock pools (Fig. 12.5; Plate 14c).

The study of lake margins received a considerable stimulus in April 1981, when the all-round botanist, Roderick Corner, discovered *Collema dichotomum* and *Lecanora achariana* growing abundantly around a high-level tarn in the Lake District. This was particularly exciting because a lichen checklist had just appeared (Hawksworth et al., 1980), which listed the latter species as extinct in the British Isles. This conspicuous, pale yellow-green lichen, looking like a cross between *Squamarina cartilaginea* and an exuberant *Lecanora muralis*, has proved to be something of a catalyst to freshwater studies in the way that

Fig. 12.5 Flat-topped boulders standing a little way off shore are a rich habitat; these support *Massalongia carnosa, Polychidium muscicola* and *Umbilicaria deusta*; Loch Bà, Rannoch Moor (V.J. Giavarini).

Lobaria pulmonaria was to woodland exploration in the 1970s. It is a 'lead in' species to a generally rich habitat; many lichenologists, when in the uplands, now deviate to take in water bodies in the hope of finding it. As a consequence of this popularity it has now been discovered at single sites in Scotland, Ireland and Wales, in every case above 610 m and where there is some basic influence in the catchment. The original nineteenth-century station, by Llyn Bodlyn in North Wales, was lost when the lake level was artificially raised, but a recent visit to the site revealed it to be a highly acidic water, quite unsuitable for the species, so a slight mystery surrounds the earliest record. When it was eventually re-found in Wales, in 1997, postcards were sent off bearing the message:

'Snowdonia drama,
We've found *L. achariana!*'

In common with streams, the ecology of lake margins is determined largely by water level fluctuations, which vary over a season from an estimated 50 cm in small water bodies to nearly 2 m in larger ones, like Lake Windermere, the level of which is regularly monitored. Wave splash extends the zonation on lee shores. Superimposed on the conspicuous vertical zonation are the effects of eutrophication, shade, suspended sediment, rock type and water chemistry. These factors combine to render each lake highly individual, but a number of general principles are emerging that help to make sense of the patterns observed in the field.

Rock type, water chemistry and eutrophication

Water bodies in the Lake District have been studied for many years by the staff of the Freshwater Biological Station at Windermere who have arranged them in a series according to their pH and alkalinity (Table 12.1). Those at low altitude or with basic rock in the catchment tend to be the most alkaline, are productive with regard to fish, rich in invertebrates, and support beds of the reed grass *Phragmites australis*, while the acid ones are smaller, more upland, in siliceous rock areas, support diminutive trout, an impoverished fauna, and contain stands of the bottle sedge (*Carex rostrata*). Mean numbers of aquatic macrophytes are five in permanently acid upland tarns and 17 in those with medium-hard water (Stokoe, 1983). Lichen surveys of upland tarns also show a striking correlation between alkalinity and lichen richness, which varies from 13 around the most acid examples to 54 at the most alkaline (Table 12.1). Acid waters are characterised by *Ephebe lanata, Micaria lignaria, Placopsis gelida, Porpidia* spp., *Stereocaulon pileatum* and *Trapelia* spp., while more alkaline ones always contain a proportion of cyanophilic lichens and a wider range of *Dermatocarpon* and *Verrucaria* spp. The majority of aquatic lichens, however, are reasonably widespread, responding to differences in water chemistry by a change in abundance.

Table 12.1 Lichen diversity of tarns in Cumbria and North Wales related to mean alkalinity and pH (water quality data from Edwards *et al.*, 1990; Rimes, 1992; Sutcliffe & Carrick, 1986; 1988).

Water type	Permanently acid	Very soft water	Soft water	Medium hard water
Range of alkalinity*	-60 to 0	1 to 100	101 to 200	201 to 2000
Range of pH	4.2 to 5.3	5.7 to 6.7	6.8 to 7.0	> 7.0
Tarns/lichen diversity	Red Tarn[+] 18	Dock Tarn 28	Angle Tarn[+] 47	Ullswater 43
	Llyn Dublyn 13	Ffynnon Lloer 34	Small Water 46	Windermere ?
	Haystacks 15	Blackbeck 25	Brown Cove 57	

* Alkalinity expressed as micro-equivalents per litre.
[+]Red Tarn, Langdale; Angle Tarn, Hartsop.

It is not known whether the alkalinity of the water or the alkalinity of the substrate is the more important, though both are linked. In the absence of water quality data an indirect guide to potential lichen richness is the status of the water body as a fishery; the more highly it is valued by anglers, the more diverse the lichens are liable to be. For example, the newly discovered Welsh site for *Lecanora achariana* has been described as the best upland lake for trout in the principality.

Around most lakes and tarns the largest flat-topped boulders, in particular those standing a little way out into the water, support lichen communities that indicate eutrophication. There is a strong link between the productivity/alkalinity of the lake and the development of these communities. The most acid water bodies have just the occasional boulder sufficiently eutrophicated to bring in a low cover of weak nitrophiles such as *Aspicilia caesiocinerea, Lasallia pustulata, Parmelia conspersa* and *Scoliciosporum umbrinum*. Soft water lakes may have a dozen or more boulders coloured white, yellow and green by a dense growth of *Caloplaca arenaria, Candelariella vitellina, Lecanora muralis, Physcia caesia* and *Tephromela atra*. Mildly eutrophicated boulders by alkaline upland lakes are highly exciting places, being the main habitat in Britain for rarities such as *Aspicilia melanaspis, A. recedens, Lecanora achariana, Phaeophyscia endococcina, P. sciastra* and *Rinodina fimbriata*.

The margins of large lakes lying in main valleys, for example, Bassenthwaite, Derwent Water, Loch Lomond and Windermere, are often fringed by pebble beaches or reed beds which form a poor habitat for lichens, but where outcrop or stony margins occur, eutrophication is the norm, and nitrophiles, particularly *Physcia* and allied genera, are abundant. Field observation suggests that this enrichment originates from insectivorous birds feeding along the shoreline; the more productive the lake, the more food there is available and therefore the greater the guano deposition.

Shaded lake margins

Under natural conditions, water bodies below 610 m altitude would have margins shaded by woodland or scrub. Low illumination severely restricts lichen diversity, so the past distribution of woodland may explain why tarns and lochans above the postglacial tree line support the richest lichen flora. Shaded lake margins have been investigated at several sites in the Lake District, an undertaking that is most conveniently achieved from a rowing boat as the widely spaced lakeside crags often have poor public access. The results of a transect

Fig. 12.6 Transect illustrating the lichen zonation on a shaded mossy bluff beside Lake Windermere. *Endocarpon adscendens* is a Red Data List species.

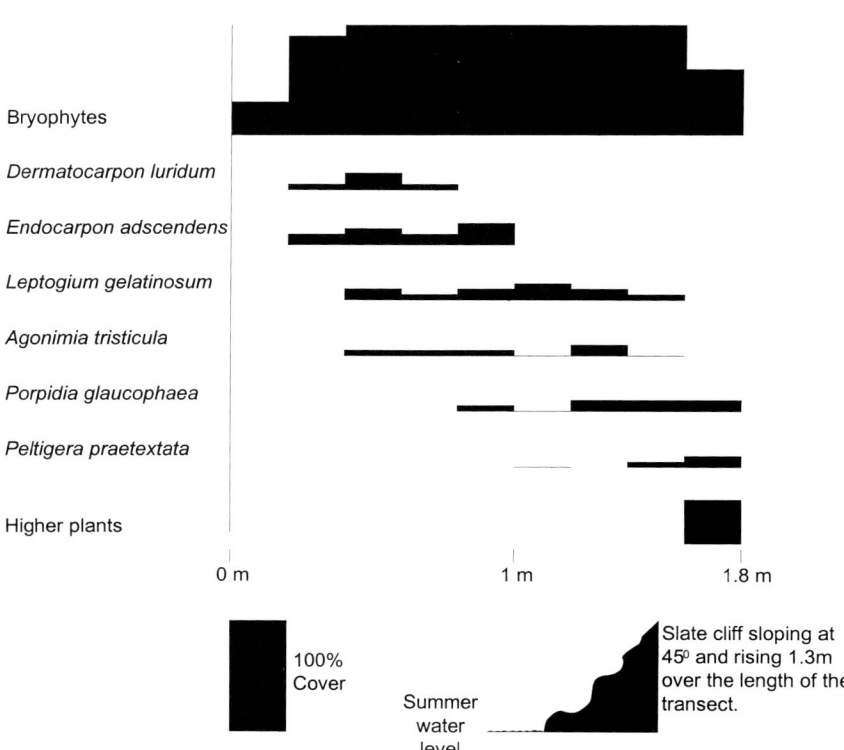

study up a rocky bluff overhung by oak trees at the north end of Lake Windermere are shown in Figure 12.6. The rocks are almost entirely covered with bryophytes, and the few lichens present, apart from *Porpidia glaucophaea*, are muscicolous. Of particular interest is the presence of the Red Data List lichen *Endocarpon adscendens*, which occurs low down in the zonation with *Dermatocarpon luridum*. When wet, the *Endocarpon* appears as a distinctive sward of apple-green, overlapping scales (Plate 15a); as these dry out they change to pale brown in which condition the species could be mistaken for a diminutive *Dermatocarpon*. Zoned above it are *Agonimia tristicula*, large foliose species of *Leptogium* and *Peltigera praetextata*.

Shady niches also occur around the margins of upland tarns in association with large boulders, where scree descends into the water, and on islands where ungrazed vegetation overhangs the margins. A search of such nooks and crannies, which often involves pulling loose material out from the back of deep crevices, will often yield fragments of rock covered with the tiny white fruits of *Absconditella delutula*, the green thalli of *Bacidia carneoglauca*, *B. inundata* and *B. fuscoviridis*, *Opegrapha gyrocarpa*, *Porina chlorotica* and various forms of *Trapelia coarctata*.

A feature of the margins of rocky lakes in the west of Great Britain and Ireland is the presence of large foliose lichens that are normally regarded as old woodland indicators. They include *Degelia atlantica*, *Dimerella lutea*, *Nephroma laevigatum*, *N. parile*, *N. tangeriense*, *Leptogium burgessii*, *Pannaria pezizoides*, *Parmeliella triptophylla* and *Pseudocyphellaria intricata*, which are found colonising the vertical sides of boulders projecting from the water. Their first discovery in these semi-alpine aquatic habitats caused astonishment. It is assumed that this humid, sheltered habitat is a natural alternative to the mossy tree trunks where they are usually found.

Lake margin zonations

In small water bodies the lichen zonation associated with length of submergence is most easily observed on the sides of boulders projecting from the water. The zones have not been formally named, but can be designated the submerged, lower splash and upper splash zones. Their arrangement at one unusually rich Lake District tarn are shown in Figure 12.7. Regularly submerged communities typically contain species such as *Dermatocarpon luridum*, *Hymenelia lacustris*, *Placynthium flabellosum*, *Porpidia hydrophila*, *Staurothele fissa*, *Verrucaria hydrela*, *V. pachyderma* and *V. rheitrophila*. Above this is a distinctive 'lower splash zone' of semi-aquatic lichens that include *Bacidia inundata*, *Collema flaccidum*, *Dermatocarpon meiophyllizum*, *Ephebe lanata*, *Porina interjugens*, *Pyrenopsis impolita*, *Rhizocarpon lavatum* and *Verrucaria aethiobola*, these forming a band extending from around summer water level to 15 cm above it. The 'upper splash zone' is composed of lichens that have a preference for damp rock, mild eutrophication or both, such as *Aspicilia caesiocinerea*, *Lecanora muralis*, *Massalongia carnea*, *Placopsis gelida*, *Polychidium muscicola*, *Porpidia soredizodes*, *Scoliciosporum umbrinum* and *Umbilicaria deusta*. These grade into the typical acidophilous lichen flora of the surrounding hillside; the presence of *Cladonia* spp., *Parmelia omphalodes* or *Rhizocarpon geographicum* indicates that the aquatic influence has become minimal.

The zonation around large lakes, such as Bassenthwaite, Derwent Water, Ullswater and Windermere, extends vertically for 2 m and is most

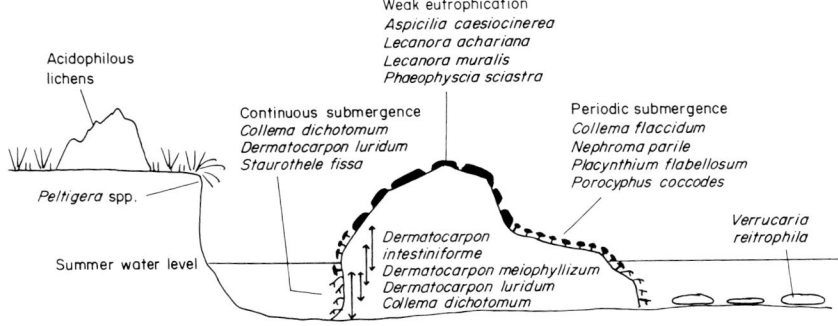

Fig. 12.7 A boulder at the margin of a Lake District tarn showing the distribution of the main lichen assemblages in relation to summer water level.

conveniently displayed where well-illuminated, vertical rock faces drop straight into the water. Transect work has enabled the following six zones to be identified.

1) The lowest zone, replacing that dominated in clear upland tarns by species of *Verrucaria*, is a layer of fine silt impregnated by algae and cyanobacteria.

2) Similarly silt-impregnated, but *Dermatocarpon luridum* present.

3) A band with *Dermatocarpon leptophylloides*, the algal cover starts to break up to expose rock allowing the entry of *Hymenelia lacustris*.

4) *Dermatocarpon intestiniforme* and/or *D. miniatum* (Plate 15b) are the most conspicuous lichens but, owing to the cleaner nature of the rock, they are accompanied by up to a dozen semi-aquatic lichens such as *Catillaria chalybeia, Porina lectissima* and *Rhizocarpon lavatum*, including weak calcicoles like *Catapyrenium squamulosum, Lecidella stigmatea* and *Porpidia speirea*.

5) This belt is dominated by nitrophilous species, which at the sites studied include *Aspicilia caesiocinerea, Caloplaca arenaria, C. crenularia, Lecanora dispersa, L. muralis, Pertusaria albescens, Physcia dubia* and *Scoliosporum umbrinum*.

6) A band of species favoured by very slight enrichment, for example *Lecanora gangaleoides, Ochrolechia parella, Parmelia verruculifera* and *Pertusaria pseudocorallina*; these grade into typical acidophilous communities.

Doubtless there are several variations from the zonation described above which is widespread in the Lake District.

Reservoirs

Reservoir drawdown zones are well known to naturalists for supporting rare ephemeral mosses and liverworts, but they have been largely neglected by lichenologists. When reservoirs in the Millstone Grit region of the Southern Pennines are drawn down, one lichen, *Trapelia coarctata*, is seen to be present in great abundance, extending to 5 m (vertical) below top water mark. It becomes less frequent with increasing distance from the shoreline and at the lowest level is represented by scattered sterile areoles. The only lichen associated with it is occasional *Verrucaria hydrela*. The main feature of more productive lowland reservoirs in the same area is a metre-wide band of *Lecania erysibe* just below top water mark. During the exceptionally dry period, 1995–6, a local naturalist, Paul Ardron, found the rare endemic lichen, *Thelocarpon magnussonii*, growing on a ruin that normally lies 10 m beneath the surface of Ladybower Reservoir, Derbyshire. At the same site there was evidence that nettles (*Urtica dioica*) and various fungi such as the sulphur tuft (*Hypholoma fasciculare*) had survived several decades of continuous submergence. Whether the lichen had, remains an open question.

13

Coastal Habitats

Coastal habitats are formed by the interaction between sea and land; in places the sea cuts into the land to create cliffs, often separated by rocky beaches, while elsewhere deposits of shingle, sand or mud build up. Whatever the type of shore there is something to interest lichenologists, though one is constantly aware that lichens are just one facet of the astounding variety of life to be found there. Survey work may be carried out with screaming gulls overhead, slippery seaweed underfoot and be interrupted by encounters with attractive shells. This narrow strip of land holds interest out of all proportion to its extent: it is the most complex of all environments and a wonderful training ground for the young lichen ecologist as it is impossible to remain unaware of the influence of the tides that divide the habitat into a series of zones, each with its characteristic species. Here is not one environment but many.

Figure 13.1 is a distribution map of the well-known saxicolous coastal lichen *Verrucaria maura*, sometimes called the tar-spot lichen, as its black thallus has frequently been mistaken for oil pollution on rocks around high tide mark. Dozens of coastal lichens have a similar distribution. The scarcity of many species along the east and southeast coasts of England is notable and has a lot to do with the Exe-Tees line that divides Highland and Lowland Britain. Hard acid rock is regarded as the prime coastal habitat for lichens and this is largely absent in the lowlands. The occasional records along the east and southeast coasts are of isolated thalli on shingle, chalk, artificial substrata such as sea defences, or concrete tank traps dating from the Second World War.

Nonetheless, it would be a mistake to think that coastal habitats in the lowlands can be disregarded by lichenologists; just the opposite. Rocky shores are only one of nine potential habitats, and have been overworked in comparison to the others. Dunes and shingle are fairly well studied, but the following six have been either neglected or severely neglected with regard to their lichens: salt marshes, intertidal mud flats, saline lagoons, soft cliffs, maritime cliff grassland and coastal heath. It will be demonstrated that all are ripe for more detailed work.

Lichen zones on rocky shores

A Frenchman, H. A. Weddell (1875), studying the Isle d'Yeu off the Vendée coast of France, appears to have been the first person to recognise that lichens on the shore occur in distinct zones. He noted 1) a marine zone, 2) a semi-marine zone subject to wave splash and 3) a maritime zone beyond the reach of the waves, and listed species typical of each. The next significant advance was achieved by Matilda Knowles who worked in Dublin at the National Museum of Ireland and made a detailed investigation of the lichens of the local rocky coast (Knowles, 1913). She identified three colour belts – grey-green, orange and dark – 'that were found all round the coast occurring with beautiful regularity' (Plate 15c). Her classification is shown in Table 13.1.

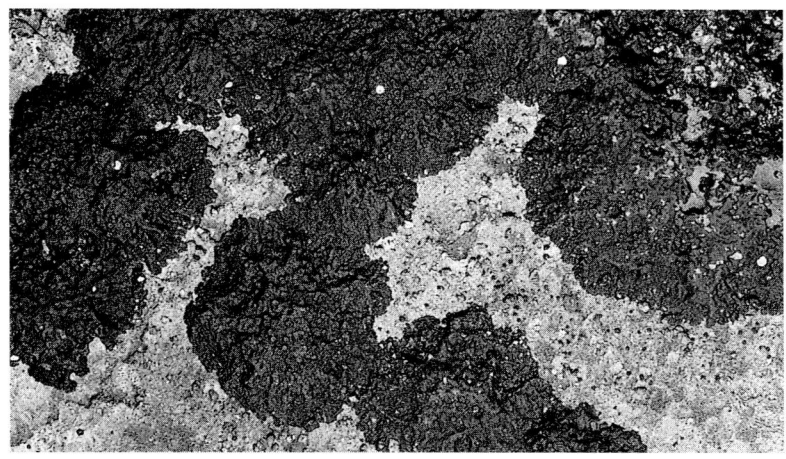

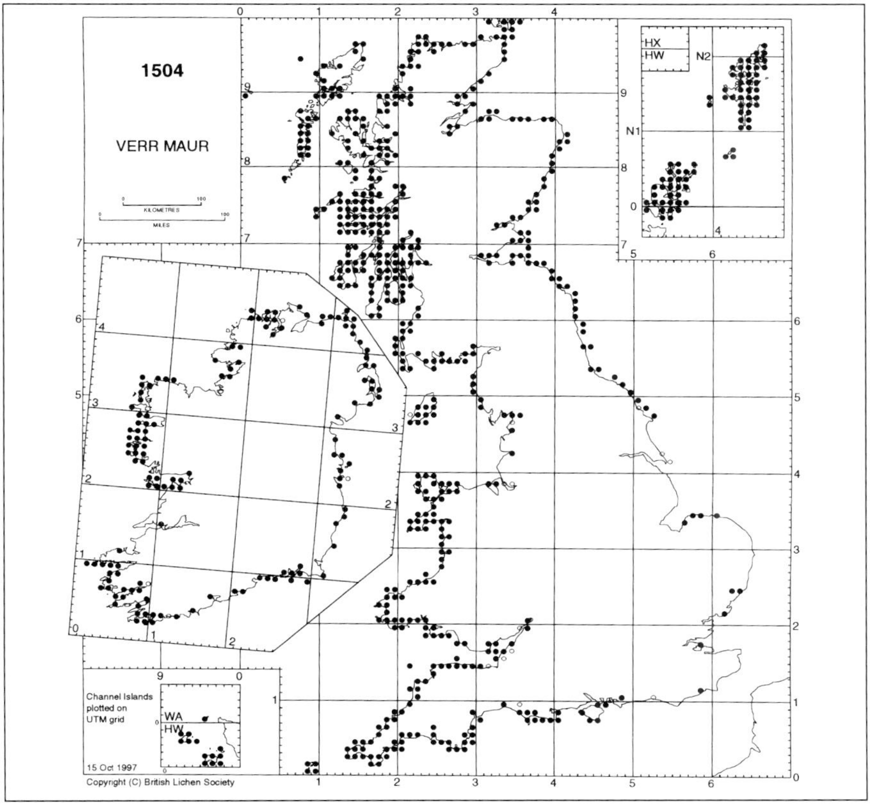

Fig. 13.1 The tar-spot lichen (*Verrucaria maura*) together with a distribution map that highlights the scarcity of suitable habitat along the east and southeast coasts. (Photo J.M. Gray; Map BLS Mapping Scheme).

Table 13.1 The lichen zones proposed for rocky coasts by Knowles (1913; 1915).

White belt	*Ochrolechia parella, Tephromela atra*, etc.	Top of shore
Grey belt (1)	1) *Ramalina* zone	
Orange belt (2)	2) *Caloplaca, Xanthoria* zone	
Black belt (3–5)	3) *Lichina confinis* zone	
	4) *Verrucaria maura* zone	
	5) Marine *Verrucaria* zone	Bottom of shore

Two years later, when working on the west coast of Ireland, she described a still higher 'white belt' formed by *Ochrolechia parella* and *Tephromela atra* where the rocks appeared as if whitewashed. However, it is her three zone colour system, now generally known as grey, orange and black for which she is remembered, as it conveniently extends the red (lower), brown (middle) and green (upper) marine algal zones into the wave-splashed region. The adaptive significance of the colours is not known.

Following useful studies in the west of Ireland and Pembrokeshire (Sheard, 1968; Ferry & Sheard, 1969), the next major advance in our understanding of coastal lichen zonation was made by Fletcher (1973 a,b; 1975 a,b) who carried out detailed work on the siliceous shores of Anglesey. His classification of the lichen vegetation is based on the system of Lewis (1964). The trouble with pre-Lewis classifications was that they contained an anomaly. The bench mark on the shore was extreme high water spring tides (EHWST). However, species characteristic of the marine zone i.e. that part of the shore inundated by the tides, are also, on exposed coasts, regularly present in the maritime zone i.e. that part of the shore never inundated by the tide. Lewis's breakthrough was a

Fig. 13.2 Frond of the brown seaweed *Pelvetia canaliculata* bearing black, hemispherical fruit bodies of the lichen *Pyrenocollema pelvetiae* (J. Benn).

classification independent of tide levels. The major subdivision between the Littoral (marine) and Supralittoral (maritime) zones is considered by Lewis to be the upper limits of the lichen *Verrucaria maura* and the periwinkle *Littorina*. On sheltered shores this corresponds to EHWST, but the boundary can be elevated many metres on exposed coasts. Owing to the diffuse upper boundary of *Verrucaria maura*, Fletcher has suggested that a better lichen marker is provided by the lowest thalli of *Caloplaca marina, Lecanora actophila* or *L. helicopis*. Fletcher then fixed the boundary between the Supralittoral zone and the Terrestrial region as being where the dominant species obtain most of their water and nutrient requirements from rain and soil solution rather than windborne spray. His classification, which has been accepted world wide, is shown in Table 13.2. In general, rocky shores in western Britain are richer than comparable ones in the east. For example, a shore in Wales might have 150 lichens, while only 40–50 can be expected in Northumberland.

Table 13.2 Summary of the internationally accepted zonation pattern for marine and maritime lichens with indicator species for British shores (from Fletcher, 1973 a,b).

Terrestrial	Halophobic		*Parmelia omphalodes, Sphaerophorus*
	Halophilic		*Parmelia saxatilis, P. caperata*
Supralittoral	Xeric		*Anaptychia runcinata, Parmelia pulla,*
			Ramalina siliquosa, Rhizocarpon richardii,
	Submesic		*Xanthoria parietina*
	Mesic		*Caloplaca marina, Lecanora actophila* (sun),
			L. helicopis (shade), *Lichina confinis*
Littoral	Littoral fringe		Periwinkles. *Lichina pygmaea, Verrucaria amphibia,*
			V. maura
	Eulittoral		Barnacles. A few lichens such as *Pyrenocollema*
			halodytes and *V. mucosa* penetrate into this zone
Sublittoral			No lichens recorded below low tidemark on
			British shores

The black belt (Littoral)

As implied above, the Littoral zone broadly corresponds to the intertidal area or marine zone of the shore, but its upper limit can be uplifted 50 m by wave action on exposed Atlantic coasts. It is divided into two sections, an upper Littoral fringe without barnacles and a lower Eulittoral where barnacles occur. The highly specialised lichen flora is composed of 16 species: *Lichina* (2), *Pyrenocollema* (5) and *Verrucaria* (9); this includes an undescribed species of *Lichina* found in sheltered sea lochs in Scotland and *Pyrenocollema pelvetiae*, which grows on fronds of the brown seaweed *Pelvetia canaliculata* (Fig. 13.2). The latter species was thought to be confined, in Britain, to the Orkney Isles, but it has now been discovered at several sites in Devon. Until fairly recently the marine species of *Verrucaria* were poorly known; in consequence, during work for his doctorate Fletcher (1973a) found four species new to Britain. Most of them have thin brown thalli and sessile perithecia < 0.2 mm diameter so are easily overlooked. The dominant marine species are present as a series of overlapping bands related to sea level, though the presence of drier ridges and wetter depressions renders the boundaries irregular. Position on the shore acts as a guide when identifying the bands; from the top of the shore downwards they run: 1) *V. maura*, 2) *V. amphibia*, 3) *V. striatula*, 4) *V. halizoa/ V. mucosa*. The lowest is often present as olive to dark green patches between barnacles and the red, crustose alga *Hildenbrandia*. If difficulty is experienced finding all these, scan the damp rock under seaweed with a hand lens. Identification should normally be confirmed by making a small collection.

No-one should have difficulty recognising the intertidal *Lichina pygmaea*. Its loose tufts of dark, flattened, palmately divided branches are regularly to be seen forming large patches on sunny exposed rock faces in mid-shore situations. Examination with a lens will reveal that the tufts support large numbers of a minute reddish bivalve *Lasaea rubra*, which is so constant that it constitutes a subsidiary identification character (Fig. 13.3). *L. pygmaea* requires more regular submersion than any other macrolichen. Perhaps for this reason it was for a long time regarded as a diminutive brown alga and went under the name *Fucus pygmaeus*, so in the older floras records must be sought in the algal section. A mid-shore specialist is *Pyrenocollema halodytes*; it can be separated from the marine *Verrucarias* by its one-septate spores and its habit of penetrating onto acid shores by growing on barnacles (Plate 15d), limpets and mussel shells. The perithecia, 0.25 mm in diameter, are just visible to the naked eye. *P. sublitorale* has been found on barnacles and limestone down to the level of the *Laminaria* belt. Only one lichen has been recorded growing below low tide mark, namely *Verrucaria serpuloides*, which was dredged from a depth of 30 m in Antarctic waters (Lamb, 1973). If the opportunity arises to examine the contents of dredgers it should be grasped as the seabed around Britain is unknown territory to lichenologists.

The upper limit of the Littoral zone is largely determined by wave action. On very sheltered shores all the marine *Verrucaria* spp., except for *V. maura*, are confined to below EHWST and *Lichina confinis* to no more than 1 m above this benchmark. As exposure increases the height to which individual lichens occur increases until on peninsulas projecting into the Atlantic many Littoral species are never submerged by the tide. In Britain exposure reaches its maximum on remote islands in the Hebrides. On the west coast of North Rona, for example, the belts are uplifted 15 m, the total cover of lichens is much lower, and zones which on a sheltered shore might be less than 1 m wide (vertical height) extend over 7 m. Each species responds to wave exposure in its own way; *V. mucosa* becomes rare, *L. pygmaea* is absent from both very sheltered and very exposed sites, while poor competitors, such as the *Pyrenocollema* spp., become progressively more abundant as bare rock increases. On very sheltered shores *V. maura* may be the only representative of the Littoral zone.

A feature of the Littoral zone in the north of Scotland is the presence of a

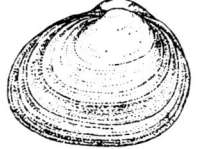

Fig. 13.3 This minute bivalve mollusc *Lasaea rubra* is constantly associated with the lichen *Lichina pygmaea*. Actual size = 1 mm.

prominent belt of cyanobacteria (*Aphanocapsa, Lyngbya, Nostoc* and *Pleurocapsa* spp.) their black, felt-like growth largely replacing lichens in the upper Littoral. Cyanobacteria are usually most abundant on the shore during the winter, northern Scotland being the only part of Britain where they persist through the summer months. In 1972 this belt on North Rona extended vertically for 7 m, but its extent is likely to vary both from year to year and season to season so it could be responsible for reports of broad bands of bare rock in the upper Littoral zone.

Grazing by invertebrates in this zone is an important and largely uninvestigated factor. Limpets browse in a rough circle around their home base, creating a territory devoid of lichens and algae, although if algae are abundant they may ignore the lichens. Periwinkles (*Littorina* spp.) shelter in crevices when the tide is out, but once submerged, they clear circular areas around their refuges which then look like bull's-eyes (Fig. 13.4). A cluster of 20 periwinkles might graze a circle 30 cm in diameter. This factor could be investigated experimentally by removing shells from an area of rock and recording what happens. Algologists report that rocks protected from grazing quickly become covered with seaweed. Many lichen thalli appear able to resist permanent settlement by algae and barnacles.

The orange and grey belts (Supralittoral)

This zone is never inundated by the sea, which influences it entirely through salt spray. It is current practice to divide the lichens of the Supralittoral into three groups; the Xeric-, Submesic-and Mesic-Supralittoral.

Fig. 13.4 A cluster of periwinkles showing the area around their refuge grazed free of lichens and algae (M.J. Lindley).

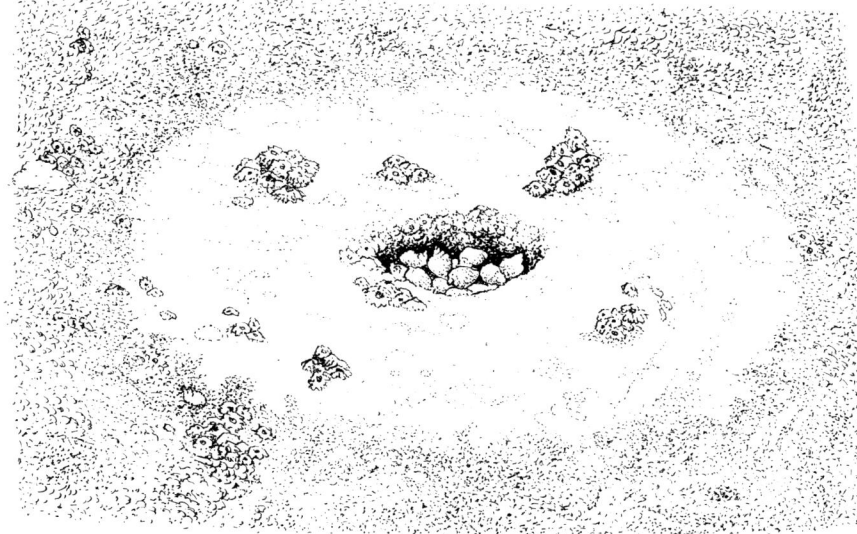

The orange belt or Mesic-Supralittoral supports a small, well-defined group of specialist lichens with a high requirement for salt spray. Around 20 species are involved, half of them exclusively maritime. Examples are *Caloplaca marina*, *C. thallincola*, *Catillaria chalybeia*, *Leconora actophila*, *L. helicopis*, *Lecania erysibe* and *Lichina confinis*. *Verrucaria maura* is usually present in damp runnels and depressions, its dark thallus frequently bearing the minute pale yellow squamules of *Caloplaca microthallina*. In eutrophicated sites *Aspicilia leprosescens*, *Caloplaca verruculifera* and *Lecanora poliophaea* are frequent. A further element of variation is shelter; along the west coast recesses near the top of this zone should be searched for the attractive orange, densely isidiate *Caloplaca littorea* and the conspicuous perithecia of *Verrucaria prominula*. On shaded shores the cover of yellow *Caloplaca* spp. is greatly reduced, so the overall colour becomes leaden-grey; consequently the Orange zone of Knowles is only conspicuous on sunny shores.

The top of the orange belt includes the Submesic-Supralittoral zone. It was separated by Fletcher because the well-known, conspicuous yellow lichen *Xanthoria parietina* forms a distinctive belt at this level. It is often accompanied by *Physcia adscendens*, *P. tenella* and *X. candelaria*, which suggests that this is the lowest level on the shore at which foliose lichens can exist in the face of wave action. Where there is significant eutrophication *X. parietina* is more widespread so the zone is not as well marked.

The grey zone or Xeric-Supralittoral is where the majority of the 70 or so exclusively maritime species in Britain are to be found, but if generalist lichens are included the total in this zone is doubled. On exposed coasts it invests entire headlands and cliffs, and in the Outer Hebrides whole islands. The zone is typified by species belonging to the *Ramalina siliquosa* complex, which form dense swards on exposed rock. This species aggregate has had a highly complicated taxonomic and nomenclatural history, which is related by Sheard & James (1976). Lichenologists received a major shock in the late 1960s when, after studying populations in North Wales, Culberson identified six chemical races. Because they also had slight niche preferences, he proceeded to recognise them at the species level (Culberson, 1967; Culberson & Culberson, 1967). This chemical taxonomy was adopted for a few years, but gradually a simpler one was introduced. Now just two species are recognised; *R. cuspidata*, which is concentrated in the lower part of the grey zone, and *R. siliquosa*, found mostly above it (Fig. 13.5).

When exploring this zone, other species likely to be encountered are brown rosettes of *Anaptychia runcinata*, *Lecanora fugiens*, *L. gangaleoides*, *Lecidella subincongrua*, *Parmelia delisei*, *Ramalina subfarinacea*, *Rhizocarpon richardii*, *Rinodina atrocinerea*, *R. luridescens* and *Verrucaria fusconigrescens*. These will be growing alongside lichens that are not seaside specialists, such as *Caloplaca crenularia*, *Lecanora rupicola*, *L. sulphurea*, *Ochrolechia parella* and *Parmelia saxatilis*. The composition of the assemblage differs according to whether the rocks are vertical or horizontal, the level of eutrophication, degree of shelter, rock type, light intensity and position within the British Isles. It always supports a colourful array of lichens (Plate 16a). A rich crevice and underhang flora often develops containing a selection of easily identifiable seaside specialists such as *Bacidia scopulicola*, *Leprocaulon microscopicum*, *Moelleropsis nebulosa*, *Solenopsora holophaea*, *S. vulturiensis* and *Trapeliopsis wallrothii*. North-facing situations have a considerably reduced flora lacking orange species especially.

Fig. 13.5 Ramalina siliquosa (left) and *R. cuspidata* (right) drawn to emphasise their differences. The latter has a blackened base, bears dark pycnidia, and has main branches that are more or less rounded with a smooth surface (M.J. Lindley).

What makes for excitement is that in this zone every shore is different. For example, the furthest Hebrides, embracing the Flannan Isles, St Kilda and North Rona, border the Arctic Ocean and support enormous colonies of sea birds. Here, large isolated guano-spattered boulders carry up to 25 lichens each, including such rarities as *Buellia coniops, Caloplaca scopularis* and *Lecanora straminea* (Plate 16b). The latter, handsome, Red Data List species, like a lavish *L. muralis* but C+ orange, is not substrate limited as it has colonised the aluminium body of a crashed Wellington bomber on St Kilda.

At the other end of the country in Cornwall, the Isles of Scilly and the Pembrokeshire Islands, species more characteristic of the Mediterranean area grow. Here, in favoured localities where warm humid air blows over acid boulders, *Pertusaria chiodectonoides, P. gallica, P. monogona, Ramalina chondrina, R. portuensis* and *Teloschistes flavicans* occur (Plate 16c), while in the surrounding turf *Heterodermia leucomelos* and *H. obscuratum* may be found. In this region dry, often sheltered recesses on north and east facing cliffs should be sought out as they are particularly rewarding. This is the habitat of the frequently elegant *Roccella fuciformis* and *R. phycopsis*, though they are equally likely to be present as small, gnarled, easily overlooked individuals. Surprisingly the genus has not been recorded from Ireland, where there is a bottle of whisky on its head.

Animals play some part in determining the lichen flora of this zone. The hardier breeds of sheep, such as Soya and Scottish Blackface, greedily graze on *Ramalina*. This may explain a curious observation made by MacCulloch (1819) when he visited Eilean Mor, the largest of the Flannan Isles. He remarked on 'the utter absence of lichens which gave a sharpness to the rock scenery, so everything appeared as if it had been cut and polished by a lapidary'. A visit by chartered trawler in 1975 (Gilbert & Wathern, 1976) revealed that the resident sheep had nibbled *Ramalina* off every accessible boulder and outcrop, so greatly reducing the biomass of lichen on the island. The 250 strong herd of feral goats on the Isle of Rhum have been observed grazing on seashore lichens. The species involved were *Anaptychia runcinata* and *Xanthoria parietina*, which were removed by rasping with the bottom incisors; goats only consume lichen when other sources of food are in short supply.

At a different level it was observed that foliose lichens on the Flannan Isles were heavily infested with arthropods. The most important grazers among these were the tiny oribatid mites *Ameronothrus maculatus* and *Phauloppia lucorum*, both well known as lichen feeders. They live under the lichen mat. Existing among and preying on this large colony of mites were considerable numbers of the carnivorous anthocorid bug *Temnostethus pusillus*. The food chain is completed by insectivorous birds such as rock pipits (*Anthus spinoletta*), which eat the bugs.

Terrestrial halophilic and halophobic zones

The Terrestrial zone commences where species receive most of their water and nutrients as rain, dew or from the soil rather than from sea spray. This change is accompanied by a sharp drop in pH as sea water is strongly alkaline (pH

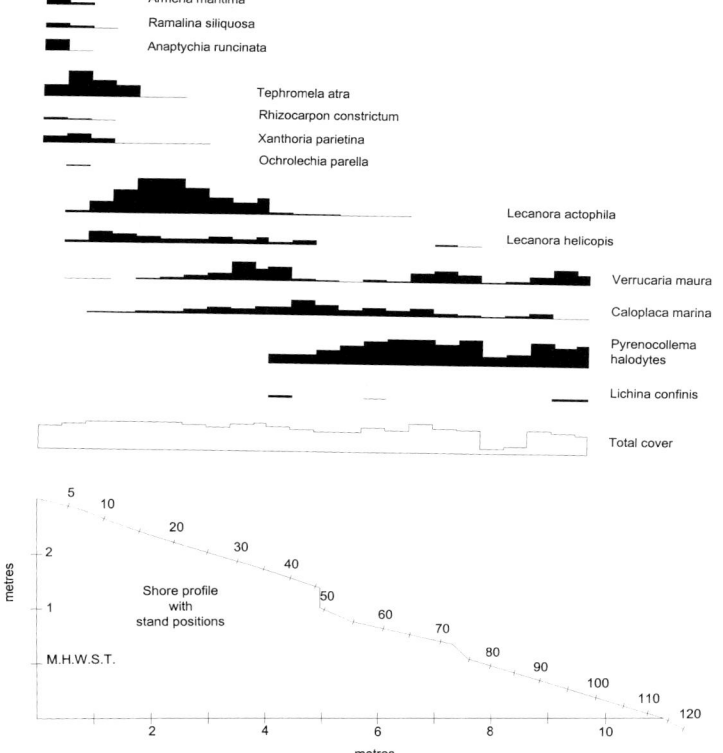

Fig. 13.6 Transect showing how species progressively replace each other on a rocky shore.

8.0–8.3). Terrestrial halophilic lichens are tolerant of occasional inundation by sea water, though some are discoloured after prolonged exposure. Typical species include *Fuscidea cyathoides, Parmelia glabratula, P. saxatilis, P. sulcata* and *Rhizocarpon obscuratum*. The Terrestrial halophobic zone is occupied by a normal inland lichen flora that commences first in sites conspicuously sheltered from sea spray such as landward facing outcrops. The sequence from eulittoral to halophobic is an outstanding example of an ecotone (ecological gradient) which can be investigated through transect work (Fig. 13.6).

A feature that reaches its finest development in Europe along the strongly oceanic west coast of Britain is the occurrence of rich saxicolous Lobarion communities close to the sea. These have been reported from islands that were never wooded such as Bardsey, The Blaskets and St Kilda, so the coast is a primary habitat. Their typical niche is a near vertical face below a vegetation-capped ledge, an arrangement that ensures the rock face is periodically flushed by water which has drained through stands of tall herb vegetation. On St Kilda the community is present on rock types ranging from acid granophyre to eucrite and gabbro, but the most species-rich stands occur on the more basic rocks (Gilbert *et al.*, 1979). Genera involved include *Degelia, Lobaria, Nephroma, Pannaria, Parmelliella, Pseudocyphellaria* and *Sticta*, they are found growing alongside maritime species such as *Anaptychia fuscata* and *Ramalina siliquosa*.

Limestone shores

Apart from work in Fletcher's doctoral thesis (1972) there have been few studies of limestone shores in Britain despite the presence of fine Carboniferous limestone headlands at Brean Down, on the Gower peninsula, the Great and Little Ormes and Humphrey Head. One of the most detailed published accounts is from Steep Holm island in the Bristol Channel where I spent a week in 1980 (Gilbert, 1984c). The shores here are very sheltered, eutrophicated by gulls and affected by silt deposition so may not be entirely typical. In the Littoral zone *Pyrenocollema halodytes, Verrucaria ditmarsica* and *V. mucosa* are locally present occupying crevices and hollows but always with a low cover, and reaching their fullest development on mineral veins containing chert. Towards the top of the zone they are replaced by *V. maura*. In the Mesic-Supralittoral zone *Caloplaca marina* and *Lecanora helicopis* form a discontinuous belt, being joined slightly higher up by *Caloplaca flavescens, C. thallincola* and *Xanthoria calcicola*. Instead of grading into a rich Xeric-Supralittoral zone the orange zone passes directly into the normal limestone lichen flora of the island. The only grey zone lichen encountered was a small patch of *Ramalina siliquosa* on an acid mineral vein near the cliff top. Only seven exclusively coastal species were present in this limestone zonation. A similarly attenuated flora is present on sheltered limestone along the north side of Morecambe Bay. Here just three marine and two maritime species are present including *Lichina confinis*. Again the orange zone passes directly into a terrestrial limestone flora of *Aspicilia calcarea, Caloplaca alociza, C. variabilis, Lecanora crenulata, Porina linearis* and *Rinodina bischoffii*.

Both the above shores are sheltered, and sheltered shores often have common species missing, so a comparison with exposed limestone shores along the Gower coast is instructive. Here 13 maritime species occur in the black (Littoral) zone, which is the entire British list with the exception of those characteristic of northern Scotland. In contrast, maritime species were rare in the Mesic- and Submesic-Supralittoral and absent from the Xeric-Supralittoral. So limestone shores support a highly distinctive zonation; the black zone can be as rich as anywhere, the orange zone is severely impoverished, while the grey belt is missing, being largely occupied by a normal terrestrial lichen flora.

Sand dunes

Sand dunes are taken to include all areas of coastal windblown sand whether forming true dunes, undulating links, sand sheets or machair. The largest and richest dunes are all in Scotland where magnificent examples occur both on the east and west coasts. Those on the east coast tend to be dour, acid and sufficiently extensive to get lost in, while the west coast examples are machair produced where shell-sand has blown inland to cover the acid landscape with a calcareous blanket. To botanise the machair on a sunny day with a white edge to the surf, surrounded by a profusion of flowers and blue butterflies is unforgettable.

Early work on the ecology of sand dune lichens was carried out by Watson (1918) at Braunton Burrows, Devon. He noted that lichens were absent from

Fig. 13.7 Histograms showing the distribution of the main lichen species along an 800 m transect crossing acid dunes at Studland Heath, Dorset (Alvin, 1960).

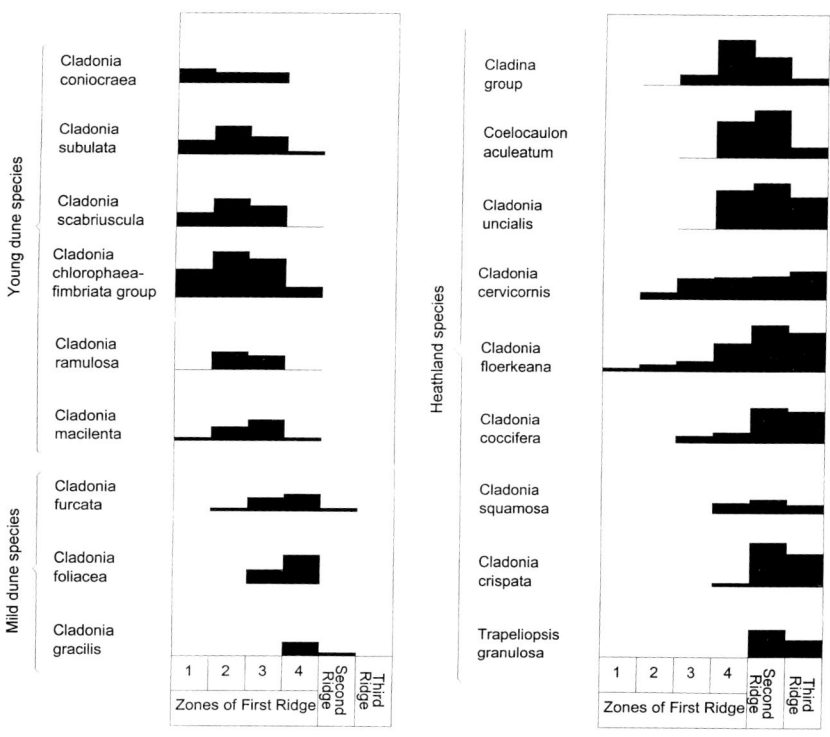

'mobile dunes' on the seaward side of the system, but three species of *Peltigera* occurred on the second line of sand-hills. Further back, the fixed dunes supported 21 terricolous lichens including *Bacidia sabuletorum, Cladonia foliacea, C. furcata, C. pocillum, Diploschistes muscorum, Peltigera* spp. and a range of Collemataceae. These were accompanied by several normally epiphytic species such as *Evernia prunastri, Parmelia caperata, Ramalina farinacea* and *Usnea hirta* growing on the sand.

Forty years elapsed before Alvin (1960) made the next advance when studying the very acid dunes at Studland Heath, Dorset. Here, three main ridges run parallel to the shore forming a system some 800 m wide which he divided into six sections, typifying the lichens of each by determining their frequency in a 100 random quadrats (Fig. 13.7). Huge differences occurred between the youngest marram grass (*Ammophila*) zone dominated by *Cladonia chlorophaea, C. coniocraea, C. humilis, C. polydactyla* and *C. scabriuscula*; the middle zone where heather starts to replace marram, which has as the leading species *Cladonia arbuscula, C. ciliata, C. furcata, C. gracilis, C. portentosa* and *Coelocaulon aculeatum*; and the oldest heather-dominated zone with *C. cervicornis, C. coccifera, C. crispata, C. floerkeana, Hypogymnia physodes* and *Trapeliopsis granulosa*. The succession appeared to be related to increases in acidity, from pH 6.5–7.0 on the seaward dunes to 3.8–4.5 at the oldest site, and also to increasing organic content of the soil. Dune systems offer considerable scope for transect work coupled with careful soil analysis to elucidate the comparative ecology of the different lichens present.

Scottish east coast dunes

A series of vast, remote dune systems is present along the northeast coast of Scotland; to tour them is a most exciting undertaking. As a bonus, banks of shingle occur among the sand-hills. Most of the dunes are covered with a well-developed heath of the northern, heather-crowberry (*Calluna-Empetrum nigrum*) type, which is found only along the coasts of northern Scotland, Denmark and southwest Sweden. Its characteristic features will be described by reference to Culbin Sands and Cuthill Links.

Culbin Sands (3,180 ha), the largest dune system in Britain, lies on the south side of the Moray Firth and is owned by the Forestry Commission who planted it with pines in the 1930s. At four localities the trees have remained stunted, providing open yet sheltered conditions; here one of the finest examples of lowland, acid lichen-heath in the UK has developed. Reindeer lichens form a deep white carpet spreading into the distance under the trees and suppressing all other growth (Fig. 13.8); it reminded a group of Russian visitors of home. Over 30 species of *Cladonia* have been recorded, which, in the absence of fire and trampling, grow with great exuberance. Among the commonest are *C. arbuscula, C. ciliata, C. gracilis, C. portentosa, C. uncialis* and huge dark 'plates' of *Coelocaulon aculeatum*. Locally abundant are *Cladonia carneola, C. cornuta, C. foliacea, C. sulphurina* (to 10 cm tall), and an upland component that includes *C. bellidiflora, C. rangiferina* and *Ochrolechia frigida*. A boreal-continental element, better represented here than anywhere else in Britain, is composed of *Cladonia mitis, C. uncialis* subsp. *uncialis* (which differs from the oceanic subsp. *biuncialis* in the ends of the podetia which branch in a star-like fashion, (Fig. 13.9)) and *C. zopfii*. This would be a suitable site to search for *C. stellaris*, perhaps originating from Scandinavia and establishing itself as a temporary population.

Fig. 13.8 Reindeer lichens form a continuous carpet among planted pines at Culbin Sands.

A network of unsurfaced roads traverses the forest. Intermittent disturbance along their margins exactly meets the requirements of *Peltigera malacea* and *Stereocaulon condensatum*, which are locally abundant in this niche. This is one of only three extant sites in Britain for *P. malacea*; it can be separated from related species by examining the underside, which is unveined and covered with a continuous grey-brown tomentum; when wet the upper surface has aquamarine tints. Culbin is one of the best places in Britain to see lichens taking on a vagrant life form, species such as *Coelocaulon muricatum*, *Evernia prunastri* and *Hypogymnia physodes* grow wholly unattached as 'lichen balls' that blow about in the wind. This is a characteristic of desert species. It is believed that the manna of the bible, which sustained the Israelites during their exodus from Egypt, was the edible vagrant lichen *Sphaerothallia esculenta*, large amounts of which unexpectedly blew into their camp.

In contrast to Culbin, which takes several days to explore, Cuthill Links (100 ha) is more domestic in scale, close to a main road, and can be investigated in an afternoon. The lichen interest is concentrated on the floors of three valleys that lie between gorse-covered dune ridges. The valleys are lined with an open sandy, heathland composed of cushions of heather and bell heather (*Erica cinerea*); there are lichens everywhere; on the ground, on stones, on the heather, on shells, bones, fence posts, etc. Nearly all the Culbin species are present, including *Cladonia mitis*, *C. uncialis* subsp. *uncialis*, *C. zopfii*, *Peltigera malacea* and *Stereocaulon condensatum*. In addition it is the best site in the country to see normally montane species growing at sea level. *Alectoria sarmentosa*, *Cladonia phyllophora*, *Ochrolechia frigida*, *Stereocaulon saxatile*, *Thamnolia*

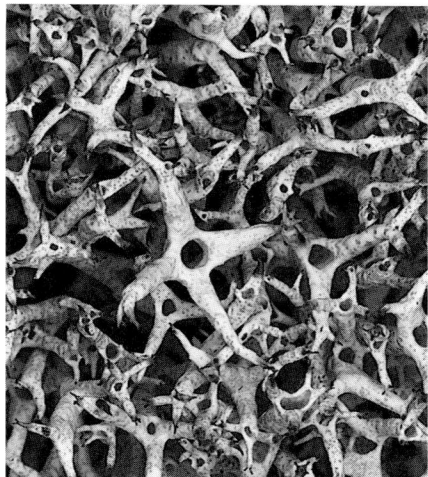 

Fig. 13.9 The rare *Cladonia uncialis* subsp. *uncialis* (left) differs from the widespread *C. uncialis* subsp. *biuncialis* (right) in the ends of the podetia which branch in a star-like fashion; Culbin Sands.

vermicularis and the woolly hair-moss *Racomitrium lanuginosum* all grow luxuriantly among the prostrate heather, recalling high ground in the Cairngorms. Cuthill Golf Course must be the only one in the country to have *Alectoria sarmentosa* in the 'rough'! As at Culbin, certain species are favoured by slight disturbance; the best place to see the dark grey, tube-like thalli of *Leptogium corniculatum* is on the verges of the new road that now bisects the site. The shingle communities at both localities are an added bonus. Cuthill provides a glimpse of what Culbin must have been like before it was planted up.

Machair

Machair, a habitat unique to Scotland, develops where calcareous shell-sand blows inland converting the wet, acid, peaty landscape into a well-drained, flowery pasture. On the seaward side the machair grades into mobile dunes at the edge of the strand while the more inland stretches are often fenced and improved. In between there is usually a belt of natural machair pasture where the vegetation is a short, species-rich turf. The habitat has not received much attention from lichenologists, though it is known to be deteriorating under the impact of ploughing, reseeding, heavy trampling and manuring. The commoner species, found widely down the west coast (Coll, Colonsay, Mull, South Uist, Tiree, Rhum), are *Agonimia tristicula*, *Bacidia muscorum*, *B. sabuletorum*, *Cladonia pocillum*, *C. rangiformis*, *Collema tenax*, *Peltigera canina*, *P. neckeri*, *P. rufescens*, *Toninia lobulata* and *T. sedifolia*.

Machair backing isolated sandy bays on the west coast of Coll was surveyed during a British Lichen Society field meeting (Gilbert *et al.*, 1984) and found to be in good condition. The richest communities occurred in association with low rock outcrops, the presence of which discourages intensive agriculture and also opens up the sward. In such sites up to 27 terricolous lichens may be present including *Catapyrenium cinereum*, *C. squamulosum*, *Cladonia firma*,

Chromatochlamys muscorum, Placidiopsis cartilaginea, Polyblastia gelatinosa, P. wheldonii, Solorina saccata, S. spongiosa and *Verrucaria muralis*. The best sites are on steep slopes where the blown sand drifts over outcrops and a complex niche structure develops that includes turf edges. In addition to the terricolous interest, acid rock among the machair supports unusual assemblages as it is affected by dry flushing from shell-sand and wet flushing by base-rich water during rain. Characteristic saxicolous species on the flushed Lewisian gneiss are *Acrocordia macrospora, Caloplaca teicholyta, Collema multipartitum, Lecanora andrewii, Leptogium plicatile, L. turgidum, Rhizocarpon concentricum, Sarcogyne privigna* and *Verrucaria glaucina*.

On Colonsay and Tiree, where grant-aided agricultural intensification has speeded the decline of the machair, golf courses now hold the richest communities. At Invernaver NNR the machair is interesting for the abundance of mountain avens (*Dryas octopetala*), bearberry (*Arctostaphylos uva-ursi*) and juniper growing at sea level. The lichens are equally notable; terricolous additions to those on Coll are *Arthrorhaphis alpina* (usually found at over 1,000 m), *Diploschistes muscorum, Gyalecta jenensis, Pannaria pezizoides, P. leucophaea, Peltigera aphthosa, Polyblastia agraria, Psoroma hypnorum* and *Squamarina cartilaginea*.

Dunes in southwest England and Wales

Dunes along the coasts of north Cornwall, Devon, Pembrokeshire and the Channel Islands support examples of the *Fulgensietum fulgentis* association, which here is at the northern edge of its range. These small disjunct populations occur in warm, sunny, sheltered sites on sandy substrata that are very freely drained. This combination of factors gives the localities a thermal similarity to sites several degrees of latitude further south. The effect is reinforced by topography, the majority of sites occupying steep south-facing slopes where high insolation helps to maximise soil temperatures. The community is calcicolous, requiring soil in the pH range 7–8. Its preference is for a well-consolidated substratum and the largest colony, at Penhale, Newquay, is on sandy raised beach deposits behind the dunes. The community is rich in squamulose and placodioid species; in addition to the bright yellow rosettes of *Fulgensia*, constant species include *Catapyrenium squamulosum, Cladonia pocillum, C. rangiformis, C. tenax, Diploschistes muscorum, Psora decipiens, Squamarina cartilaginea* and *Toninia sedifolia*. They occur in open heavily grazed sheep's fescue-wild thyme turf. At Penhale a relaxation of grazing is allowing gorse to invade. The Caldy Island population appears to have been lost as a result of this as the site at Priory Bay, where P.G.M. Rhodes collected *Fulgensia* in 1916, now carries dense scrub.

The occasional presence of normally epiphytic lichens on fixed dunes, noted by Watson (1918), is well illustrated by *Usnea articulata*. In Glamorgan this pollution-sensitive lichen is confined to three dune systems including Pennard Burrows; it is also known from dune systems in Norfolk and the Isle of Man.

Shingle

Shingle forms active beaches, relic beaches, spits and forelands. Britain and Ireland have the greatest length of shingle beach in northern Europe, much of it concentrated on the south and east coasts. Shingle has been thoroughly surveyed for lichens recently at Dungeness, Kent, on Holy Island,

Fig. 13.10 Stabilised shingle supports a wide range of saxicolous species; as a bonus the terricolous element may also be well developed; Cuthill Links.

Northumberland, at Ravensglass, Cumbria, and at a number of sites in Scotland such as Western Galloway, Culbin Forest, Moray, and Cuthill Links and Ferry-Coul Links in Sutherland. The best lichen sites are where the shingle occurs as banks on the floor of dune blowouts.

Walking across spreads of rounded stones, their upper surfaces covered with colourful lichen mosaics is an exhilarating experience; it is all too easy to collect over-zealously. From existing studies it has proved possible to identify a number of assemblages that are typical of acid shingle. A bank of stabilised acid shingle at the landward margin of a beach anywhere in Britain can be expected to yield 30 to 40 species (Fig. 13.10). A widespread element includes *Acarospora fuscata, Buellia aethalea, Catillaria chalybeia, Lecanora campestris, L. muralis, L. rupicola, Lecidea fuscoatra, Lecidella viridans, Micarea erratica, Ochrolechia parella, Porpidia tuberculosa, Tephromela atra, Tremolechia atrata* and *Xanthoria parietina*. This assemblage has a somewhat urban flavour as the conditions of intermittent disturbance and mild eutrophication found along strand lines have a counterpart in towns and villages. A small maritime element of *Acarospora impressula, Lecanora helicopis, Rhizocarpon richardii, Verrucaria fusconigrescens* and *V. prominula* may be present. Where the shingle is being invaded by grasses, species of *Peltigera* and *Cladonia* appear, later to be replaced by blackthorn or gorse as succession proceeds.

Where the shingle includes limestone pebbles, additional species can be expected. On Holy Island the deeply weathered, often honeycombed surface of Carboniferous pebbles are almost entirely covered with an intricate mosaic of microlichens, many of which are rare in northeast England. They include *Acarospora veronensis, Arthonia lapidicola, Caloplaca lactea, C. saxicola, Diplotomma epipolia, Eiglera flavida, Lecanora crenulata, Rinodina bischoffii, Sarcogyne regularis* and many pyrenocarps. Pebbles of whinstone (basalt), Triassic sandstone and seashells bring in further species. Due to the variety of substrates present, shingle and churchyards are both ideal habitats to study the ecology of saxicolous lichens; one scores for convenience the other for ambience. A few pebbles should be taken home and examined under a binocular microscope; it is surprising how many extra species can be discovered, some represented by a single fruit body.

Shingle sites of national importance are Dungeness (2,200 ha), thought to be the largest shingle structure in Europe, Chesil Bank, which is 14 km long, and the extensive dune/shingle complexes on the east coast of Scotland. Their sheer size ensures a wide range of conditions so that all are important for a combination of reasons. Among the stunted pines at Culbin Sands, ridges of shingle and shingle/sand mixtures provide an open but sheltered environment. The individual pebbles, many of which are quite large (to 30 cm), are uniformly acid and between them support a lichen flora of 40–50 species. These include fine colonies of *Lecidea plana, Parmelia mougeotii* and *Rhizocarpon lecanorinum*; many species have a well-developed border of dark fungal tissue extending beyond the limits of the normal thallus. A group of mainly coastal crustose species are shingle specialists, for example *Lecidea auriculata, L. diducens, L. brachyspora,* and *Rhizocarpon cinereovirens. L. brachyspora* shows a preference for well-rotted sandstone. A number of *Acarospora* spp. also flourish on the smooth, hard, slightly nutrient-enriched pebbles, for example *A. impressula, A. rufescens, A. smaragdula* and *A. veronensis*. Shells regularly support *Acarospora heppii* and *Caloplaca lactea*, which has bright orange fruits on a thin white thallus. Similar sites to Culbin are Cuthill Links and Ferry-Coul Links, Sutherland, Sands of Forvie, Aberdeen, and Ravensglass, Cumbria.

Dungeness is important for its wide range of habitats, which vary from open shingle, through lichen heath, to scrub. It is considered that the northern species *Bryoria fuscescens* and *Cladonia mitis* may have persisted here as postglacial relics, having survived the forest maximum on open areas of the beach. This frequently studied site (Ferry & Pickering, 1989b; Laundon, 1989) has deteriorated under the impact of gravel extraction and the construction of a nuclear power station. The apparently bare flint pebbles support several rare species such as the sorediate *Rinodina aspersa*, but *Caloplaca atroflava*, present on shingle elsewhere in the south, has not been reported. Where a little surface humus has accumulated, a lichen heath develops in which *Cladonia cervicornis, C. foliacea, C. gracilis, C. portentosa* and *C. rangiformis* are abundant. Their distribution is related to distance from the sea, the inland shingle being more acid.

The shingle ridges at Dungeness are famous for their stands of prostrate blackthorn scrub, the lichens of which have been described by Ferry & Lodge (1996). Fruticose species such as *Bryoria fuscescens, Usnea glabrata* and *U. rubicunda* are locally present. This was also once the most easterly site in the UK for *Teloschistes flavicans*. A unique vegetation cycle is associated with broom on the lichen heath. Plants die from the centre and their rotting wood forms a layer over the shingle. This rotting wood is colonised by cup lichens of the *Cladonia chlorophaea* complex. As the wood decays away the site is colonised by *C. portentosa* and various higher plants. Recolonisation by broom may occur and the cycle is repeated.

Despite the list of sites just described, there are major shingle structures that have never seen a lichenologist.

Salt marshes and intertidal mud flats

Salt marshes are areas of salt-tolerant vegetation that extend from mean high water neap tides to slightly above the mean high water level of spring tides; to seaward there are extensive areas of algal-covered mud. Salt marsh is well distributed around the coasts of Britain and Ireland. The habitat is valued as a roosting and nesting site for birds and for its role in coastal defence. Intertidal mud flats, which are natural seaward extensions of the marshes, cover around ten times the area, support high densities of invertebrates, and are internationally important for birds.

The lichenological significance of these habitats will always be subordinate to their zoological value, and at first sight appears minimal. However, a walk through the upper marsh will often reveal small stones, pieces of brick, or cement, that on close inspection are covered with tiny perithecia. These are likely to belong to either *Pyrenocollema orustense* or *Verrucaria ditmarsica*, both of which favour sheltered muddy shores. Miscellaneous debris along the strandline is always worth examining. On a field meeting at Girvan, 19 species were found colonising old shoes on the beach (Coppins, 1977). Woody salt marsh species, such as shrubby seablite (*Suaeda verna*) should be searched for epiphytes.

The prime lichen habitat of salt marshes and mud flats is wooden posts, these are frequently the remains of fences or jetties. They carry communities of lichens and algae that are zoned in relation to height above the mud (Fig. 13.11). The algal bands are submerged at high tide, while the lowest lichen belt, which is dominated by common crustose species including the halophytes *Caloplaca marina*, *Lecanora helicopis* and *Verrucaria maura*, is influenced by wave-splash. Above the lowest lichen zone there is an abrupt change to a calcifuge community of *Hypocenomyce scalaris*, *Lecanora conizaeoides* and *L. expallens*. The tops of posts used as bird perches support *Physconia grisea* and *Xanthoria candelaria*. Worked timber structures in certain salt marshes in Essex and Norfolk carry pale grey, sterile, coarsely warted sheets of the Red Data List species *Cliostomum corrugatum*. The 'calcifuge zone' is home to the equally rare, vivid yellow-green *Cyphelium notarisii*; its headquarters currently appear to be salt marshes in East Sussex, but it is seldom searched for. These two lignicolous species are thought have the most strongly eastern distributions of all British bark- or wood-inhabiting lichens.

Saline lagoons

Another coastal habitat hardly explored by lichenologists is saline lagoons. These are bodies of salt water partially separated from the sea by barriers of sand or shingle; their water level is retained when the tide is out. Around 130 are known in England, concentrated on the eastern and southern coasts. These lagoons are highly valued by conservationists for their specialist flora and fauna, which includes rare sea anemones, shrimps, sand worms and the foxtail stonewort (*Lamprothamnium papulosum*). Two of the best known are The Fleet, behind Chesil Bank, Dorset, and Minsmere Lagoon in Suffolk. Those aspiring to break new ground and survey the lichen flora of these sites will find all English examples listed with size and grid reference in Pye & French (1993).

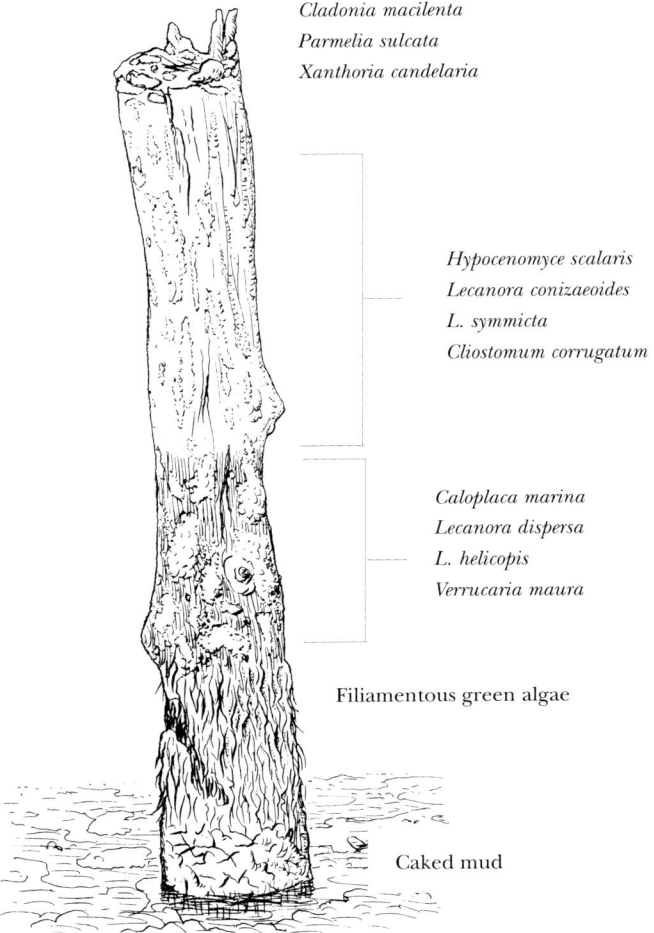

Fig. 13.11 Posts in salt marshes and mud flats support a zoned sequence of lichen communities that include rare species (M.J. Lindley).

Unprotected soft cliffs

This habitat includes sea cliffs of poorly consolidated material such as glacial drift, head or outwash deposits, but not chalk or other Mesozoic limestones. Where unfronted by any protection, such as a sea wall or gabions, the rate of cliff recession is moderate to high. The extent of these cliffs has been surveyed in England where they total 256 km (Pye & French, 1993), with the greatest lengths on Humberside, in Yorkshire, Hampshire, Norfolk, Suffolk and the Isle of Wight. This habitat has been largely ignored by lichenologists despite its concentration in lowland Britain where most of them live. The semi-consolidated landslipped cliffs east of Hastings, Sussex, support many small, abundantly fertile colonies of *Caloplaca teicholyta*. In this condition, they look very different from the sterile churchyard plants, being more crustose without radiating lobes. Further study may show they are different taxa. The rare *Tornabea scutellifera* used to occur on these cliffs, but has not been seen this century.

Soft cliffs near Beer in south Devon have recently produced a few surprises. At one spot, where Greensand has slumped down over the chalk, a second extant UK population of the Red Data List species *Endocarpon pusillum* has been discovered (Fig. 13.12). Its small red-brown squamules are rather inconspicuous so it may be more widespread than we realise. The older lichenologists knew it from cliffs at Rottingdean in East Sussex and Alum Bay on the Isle of Wight.

Coastal heath

Coastal heath communities are dominated by ericaceous plants. A distinction is usually made between maritime heaths influenced by salt spray, and other heath communities close to the shore, but less influenced by the sea and containing gorse. Both are quite scarce habitats, which in England extend to only 460 ha (Pye & French, 1993). Cornwall is the county with the largest total followed by Devon, Norfolk and Dorset. It is more widespread in Scotland where crowberry (*Empetrum nigrum*) and cross-leaved heather (*Erica tetralix*) are frequent associates of the dominant heather (*Calluna vulgaris*). To seaward, maritime heath grades into maritime cliff grassland, which is treated in the following section.

Lichenologists have largely ignored coastal heathland. There have been no dedicated surveys, though some time ago a working party (Fletcher, 1984)

Fig. 13.12 Barbara Benfield examines the population of *Endocarpon pusillum* that she discovered on the Greensand cliffs of South Devon.

assessed the known sites for conservation purposes and concluded that none were of international importance. Since the UK has more of this habitat than any other European country, it must simply be a case that the top sites have not been recognised.

An introduction to the habitat can be gained by examining windy headlands projecting into the Atlantic off the north coast of Scotland. Here, a wide range of terricolous lichens occur on the acid peaty soil including *Baeomyces rufus, Coelocaulon aculeatum, Icmadophila ericetorum, Ochrolechia frigida, Pycnothelia papillaria, Sphaerophorus globosus, Trapeliopsis wallrothii* and numerous *Cladonia* spp. Compared with inland heaths, red-fruited *Cladonia* and *Micarea* spp. (except *M. incrassata*) are rare, while *Sphaerophorus* is more abundant and regularly grazed to a stump. Rarer species that may be present include *Pertusaria xanthostoma* on old heather stems, *Porina mammillosa* on peat and *Thrombium epigaeum* on mineral soil. Baligill, 14 km east of Bettyhill, is said to be one of the best examples of maritime heath in Sutherland. Here 21 lichens occur among the heather, including *Pannaria pezizoides, Psoroma hypnorum* and several *Peltigera* spp., their presence suggesting that the site receives a substantial nutrient input from salt spray.

Some of the best studied areas of coastal heath, and quite different from the examples described above, are on the Lizard Peninsula in Cornwall, where closed-canopy stands of Cornish heath (*Erica vagans*) and gorse occur close to the sea. Their interest is almost entirely epiphytic and includes extensive swards of *Cladonia portentosa* and *C. ciliata* growing a metre above ground level in wind-clipped gorse bushes where they look decidedly out of place. When the 1 m tall bushes of Cornish heath start to loose their vigour, the canopy opens out and becomes colonised by conspicuous tufts of *Evernia prunastri, Parmelia* and *Usnea* spp. and occasionally *Teloschistes flavicans*. To discover the principal lichen riches of this habitat it is necessary to burrow into the shady depths where conditions of high humidity prevail. At first the heather stems appear bare, but closer examination reveals thin, pale green or grey crusts, or sometimes groups of fruit bodies with no apparent thallus. The following species are present on Lizard Down: *Athopyrenia antecellans, Byssoloma leucoblepharum, B. subdiscordans. Dimerella lutea, D. pineti, Fuscidea lightfootii, Gyalideopsis anastomosans, Lecanora confusa, Micarea peliocarpa, M. nitschkeana, M. prasina, M. denigrata, Porina leptalea* and *Thelotrema lepadinum*. The fruits of *Byssoloma* have a furry margin and are objects of great beauty.

Further west, the heather on the Isles of Scilly is even richer. Rocks and heather stems on Wingletang Down, St Agnes, were once home to the handsome, large, foliose lichen *Pseudocyphellaria aurata*, but a recent search failed to re-find it. Short heather on Tresco is the only site in Europe for *Heterodermia propagulifera*.

Leggy coastal heathland clearly represents an alternative, largely unrecorded habitat for many rare and local epiphytes of dense woodland. It also has its own specialities and is worthy of closer study. Gilbert *et al.* (1984), McCarthy *et al.* (1985) and Coppins & Gilbert (1990) have produced long lists of epiphytes from deep within coastal stands of unburnt heather, which suggests this is a general feature of strongly Atlantic heathland.

There is a striking similarity between the 'wave' Callunetum of exposed ridges in the Cairngorm Mountains and wind-clipped Callunetum on the north Cornish coast. In both communities the canopy of the prostrate heather is decked with yellow fruticose lichens, *Usnea* spp. in Cornwall, *Alectoria* and *Cetraria nivalis* in the Cairngorms, while bare ground between the bushes support a similar range of yellow, usnic acid containing species. Why they should become dominant in windswept sites is not known.

Maritime cliff grassland

Along cliff tops, the dark maritime heath grades into a green fescue grassland which, periodically drenched by sea spray, has an alkaline soil pH and contains thrift (*Armeria maritima*) and buck's-horn plantain (*Plantago coronopus*). The extent of this habitat in England is less than 2,000 ha, half of it in Cornwall. Maritime grassland is usually poor in lichens except where rocks protrude which bring in crevice and ledge species. At a few sites, however, conditions combine to favour a rich terricolous flora; a noted example occurs on the remote island of St Kilda (Fig. 13.13).

The special conditions required on this exposed, cliff-girt island, are geos (gullies) up which the sea rushes during stormy weather hurling clouds of spray onto the close-cropped sward 30 to 40 m above. Around the head of Geo Chille Brianan the steep red fescue-buck's-horn plantain (*Festuca rubra-Plantago coronopus*) grassland contains up to 12 lichens per square metre. They include *Bacidia arceutina, Caloplaca cerina, C. holocarpa, Collema tenax, Leptogium teretiusculum, L. tremelloides, Moelleropsis nebulosa, Pannaria pezizoides, Parmeliella triptophylla, Solenopsora holophaea* and *S. vulturiensis*. From grassland at the east end of Village Bay *Degelia atlantica, D. plumbea, Dermatocarpon cinereum, Lecidea*

Fig. 13.13 An unusually calm day on St Kilda. Atlantic gales regularly hurl clouds of spray onto the cliff tops which, together with a thousand years of sheep grazing, has produced maritime grasslands rich in lichens (H. Balharry).

hypnorum, *Pannaria leucophaea* and *Polyblastia gelatinosa* have been recorded. Many of these species are essentially calcicoles and owe their presence to the alkaline influence of the salt spray, which raises the pH of the acid soil to over 6. The factors that render these grasslands so favourable to lichens are, first, the unbroken history of close grazing by the Soay sheep, possibly over 1,000 years, which has produced a stable habitat in which abundant light reaches the soil surface. Secondly, there is a complex niche structure, many of the smaller lichens growing on dead vegetation and the haulms of red fescue which, under the influence of salt spray, decay only very slowly. On the adjacent island of Dun, in contrast ungrazed since 1931, a rank growth of red fescue has all but eliminated the terricolous lichens.

Other places where a lichen-rich maritime cliff grassland has been reported are all islands or exposed coasts that regularly become enveloped in spray. Most sites in the north of Scotland, such as Ard Mor on the east side of Torrisdale Bay, are headlands the end and flanks of which carry a grassland containing *Agonimia tristicula*, *Bacidia muscorum*, *Brigantiaea fuscolutea*, *Catapyrenium squamulosum*, *Opegrapha multipunctata*, *Peltigera rufescens* and a selection of the species found on St Kilda. Further south, off the Welsh coast, islands such as Bardsey and Skomer (Wolseley *et al.*, 1996) carry extensive swards of thrift, which are worth examining on hands and knees as they are home to *Caloplaca cerina*, *Heterodermia obscurata*, *Lecanora dispersa* f. *zosterae*, *Opegrapha ochrocheila*, *Rinodina confragosa* and *R. subglaucescens*. The cliff tops on Lundy are known for their grassland populations of *Cladonia firma* and *Teloschistes flavicans*; those at Land's End are the place to see *Heterodermia leucomelos*; while thrift swards at Hurlstone Point, Porlock, support *H. obscuratum*, *Parmelia endochlora* and *P. taylorensis*. All the best sites are readily accessible from long-distance coastal footpaths.

Throughout this volume I have indicated which habitats have been neglected during the modern period, and pointed out topics where further field work or systematic observation is required. Many of these ventures are suitable for investigation by the amateur naturalist armed with no more than a hand lens, paper, pencil, tape measure, application and enthusiasm. The time commitment would be minimal, perhaps one day a year for permanent quadrats or two 'holidays' a year over several years for national surveys. I call this type of endeavour 'focused field work' and can recommend it alongside more relaxed general days out, site visits and the filling in of mapping cards. Of all the environments in the UK, maritime ones, with the exception of rocky shores, are ripe for committed investigation to bring them into line with woodland, montane, churchyard and other better-studied habitats.

Appendices

APPENDIX 1

The new approach to lichenometry developed by Vanessa Winchester

Lichenometric dating depends primarily upon accurately constructed growth curves. When compiling them, care is needed to measure species growing under comparable environmental conditions, and the help of an experienced lichenologist may be needed to check species identification. The size of the largest thalli can most conveniently be measured using a clear, flexible, plastic or acetate film inscribed with concentric circles at 5 mm intervals (Fig. A.1). Using this method, not only orbicular but also irregular or imperfect growths on uneven surfaces can be measured as long as one well-developed arc is present.

The radii of the largest individuals on each dated surface at a site (churchyard, moraine) are measured and plotted on a graph, size against age, one graph for each species. A curve can then be drawn by eye through maximum growths (Fig. A.2). With crustose growth forms, which often persist longer than foliose or placoid ones, it may be permissible to project the steady, slow growth period beyond the last point on the graph using a dotted line; this has been done in Figure A.2.

The strength of Winchester's method is that, wherever possible, she uses a multi-species approach to improve accuracy. Figure A.3 shows how, by using six species to date boulders by the River Kennet, a range of dates from 1961 to

Fig. A.1 A clear acetate sheet, inscribed with concentric circles at 5 mm intervals, can be used to measure even irregular growths. In this example thallus diameter was estimated at 27 mm. From Winchester, 1984.

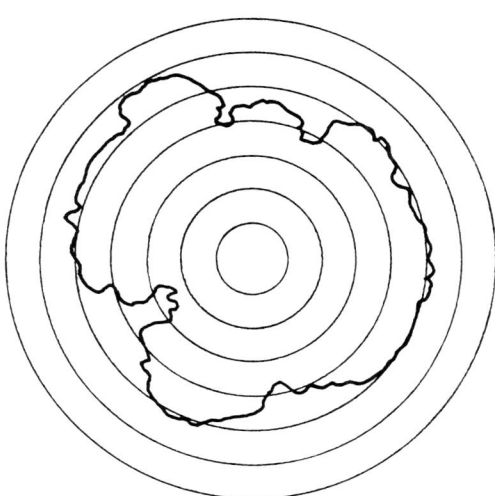

1964 was obtained with a standard deviation of 1.7 years and a mean of 19.66 years, which was within a month of the date the farmer recalled dragging the boulders from his field.

Despite theoretical drawbacks, such as the presence of anomalous thalli, fused thalli, thalli trapped in mosaics, the uncertain time-lag before colonisation commences, and reliance on field identification (especially with *Rhizocarpon geographicum* agg.), the technique outlined above consistently produces meaningful results. (The above is summarised from Winchester, 1984).

Fig. A.2 (above) Growth curve for *Lecanora campestris* fitted to the largest colonies on tombstones in Avebury churchyard. From Winchester, 1984.

Fig. A.3 (below) Growth curves for six lichens, derived from Avebury churchyard, were used to estimate the date of exposure of boulders by the river Kennet 5 km away. From Winchester, 1984.

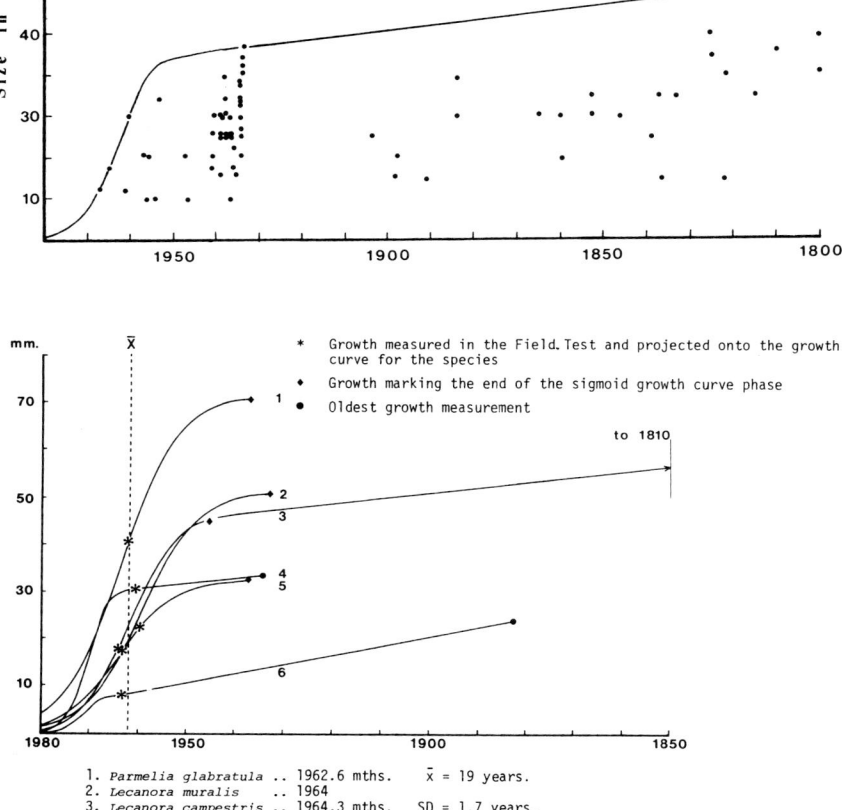

APPENDIX 2

Biological estimation of air pollution

By working in areas where the concentration of sulphur dioxide (SO_2) is known, lichenologists have been able to produce semi-quantitative scales from which levels of this pollutant can be estimated. Most widely used is the ten point Hawksworth-Rose scale (1970; 1976) that employs epiphytes, separate lists are provided for eutrophicated and noneutrophicated trees (Table A.1). It can be used to calculate mean winter SO_2 levels in England and Wales. The scale works best in areas where levels of pollution are reasonably constant over several years, where there has been rapid amelioration it is unreliable.

Table A.1 The Hawksworth-Rose pollution scale.

Zone	Moderately acid bark	Eutrophicated bark	Mean wint. SO_2 µg/m3
0	Epiphytes absent.	Epiphytes absent.	?
1	*Desmococcus viridis* confined to the base the trunk.	*Desmococcus viridis* extends up the trunk.	>170 .
2	*D. viridis* extends up the trunk *Lecanora conizaeoides* confined to the base.	*Lecanora conizaeoides* abundant; *L. expallens* occasional on base.	~150
3	*L. conizaeoides* extends up the trunk; *Lepraria incana* frequent on base.	*L. expallens* and *Buellia punctata* abundant; *Diploica canescens* appears.	~125
4	*Hypogymnia physodes* and/or *Parmelia saxatilis*, or *P. sulcata* on base. *Hypocenomyce scalaris*, *L. expallens* and *Chaenotheca ferruginea* often present.	*Diploica canescens* common; *Physcia adscendens* and *Xanthoria parietina* on base; *Physcia tribacea* appears in the south.	~70
5	*H. physodes* or *P. saxatilis* extend up trunk to 2.5 m or more; *P. glabratula*, *P. subrudecta*, *Parmeliopsis ambigua* and *Lecanora chlarotera* appear; *Calicium viride*, *Chrysothrix candelaris*, *Pertusaria amara* may occur; *Ramalina farinacea* and *Evernia prunastri* if present largely confined to base: *Platismatia glauca* may be present.	*Physconia grisea*, *P. perisidiosa*, *Diplotomma alboatrum*, *Phaeophyscia orbicularis*, *Physcia tenella*, *Ramalina farinacea*, *Haematomma ochroleucum*, *Schismatomma decolorans*, *Xanthoria candelaria*, *Opegrapha varia* and *O. vulgata* appear; *Diploicia canescens* and *X. parietina* common; *Parmelia acetabulum* appears in the east.	~60
6	*P. caperata* present at least on the base; rich in species of *Pertusaria* and *Parmelia* (except in NE), *P. tiliacea*, *P. exasperatula* (in N), *Graphis elegans* appearing; *Pseudevernia furfuracea* and *Bryoria fuscescens* in upland areas.	*Pertusaria albescens*, *Physconia distorta*, *Hyperphyscia adglutinata*, *Acrocordia gemmata*, *Xanthoria polycarpa*, and *Lecania cyrtella* appear; *Physconia grisea*, *Phaeophyscia orbicularis*, *Opegrapha varia* and *O. vulgata* become abundant.	~50

7	*P. caperata, P. revoluta*, (except in NE), *P. tiliacea, P. exasperatula* (in N) extend up the trunk; *Usnea subfloridana, Pertusaria hemisphaerica, Rinodina roboris* (in S) and *Arthonia impolita* (in E) appear.	*Physcia aipolia, Anaptychia ciliaris, Bacidia rubella, Ramalina fastigiata, Candelaria concolor* and *Anisomeridium biforme* appear.	~40
8	*Usnea ceratina, Parmelia perlata* or *P. reticulata* (S and W) appear; *Rinodina roboris* extends up trunk (in S); *Normandina pulchella* and *U. rubicunda* (in S) usually present.	*Physcia aipolia* abundant; *Anaptychia ciliaris* in fruit; *Parmelia perlata, P. reticulata* (in S and W), *Gyalecta flotowii, Ramalina canariensis, R. pollinaria* and *R. lacera* appear.	~35
9	*Lobaria pulmonaria, L. amplissima, Pachyphiale carneola, Dimerella lutea*, or *Usnea florida* present; if these absent then crustose flora well developed with often more than 25 species on large well-lit trees.	*Ramalina calicaris, R. fraxinea, R. subfarinacea, Physcia leptalea, Caloplaca flavorubescens* and *C. cerina* appear.	< 30
10	*L. amplissima, L. scrobiculata, Sticta limbata, Pannaria* spp., *Usnea articulata, U. filipendula* or *Teloschistes flavicans* (in SW) present to locally abundant.	As 9	Pure air

APPENDIX 3

Lichens as indicators of ecological continuity

Several scales have been produced that use lichens to assess the probability of a wood being ancient. They should only be used in woodland with mature trees of native species. The most useful scale is the Revised Index of Ecological Continuity (RIEC) (Rose, 1976). It contains 30 species, not all of which have identical geographical distributions in Britain, so it is considered reasonable to regard the presence of at least 20 out of the 30 species as an indication that a wood is of at least early medieval origin, and probably much older (Table 5.1, p.71). The initial choice of species was based on sites known to be very old from documentary evidence. The presence of only a few Index species signifies little; it could be a result of air pollution, of intensive management or a recent origin to the wood.

High values (>60) indicate a strong probability that a woodland, forest or park was formed at a time when fragments of the wildwood existed on or near the site. The value is calculated for a given site as follows:

$$\text{RIEC} = n/20 \times 100 \qquad \text{Twelve RIEC spp. give a value of 60.}$$

where n is the number of Index species given in Table 5.1. Exceptional sites have an RIEC value of over 100. Boconnoc Park in Cornwall, for example, has 29 of the RIEC species, with a resultant score of 145.

Improved sensitivity can be achieved by using a larger number of species and/or regional indices. With this in mind The New Index of Ecological Continuity (NIEC) (Rose, 1992) has been proposed for use in lowland Britain, reaching north to and including Yorkshire, Cumbria and Galloway. The NIEC involves 70 species. Computation of the NIEC value for a site is slightly more complex than for the RIEC. First the total number of species present on the main list (Table A.3) is counted. Then to this value the total number of bonus species present (Table A.3) is added to obtain the final Index score denoted by T.

Calculating NIEC values for a large number of sites has shown it to be highly effective at classifying woodlands in terms of their conservation value. Sites with T values <20 are of limited conservation importance; a score over 40 is exceptional. The RIEC is a complementary tool for assessing the relative age and degree of disturbance.

Rose has prepared regional indices for the rest of Britain; these can be found in Hodgetts (1992). They cover 1) euoceanic west Scotland; 2) southwest Ireland (in preparation); 3) acidic oak-birch woods in upland Britain; 4) the continental woods of east Scotland, Northumberland and the Welsh Borders; and 5) the Caledonian pinewoods (Coppins, in preparation). Comparisons between regions is difficult as the numbers of main species and bonus species differ, but within a region the scales are remarkably sensitive. To use any but the RIEC it is necessary to be a specialist in woodland lichens.

Table A.3 Lichen epiphytes used to calculate the New Index of Ecological Continuity (NIEC).

Main species

Agonimia octospora	*Megalospora tuberculosa*
Arthonia astroidestra	*Micarea alabastrites* or *M. cinerea*
A. vinosa	*M. pycnidiophora*
Arthopyrenia antecellans	*Nephroma laevigatum*
A. ranunculospora	*N. parile*
Arthothelium ilicinum	*Ochrolechia inversa*
Bacidia biatorina	*Opegrapha corticola*
B. epixanthoides	*O. prosodea*
Biatora sphaeroides	*Pachyphiale carneola*
Buellia erubescens	*Pannaria conoplea* or *P. rubiginosa*
Catillaria atropurpurea	*Parmelia crinita*
Cetrelia olivetorum	*P. reddenda*
Chaenotheca spp. (count only one sp. but not *C. ferruginea*)	*Parmeliella jamesii*
	P. triptophylla
Cladonia caespiticia	*Peltigera collina*
C. parasitica	*P. horizontalis*
Collema furfuraceum or	*Pertusaria multipuncta*
C. subflaccidum	*P. velata*
Degelia atlantica or *D. plumbea*	*Phaeographis dendritica* or *P. inusta* or *P. lyellii*
Dimerella lutea	
Enterographa sorediata	*Phyllospora rosei*
Heterodermia obscurata	*Polyblastia allobata*
Lecanactis amylacea	*Rinodina isidioides*
L. lyncea	*Schismatomma niveum*
L. premnea	*S. quercicola* or *Pertusaria pupillaris*
L. subabietina	*Stenocybe septata*
Lecanora jamesii	*Sticta limbata*
L. quercicola	*S. fuliginosa* or *S. sylvatica*
Lecidea sublivescens	*Strangospora ochrophora*
Leptogium cyanescens	*Thelopsis rubella*
L. lichenoides	*Thelotrema lepadinum*
L. teretiusculum	*Usnea ceratina*
Lobaria amplissima	*Usnea florida*
L. pulmonaria	*Wadeana dendrographa*
L. scrobiculata	*Zamenhofia coralloidea*
L. virens	*Z. hibernica*
Loxospora elatina	

Bonus species

Anaptychia ciliaris (Devon only)	*P. sampiana*
Arthonia arthonioides	*Parmelia arnoldii*
A. anglica	*P. horrescens*
A. zwackhii	*P. minarum*
A. anombrophila	*P. sinuosa*
Bacidia circumspecta	*P. taylorensis*
B. subincompta	*Parmeliella testacea*
Catillaria laureri	*Pseudocyphellaria crocata*

Caloplaca lucifuga
Collema fragrans
C. nigrescens
C. subnigrescens
Cryptolechia carneolutea
Gyalecta derivata
Leptogium burgessii
L. colcheatum
Megalaria grossa (in S. England)
Opegrapha fumosa
O. multipunctata
Pannaria mediterranea

P. thouarsii agg.
Ramonia species (any)
Schismatomma graphidioides
Sphaerophorus globosus (S. England)
S. melanocarpus (S. England)
Sticta dufourii or blue-green algal morphotype: *S. canariensis*
Teloschistes flavicans
Tomasellia lactea
Usnea articulata (New Forest and Sussex)
Zamenhofia rosei

Glossary

Acidophilous. Describing plants that are confined to, or are more common in acid habitats

Allelopathic. A form of interference competition by means of chemicals; i.e. compounds produced by one species that reduces the growth or survival of other species.

Apothecium(a). A more or less flat cup or saucer-like fruit body.

Areole. An island-like portion of a crustose thallus separated from adjacent areoles so the thallus appears like crazy paving.

Ascomycetes. One of the main divisions of the fungi characterised by the production of asci which are enlarged, commonly elongated cells containing usually eight spores.

Ascospore. Reproductive spore found within an ascus.

Ascus. A sac-like cell in which ascospores are produced (Fig. 2.6).

Ashlar. Squared stone used for facing buildings or walls.

Basidiomycetes. One of the main divisions of the fungi characterised by the possession of basidium bearing typically four basidiospores; includes toadstools, rusts and smuts.

Basiphilous. Describing plants that are confined to, or are more common in, alkaline habitats.

Calcicole. A plant that is confined to, or is more common in calcareous habitats.

Calcifuge. A plant that is limited to, or more abundant in, habitats of low calcium status, usually of pH 5 or less.

Climax community. The final stage of an undeflected plant succession. Today the term 'late successional stage' is preferred

Clint. The flat table-like surface of a limestone pavement that is divided by solution widened vertical joints known as grykes.

Cohort. A group of conspecific individuals belonging to the same generation.

Coppice-with-standards. A method of woodland management that involves regular cyclical cropping while leaving a scattering of trees to reach maturity.

Crustose. A lichen growth form in which the thallus forms a crust on the substratum.

Cryptic. Protectively concealing.

Cyanophilic. A lichen in which the 'algal partner' or photobiont is a cyanobacteria (formerly known as blue-green algae).

Dolomitisation. The process by which a considerable proportion of the calcium carbonate in a limestone is replaced by magnesium carbonate to form the rock type known as dolomite.

Epidiorite. A basic, metamorphosed, igneous rock occurring as sills in the Scottish Highland. Of great interest to botanists on account of its alkaline nature.

Epiphyte. A plant growing on another plant, without being parasitic.

Eutrophication. Biological effects of an over enrichment in plant nutrients, (usually nitrogen and phosphorus).

Euoceanic. Strongly oceanic.
Foliose. A lichen growth form in which the thallus is leaf-like and usually readily separable from the substratum.
Fruticose. A lichen growth form in which the thallus is shrub-, beard- or worm-like; often grows in tufts.
Fungi Imperfecti. A large, artificial group of fungi that have no sexual state i.e. do not produce sexual fruiting bodies.
Halophilic. A plant tolerant of salt-rich habitats.
Halophobic. A plant intolerant of salt-rich environments.
Hypertrophication. A serious over enrichment in plant nutrients; usually associated with the activities of man.
Hypha(e). A fungal thread of microscopic proportions.
Hypothallus. A layer of fungal material without photobiont cells found under certain lichens.
Graphidion. A lichen community found on smooth bark; most the crustose species involved have fruit bodies that are either lirellae or perithecia.
Isidia (Isidiate). A photobiont-containing protuberance on the upper surface of a lichen that can become detatched and grow into a new plant.
Karst. Rough limestone country with underground drainage.
Lecideine. An apothecium without a thalline margin; composed entirely of fungal tissue. (Fig. 2.6).
Leprose. The surface of a thallus that is entirely granular or powdery.
Lichenisation. The process by which a fungus and an alga or cyanobacterium come together to form a lichen.
Lignicolous. Growing on wood.
Lirellae. Long, narrow apothecia.
Lithology. The composition and texture of stone.
Lobarion. A lichen community mainly composed of large foliose lichens, including species of *Lobaria*, that appears to be the natural forest climax on mature hardwood trees in western Europe.
Mesic. Of conditions that are intermediate; in no way extreme.
Mesoclimate. A scale of climate intermediate between macro and microclimate.
Morphotype. A group of morphologically differentiated individuals of unknown taxonomic significance.
Muscicolous. Growing on moss.
Mycorrhiza. A mutualistic association between a fungal mycelium and plant roots which supplies the plant with nutrients derived from the humus.
Nitrophilous. A plant that is limited to, or more abundant in, habitats that are rich in nitrogenous compounds.
Pantile. A roofing tile whose cross section forms an ogee curve.
Perithecium(a). A flask-like fruit body opening by a pore at the top (Fig. 2.6).
Photobiont. A photosynthetic partner which may be either a green alga or a cyanobacterium (blue-green alga).
Physodion. A lichen community of acid habitats in which *Hypogymnia physodes* is abundant.
Placodioid (Placoid). A lichen growth form in which the thallus is crustose with a radiating lobed margin.
Podetium(a). A stalk-like portion of a lichen thallus usually bearing an apothecium (fruit body) at its end.

Propagule. A part of a lichen such as a spore, soredium or isidium concerned with propagation.
Prothallus. A part of the thallus without algae that is sometimes visible around the edge of the thallus or between the areoles.
Pruina. A frost-like or flour-like surface covering.
Pyrenocarpous. A lichen in which the fruit bodies are perithecia.
Rhizinae. Root-like structures on the underside of a lichen that act as attachment organs.
Saxicolous. Of lichens growing on rock.
Soralium(a). A structure or region of a thallus bearing soredia.
Soredium(a). Powdery granules composed of algal and fungal cells capable of reproducing the lichen vegetatively.
Squamule. A small scale.
Squamulose. A lichen growth form in which the thallus is composed of many small scales, at least towards the circumference.
String course. A projecting horizontal line of moulding running along the face of a building.
Symbiosis. An association between unlike organisms, generally persisting for long periods. Each is known as a symbiont.
Terricolous. Of lichens growing on the ground attached to soil or general plant debris.
Thalline margin. The margin of an apothecium which contains algal as well as fungal cells and is of the same colour and consistency as the thallus (Fig. 2.6).
Thallus. A vegetative plant body not differentiated into leaf, stem or root.
Type locality. Site from which the type specimen of an organism was collected.
Umbilicate. A lichen growth form in which the foliose thallus is button-like, being attached by a central holdfast.
Watsonian vice-county. One of the 112 provinces that Britain is divided into for the purposes of Botanical recording. The system originated in 1852.
Xanthorion. A lichen community found on well-lit, nutrient enriched trees and rocks; species of *Phaeophyscia*, *Physcia* and *Xanthoria* are usually prominent.

Bibliography

Ainsworth, G.C. 1976. *Introduction to the history of mycology*. Cambridge University Press, Cambridge

Alvin, K.L. 1960. Observations on the lichen ecology of South Haven Peninsula, Studland Heath, Dorset. *Journal of Ecology* **48**: 331–339.

Anon 1786. *Memorial of Mr Cuthbert Gordon relative to the discovery and use of cudbear and other dying wares*. London.

Anon 1905. Reports of the meetings of the Berwickshire Naturalists' Club for 1904. Kielder, Northumberland. *History of the Berwickshire Naturalists' Club* **19**: 117–122.

Anon 1975. Lichen chimeras. *Report on the British Museum 1972–1974*. British Museum of Natural History, London.

Arkell, W.J. 1947. *Oxford Stone*. Faber & Faber, London.

Arup, U., Ekman, S., Froberg, L., Knutsson, T. & Mattsson, J.-E. 1989. Changes in the lichen flora on Romeleklint, S. Sweden, over a 50-year period. *Graphis Scripta* **2**: 148–155.

Atkinson, R. 1940. *Island going*. Collins, London.

Barkman, J.J. 1958. *Phytosociology and ecology of cryptogamic epiphytes*. Van Gorcum, Assen, Netherlands. [Reissued 1969.]

Bary, A. de 1866. *Morphologie und physiology der pilze, flechten und myxomyceten*. Leipzig.

Bates, J.W., Bell, J.N.B. & Farmer, A.M. 1990. Epiphyte recolonisation of oaks along a gradient of air pollution in southeast England, 1979–1990. *Environmental Pollution* **68**: 81–99.

Benfield, B. 1994. Impact of agriculture on epiphytic lichens at Plymtree, East Devon. *Lichenologist* **26**: 91–96

Benson, A.C. 1981. Edwardian excursions from the diaries of A.C. Benson 1898 to 1904. Murray, London.

Bolton, E.M. 1960. *Lichens for vegetable dying*. Studio Press: London.

Borrer, W. 1805. In: **D. Turner & L.W. Dillwyn** (eds.). *The botanist's guide through England and Wales*, 2 vols. Phillips & Fardon, London.

Brightman, F.H. & Seaward, M.R.D. 1977. Lichens of man-made substrates. In: **M.R.D. Seaward** (ed.) *Lichen ecology*. pp.253–293. Academic Press, London.

Broadhead, E. 1958. The psocid fauna of larch trees in northern England. An ecological study of mixed species populations exploiting a common resource. *Journal of Animal Ecology* **27**: 217–263.

Brunskill, R. & Clifton-Taylor, A. 1978. *English brickwork*. Ward Lock Ltd, London.

Carrol, I. 1865. Contributions to British lichenology; being notices of new or rare species observed since the publication of Mudd's Manual. *Journal of Botany* **3**: 286–293.

Chester, T.W. 1997. The saxicolous churchyard flora of lowland England. *British Wildlife* **8** (3): 161–172.

Church, J.M., Coppins, B.J., Gilbert, O.L., James, P.W. & Stewart, N.F. 1996. *Red data books of Britain and Ireland: Lichens. Volume 1: Britain*. Joint Nature Conservation Committee, Peterborough.

Clifton-Taylor, A. 1987. *The pattern of English building*. 4th. ed. Faber & Faber, London.

Cooke, L.M., Rigby, K.D. & Seaward, M.R.D. 1990. Melanic moths and changes in epiphytic vegetation in northwest England and North Wales. *Biological Journal of the Linnean Society* **39**: 343–354.

Coppins, A.M. 1996. *Fyfield Down NNR and SSSI: saxicolous and terricolous lichens*. pp. 23. Report to English

Nature, Devizes.

Coppins, B.J. 1976. Distribution patterns shown by epiphytic lichens in the British Isles. In: **D.H. Brown, D.L. Hawksworth & R.H. Bailey** (eds.) *Lichenology: progress and problems*. pp. 249–278. Academic Press, London.

Coppins, B.J. 1977. Field meeting at Girvan, Ayrshire. *Lichenologist* **9**: 153–167.

Coppins, B.J. 1978. Lichens. In: *Breadalbane survey 1977/78, Royal Botanic Garden, Edinburgh*: 47–62, 63–73. [Manuscript.] Edinburgh: Nature Conservancy Council Scottish Headquarters.

Coppins, B.J. 1979. Lichens. In: *Breadalbane survey 1978/79, Royal Botanic Garden, Edinburgh*: 32–37. [Manuscript.] Edinburgh: Nature Conservancy Council Scottish Headquarters.

Coppins, B.J. 1983. A taxonomic study of the lichen genus *Micarea* in Europe. *Bulletin of the British Museum (Natural History). Botany* **11**: 17–214.

Coppins, B.J. 1990. Lichens of the native Scottish pinewoods. *British Lichen Society Bulletin* **66**: 11–13.

Coppins, B.J., Fletcher, A., Gilbert, O.L. & James, P.W. 1986. Field meeting in Sutherland. *Lichenologist* **18**: 275–285.

Coppins, B.J. & Gilbert, O.L. 1981. Field meeting near Penrith, Cumbria. *Lichenologist* **13**: 191–199.

Coppins, B.J. & Gilbert, O.L. 1990. Field meeting in western Galloway. *Lichenologist* **22**: 183–190.

Coppins, B.J. & Lambley, P.W. 1974. Changes in the lichen flora of the parish of Mendlesham, Suffolk, during the last fifty years. *Suffolk Natural History* **16** (5): 319–335.

Coppins, B.J. & Shimwell, D.W. 1971. Cryptogam complement and biomass in dry *Calluna* heaths of different ages. *Oikos* **22**: 204–209.

Corner, R.W.M. 1992. *Cladonia rangiferina* in northern England. *British Lichen Society Bulletin* **70**: 32–33.

Crombie, J.M. 1873. On the rarer lichens of Blair Atholl. *Grevillea* **1**: 170–174.

Crombie, J.M. 1874. On the Lichen–Gonidia question. *Popular Science Review*, **July 1874**. pp. 18.

Crombie, J.M. 1894. *A monograph of lichens found in Britain*, vol. 1. London: British Museum (Natural History).

Culberson, W.L. 1967. Analysis of chemical and morphological variation in the *Ramalina siliquosa* species complex. *Brittonia* **19**: 333–352.

Culberson, W.L. & Culberson, C.F. 1967. Habitat selection by chemically differentiated races of lichens. *Science, New York* **158**: 1195–1197.

Davidson, J.B. 1883. On some Anglo-Saxon charters at Exeter. *British Archaeological Association* **39**: 259–303.

Day, I.P. 1989. *Seatoller woodlands, Seathwaite, Borrowdale, Cumbria. Lichenological survey of pollards and other ancient trees*. Report to the National Trust, Keswick.

de Bakker, A.J. 1989. Effects of ammonia emission on epiphytic lichen vegetation. *Acta Botanica Neerlandica* **38**, 337–42.

Degelius, G. 1982. The lichen flora of the Island of Vega. *ACTA Regiae Societatis Scientiarum et Litterarum Gothoburgensis. Botanica* **2**: 1–127.

Denton, G.H. & Karlén, W. 1973. Lichenometry: its application to holocene moraine studies in South Alaska and Swedish Lapland. *Arctic and Alpine Research* **5**: 347–372.

Department of the Environment 1994. *The reclamation and management of metalliferous mining sites*. Department of the Environment Minerals Division, HMSO, London.

Design Council 1974. *Colour finishes for farm buildings*. Design Council, 28 Haymarket, London SW1Y 4SU.

Dobson, F. 1999. *Lichens: an illustrated guide*. Richmond Publishing Co. Ltd., Richmond, Surrey.

Edwards, R.W., Gee, A.S. & Stoner,

J.H. (eds.). 1990. *Acid waters in Wales.* Monographiae Biologicae 66. Kluwer Academic Publishers, London.

Ernst, G. 1995. *Vezdaea leprosa* – Spezialist am Strabenrand. *Herzogia* **11:** 175–188.

Evans, J. 1800. *A tour through part of North Wales in the year 1798, and at other times.* London.

Evelyn, J. 1818. *Memoirs illustrative of the life and writings of John Evelyn, comprising his diary 1641–1705.* London.

Farmer, A.M., Bates, J.W. & Bell, J.N.B. 1992. Ecophysiological effects of acid rain on bryophytes and lichens. In: **J.W. Bates & A.M. Farmer** (eds.) *Bryophytes and lichens in a changing environment.* pp.284–313. Clarendon Press; Oxford.

Fenton, F.A. 1960. Lichens as indicators of atmospheric pollution. *Irish Naturalists Journal* **13:** 153–159.

Ferreira, R.E.C. 1958. *A comparative ecological and floristic study of the vegetation of Ben Hope, Ben Loyal, Ben Lui and Glas Maol in relation to the geology.* Ph.D. Thesis, University of Aberdeen.

Ferry, B.W. & Lodge, E. 1996. Distribution and succession of lichens associated with *Prunus spinosa* at Dungeness, England. *Lichenologist* **28:** 129–143.

Ferry, B.W. & Pickering, M. 1989. Studies on the *Cladonia chlorophaea* complex at Dungeness, England. *Lichenologist* **21:** 67–77.

Ferry, B.W. & Sheard, J.W. 1969. Zonation of supralittoral lichens on rocky shores around the Dale peninsula, Pembrokeshire. *Field Studies* **3:** 41–67.

Fletcher, A. 1972. *The ecology of marine and maritime lichens of Anglesey.* Ph.D. Thesis, University of Wales.

Fletcher, A. 1973a. The ecology of marine (littoral) lichens on some rocky shores of Anglesey. *Lichenologist* **5:** 368–400.

Fletcher, A. 1973b. The ecology of maritime (supralittoral) lichens on some rocky shores of Anglesey. *Lichenologist* **5:** 401–422.

Fletcher, A. 1975a. Key for the identification of British marine and maritime lichens. 1. Siliceous rocky shore species. *Lichenologist* **7:** 1–52.

Fletcher, A. 1975b. Key for the identification of British marine and maritime lichens. 11. Calcareous and terricolous species. *Lichenologist* **7:** 73–115.

Fletcher, A. (ed.). 1984. *Lichen habitats – lowland heath, dune and machair.* Report to the Nature Conservancy Council, Peterborough.

Forster, T.F. 1816. *Flora Tonbridgensis.* J. Clifford, London.

Fritsch, F.E. & Salisbury, E. J. 1915. Further observations on the heath association of Hindhead Common. *New Phytologist* **14:** 116–138.

Fryday, A.M. 1996. The lichen vegetation of some previously overlooked high-level habitats in North Wales. *Lichenologist* **28:** 521–541.

Fryday, A.M. 1997. *The ecology and taxonomy of montane lichen vegetation in the British Isles.* Ph.D. Thesis, University of Sheffield.

Fryday, A.M. & Coppins, B.J. 1997. Three new species in the *Catillariaceae* from the Central Highlands of Scotland. *Lichenologist* **28:** 507–512.

Gams, H. 1962. Die halbflechten *Botrydina* und *Coriscium* als basidiolichen. *Oesterreichische botanische Zeitschrift* **103:** 164–167.

Gerard, J. 1597. *The herball or generall historie of plantes.* London.

Gerard, J. 1633. *The herball or generall historie of plantes.* [Enlarged and amended by T. Johnson]. London.

Gerson, U. & Seaward, M.R.D. 1977. Lichen–invertebrate associations. In: M.R.D. Seaward (ed.) *Lichen ecology* pp. 69–119. Academic Press, London.

Giavarini, V.J. 1986. Field meeting on Arran. *Lichenologist* **18:** 371–381.

Giavarini, V.J. 1990. Lichens of the Dartmoor rocks. *Lichenologist* **22:** 367–396.

Gilbert, O.L. 1965. Lichens as indicators of air pollution in the Tyne Valley. In: **G.T. Goodman, R.W. Edwards & J.M. Lambert** (eds.) *Ecology and the industrial society* pp. 35–47. Oxford University Press, London.

Gilbert, O.L. 1970. A biological scale for the estimation of sulphur dioxide air pollution. *New Phytologist* **69**: 605–627.

Gilbert, O.L. 1971a. Some indirect effects of air pollution on bark-living invertebrates. *Journal of Applied Ecology* **8**: 77–84.

Gilbert, O.L. 1971b. The effect of airborne fluorides on lichens. *Lichenologist* **5**: 26–32.

Gilbert, O.L. 1974. An air pollution survey by school children. *Environmental Pollution 6:* 175–180.

Gilbert, O.L. 1976. An alkaline dust effect on epiphytic lichens. *Lichenologist* **8**: 173–178.

Gilbert, O.L. 1980a. A lichen flora of Northumberland. *Lichenologist* **12**: 325–395.

Gilbert, O.L. 1980b. Effect of land-use on terricolous lichens. *Lichenologist* **12**: 117–124.

Gilbert, O.L. 1983. The lichens of Rhum. *Transactions of the Botanical Society of Edinburgh* **44**: 139–49.

Gilbert, O.L. 1984a. Lichens of the Magnesian Limestone. *Lichenologist* **16**: 31–43.

Gilbert, O.L. 1984b. Some effects of disturbance on the lichen flora of oceanic hazel woodland. *Lichenologist* **16**: 21–30.

Gilbert, O.L. 1984c. Lichen ecology on Steepholm. *Proceedings of the Bristol Naturalists' Society* **44**: 27–34.

Gilbert, O.L. 1985. Environmental effects of airborne fluorides from aluminium smelting at Invergordon, Scotland. 1971–1983. *Environmental Pollution (Series A)* **39**: 293–302.

Gilbert, O.L. 1986. Field evidence for an acid rain effect on lichens. *Environmental Pollution (Series A)* **40**: 227–231.

Gilbert, O.L. 1989. *The ecology of urban habitats.* Chapman and Hall, London.

Gilbert, O.L. 1990. The lichen flora of urban wasteland. *Lichenologist* **22**: 87–101.

Gilbert, O.L. 1992. Lichen reinvasion with declining air pollution. In: J.W. Bates & A.M. Farmer (eds) *Bryophytes and lichens in a changing environment* pp. 159–177. Clarendon Press, Oxford.

Gilbert, O.L. 1993. The lichens of chalk grassland. *Lichenologist* **25**: 379–414.

Gilbert, O.L. 1995. The conservation of chalk grassland lichens. *Cryptogamic Botany* **5**: 232–238.

Gilbert, O.L. 1996. The lichen vegetation of chalk and limestone streams in Britain. *Lichenologist* **28**: 145–159.

Gilbert, O.L. & Coppins, B.J. 1992. The lichens of Caenlochan, Angus. *Lichenologist* **24**: 143–163.

Gilbert, O.L., Coppins, B.J. & Fox, B.W. 1988. The lichen flora of Ben Lawers. *Lichenologist* **20**: 201–243.

Gilbert, O.L., Coppins, B.J. & James, P.W. 1984. Field meeting on Coll and Tiree. *Lichenologist* **16**: 67–79.

Gilbert, O.L. & Fox, B.W. 1985. Lichens of high-ground in the Cairngorm Mountains, Scotland. *Lichenologist* **17**: 51–66.

Gilbert, O.L. & Fox, B.W. 1986. A comparative study of the lichens occurring on the geologically distinctive mountains Ben Loyal, Ben Hope and Foinaven. *Lichenologist* **18**: 79–93.

Gilbert, O.L. & Fryday, A.M. 1996. Observations on the lichen flora of high ground in the west of Ireland. *Lichenologist* **28**: 113–127.

Gilbert, O.L., Fryday, A.M., Giavarini, V.J. & Coppins, B.J. 1992. The lichen vegetation of high ground in the Ben Nevis range, Scotland. *Lichenologist* **24**: 43–56.

Gilbert, O.L. & Giavarini, V.J. 1993. The lichens of high ground in the English Lake District. *Lichenologist* **25**: 147–164.

Gilbert, O.L. & Giavarini, V.J. 1997.

The lichen vegetation of acid watercourses in England. *Lichenologist* **29**: 347–367.

Gilbert, O.L., Holligan, P.M. & Holligan, M.S. 1973. The flora of North Rona 1972. *Transactions of the Botanical Society of Edinburgh* **42**: 43–68.

Gilbert, O.L. & James, P.W. 1987. Field meeting on the Lizard Peninsula, Cornwall. *Lichenologist* **19**: 319–334.

Gilbert, O.L. & Wathern, P. 1976. The flora of the Flannan Isles. *Transactions of the Botanical Society of Edinburgh* **42**: 487–503.

Gilbert, O.L., Watling, R. & Coppins, B.J. 1979. Lichen ecology on St Kilda. *Lichenologist* **11**: 191–202.

Gimmingham, C.H. 1972. *Ecology of heathlands.* Chapman and Hall, London.

Grime, J.P. 1979. *Plant strategies and vegetation processes.* John Wiley & Sons, Chichester.

Grindon, L.H. 1859. *The Manchester flora.* London.

Guest, J.P. 1995. Lichens, long-tailed tits and air pollution. *British Lichen Society Bulletin* **76**: 37–39.

Hansell, M. 1994. Lichens, long-tailed tits and velcro. *British Lichen Society Bulletin* **75**: 15–17.

Hansell, M.H. 1996. The function of lichen flakes and white spider cocoons on the outer surface of birds' nests. *Journal of Natural History* **30**: 303–311,

Harding, P.T. & Rose, F. 1986. *Pasture-woodlands in lowland Britain.* Institute of Terrestrial Ecology, Huntington.

Harting, J.E. 1880. *British animals extinct within historic times.* Trübner and Co., London.

Hawksworth, D.L. 1971. Field meeting at Leicester. *Lichenologist* **5**: 170–174.

Hawksworth, D.L. 1988. The variety of fungal-algal symbioses, their evolutionary significance, and the nature of lichens. *Botanical Journal of the Linnean Society* **96**: 3–20.

Hawksworth, D.L. & Chater, A.O. 1979. Dynamism and equilibrium in a saxicolous lichen mosaic. *Lichenologist* **11**: 75–80

Hawksworth, D.L. & Hill, D.J. 1984. *The lichen-forming fungi.* Blackie, Glasgow and London.

Hawksworth, D.L., James, P.W. & Coppins, B.J. 1980. Checklist of British lichen-forming, lichenicolous and allied fungi. *Lichenologist* **12**: 1–115.

Hawksworth, D.L., Kirk, P.M., Sutton, B.C. & Pegler, D.N. 1995. *Ainsworth & Bisby's dictionary of the fungi.* CAB International, Wallingford.

Hawksworth, D.L. & McManus, P.M. 1989. Lichen recolonisation of London under conditions of rapidly falling sulphur dioxide levels, and the concept of zone skipping. *Botanical Journal of the Linnean Society* **109**: 99–109.

Hawksworth, D.L. & Rose, F. 1970. Qualitative scale for estimating sulphur dioxide air pollution in England and Wales using epiphytic lichens. *Nature (London)* **227**: 145–148.

Hawksworth, D.L. & Rose, F. 1976. *Lichens as pollution monitors.* Edward Arnold, London

Hawksworth, D.L. & Seaward, M.R.D. 1977. *Lichenology in the British Isles 1568–1975.* The Richmond Publishing Co. Ltd: Richmond.

Henderson, A. [Vinifera] 1981. Grapevine. *British Lichen Society Bulletin* **49**: 9–10.

Henderson, A. 1984. Some memorabilia of the industrial manufacture of the lichen dyestuffs, cudbear and orchil – Part 1. *British Lichen Society Bulletin* **55**: 19–21.

Henderson, A. 1985a. Some memorabilia of the industrial manufacture of the lichen dyestuffs, cudbear and orchil – Part 2. *British Lichen Society Bulletin* **56**: 22–24.

Henderson, A. 1985b. Some memorabilia of the industrial manufacture of the lichen dyestuffs, cudbear and orchil – Part 3. *British Lichen Society Bulletin* **57**: 12–14.

Henderson, A. 1986a. Etymological notes on lichen names 1. *British Lichen Society Bulletin* **58**: 15–17.

Henderson, A. 1986b. Etymological notes on lichen names 2. *British Lichen Society Bulletin* **59**: 15–16.

Henderson, A. 1987a. Approaches to lichen aesthetics 1. *British Lichen Society Bulletin* **60**: 3–5.

Henderson, A. 1987b. Approaches to lichen aesthetics 2. *British Lichen Society Bulletin* **61**: 10–12.

Henderson, A. 1988a. Etymological notes on lichen names 5. *British Lichen Society Bulletin* **62**: 13–14.

Henderson, A. 1988b. Etymological notes on lichen names 6. *British Lichen Society Bulletin* **63**: 20–21.

Henderson, A. 1989. Approaches to lichen aesthetics 3. *British Lichen Society Bulletin* **64**: 38–39.

Henderson, A. 1990. Approaches to lichen aesthetics 4. *British Lichen Society Bulletin* **66**: 22–24.

Henderson, A. 1992. Approaches to lichen aesthetics 5. *British Lichen Society Bulletin* **70**: 26–29.

Hobbs, R.J. 1985. The persistence of *Cladonia* patches in closed heathland stands. *Lichenologist* **17**: 103–109.

Hodgetts, N.G. 1992. *Guidelines for the selection of biological SSSIs: non-vascular plants.* Joint Nature Conservation Committee, Peterborough.

Hodgetts, N.G. 1996. *The conservation of lower plants in woodland.* Joint Nature Conservation Committee, Huntingdon.

How, W. 1650. *Phytologia Britannica Natales Exhibens Indegenarum Stirpium Sponte Emergentium.* R. Cotes: London.

James, P.W. 1965. Field meeting in Scotland. *Lichenologist* **3**: 155–172.

James, P.W. 1978. Lichens. In: **A.C. Jermy & J.A. Crabbe** (eds.) *The Island of Mull: a survey of its flora and environment* pp. 14.1–14.62. British Museum (Natural History): London.

James, P.W., Hawksworth, D.L. & Rose, F. 1977. Lichen communities in the British Isles: a preliminary conspectus. In: **M.R.D. Seaward** (ed.) *Lichen ecology* pp.295–413. Academic Press: London.

James, P.W. & Henssen, A. 1976. The morphological and taxonomic significance of cephalodia. In: **D.H. Brown, D.L. Hawksworth & R.H. Bailey** (eds.) *Lichenology: progress and problems* pp. 27–77. Academic Press, London.

Johnson, W. 1879. Lichens and a polluted atmosphere. *Hardwick's Science Gossip* **15**: 217.

Johnston, G. 1831. *A flora of Berwick-upon-Tweed.* Vol. 2. Cryptogamous Plants. J. Carfrae & Son: Edinburgh.

Jones, E.W. 1952. Some observations o the lichen flora of tree boles, with special reference to the effects of smoke. *Revue Bryologique et Lichenologique* **21**: 96–115.

Kershaw, K.A. 1986. *Physiological ecology of lichens.* Cambridge University Press, Cambridge.

Kettlewell, H.B.D. 1973. *The evolution of melanism.* Oxford University Press, Oxford.

Kirby, J.E. 1998. Sunart oakwoods. *Native Woodlands Discussion Group Newsletter* **23**: 15–16.

Knowles, M.C. 1913. The maritime and marine lichens of Howth. *Scientific Proceedings of the Royal Dublin Society* **14** (N. S.): 79–143.

Knowles, M.C. 1915. Results of a biological survey of Blackrod Bay, Co. Mayo. Lichenes. pp. 22–26. Department of Agriculture Technical Institute, Ireland.

Kok, A. 1966. A short history of orchil dyes. *Lichenologist* **3**: 248–271.

Krog, H. & James, P.W. 1977. The genus *Ramalina* in Fennoscandia and the British Isles. *Norwegian Journal of Botany* **24**: 15–43.

Lamb, I.M. 1973. Further observations on *Verrucaria serpuloides* M. Lamb. the only known permanently submerged lichen. *Occassional Papers of the Farlow Cryptogammic Herbarium* **6**: 1–5.

Laundon, J.R. 1967. A study of the lichen flora of London. *Lichenologist*

3: 277–327.

Laundon, J.R. 1976. Sarsen lichens at risk. *British lichen Society Bulletin* **38**: 1–2.

Laundon, J.R. 1986. Studies on the nomenclature of British lichens 2. *Lichenologist* **18**: 169–177.

Laundon, J.R. 1989a. The species of *Leproloma* – the name for the *Lepraria membranacea* group. *Lichenologist* **21**: 1–22.

Laundon, J.R. 1989b. Lichens at Dungeness. *Botanical Journal of the Linnean Society* **101**: 103–109.

Lawrey, J.D. 1977. Adaptive significance of O-methylated lichen depsides and depsidones. *Lichenologist* **9**: 137–142.

Lawrey, J.D. 1984. *The biology of lichenized fungi*. Praeger Scientific, New York.

Lees, E. 1852. *The botany of the Malvern Hills in the counties of Worcester, Hereford and Gloucester*. (Ed. 2) David Bouge, London.

Lees, E. 1868. *The botany of the Malvern Hills*. (Ed. 2) Simpkin & Marshall, London.

Lewis, J.R. 1964. *The ecology of rocky shores*. English University Press, London.

Lindsay, W.L. 1856. *A popular history of British lichens*. L. Reeve, London.

Looney, J.H.H. 1991. Effects of acidification on lichens. In: S.J. Woodin & A.M. Farmer (eds) *The effects of acid deposition on nature conservation in Great Britain*. pp. 45–55. Focus on nature conservation No. 26, Nature Conservancy Council, Peterborough.

Mackinder, H.J. 1902. *Britain and the British seas*. Heinemann, London.

MaCulloch, J. 1819. *A description of the western islands of Scotland including the Isle of Man* (in three volumes), vol. 1. London.

McCarthy, P.M. 1983. The composition of some calcareous lichen communities in the Burren, Western Ireland. *Lichenologist* **15**: 231–248.

McCarthy, P.M., Mitchell, M.E. & Schonten, M.G.C. 1985. Lichens epiphytic on *Calluna vulgaris* L. Hull in Ireland. *Nova Hedwigia* **42**: 91–98.

McVean, D.N. & Ratcliffe, D.A. 1962. *Plant Communities of the Scottish Highlands*. HMSO, London.

Majerus, M. 1998. *Melanism: evolution in action*. Oxford University Press, Oxford.

Merrett, C. 1666. *Pinax rerum naturalium Britannicarum*. London.

Miadlikowska, J. 1997. Lichens on *Vaccinium myrtillus* in Poland. *Graphis Scripta* **8**: 1–3.

Ministry of Agriculture, Fisheries and Food 1980. *Lichen on farm roofs*. Leaflet No. 753. HMSO, London.

Morison, R. 1699. *Plantarum historiae universalis Oxoniensis*. Vol. 3. Oxford.

Moxham, T.H. 1980. Lichens and perfume manufacture. *British Lichen Society Bulletin* **47:** 1–2.

Nylander, W. 1866. Les lichens du Jardin du Luxembourg. *Bulletin Société Botanique de France* **13:** 364–372.

O'Dare, A.M. & Laundon, J.R. 1986. Field meeting in north Wiltshire. *Lichenologist* **18**: 269–273.

Parkinson, J. 1640. *Theatrum botanicum*. Sherard: London.

Pentecost, A. 1987. The lichen flora of Gwynedd. *Lichenologist* **19**: 97–166.

Pentecost, A. & Rose, F. 1985. Changes in the cryptogam flora of the Wealden sandstone rocks, 1688–1984. *Botanical Journal of the Linnean Society* **90**: 217–230.

Petch, R. 1984. Some gaelic lichen names. *British Lichen Society Bulletin* **54:** 24–25.

Pettersson, R.B., Ball, J.P., Renhorn, K.-E., Esseen, P.-A. & Sjoberg, K. 1995. Invertebrate communities in boreal forest canopies as influenced by forestry and lichens with implications for passerine birds. *Biological Conservation* **74:** 57–63.

Plomer, W. (ed.). 1938. *Kilvert's diary, selections*. Jonathan Cape, London.

Plot, R. 1677. *The natural history of Oxfordshire*. Oxford.

Popescu, C., Broadhead, E. &

Shorrocks, B. 1978. Industrial melanism in *Mesopsocus unipunctatus* Müll (Psocoptera) in northern England. *Ecological Entymology* **3**: 209–219.

Purvis, O.W. 1996. Interactions of lichens with metals. *Science Progress* **79**: 283–309.

Purvis, O.W., Coppins, B.J., Hawksworth, D.L., James, P.W. & Moore, D. M. 1992. *The lichen flora of Great Britain and Ireland.* London: Natural History Museum Publications.

Purvis, O.W., Coppins, B.J & James, P.W. 1994a. *Checklist of lichens of Great Britain and Ireland.* Natural History Museum, London.

Purvis, O.W., Gilbert, O.L. & Coppins, B.J. 1994b. Lichens of the Blair Atholl Limestone. *Lichenologist* **26**: 367–382.

Purvis, O.W. & Halls, C. 1996. A review of lichens in metal enriched environmrents. *Lichenologist* **28**: 571–601.

Purvis, O.W. & James, P.W. 1985. Lichens of the Coniston Copper Mines. *Lichenologist* **17**: 221–237.

Pyatt, F.B. 1967. The inhibitory influence of *Peltigera canina* on the germination of graminaceous seeds and the subsequent growth of these seedlings. *Bryologist* **71**: 97–101.

Pye, K. & French, P.W. 1993. *Targets for coastal habitat recreation.* English Nature Science No. 13. English Nature, Peterborough.

Rackham, O. 1976. *Trees and woodland in the British landscape.* J.M. Dent & Son: London.

Rackham, O. 1980. *Ancient woodland, its history, vegetation, and uses in England.* Edward Arnold, London.

Rackham, O. 1986. *The history of the countryside.* Dent & Sons, London.

Raistrick, A. & Gilbert, O.L. 1963. Malham Tarn House: its building materials, their weathering and colonization by plants. *Field Studies* **1** (5): 89–115.

Ramaut, J.L. & Corvisier, M. 1975. Effects inhibiteurs des extraits de *Cladonia impexa* Harm., *C. gracilis* (L) Willd. et *Cornicularia muricata* (Ach.) Ach. sur la germination des graines de *Pinus sylvestris* L. *Oecology Plantarum* **10**: 295–299.

Ratcliffe, D.A. 1968. An ecological account of Atlantic bryophytes in the British Isles. *New Phytologist* **67**: 365–439.

Ratcliffe, D.A. 1977. *A nature conservation review.* 2 vols. Cambridge University Press, Cambridge.

Ray, J. 1670. *Catalogus plantarum Angliae et insularum adjacentium.* Ed. 2. London.

Ray, J. 1686. *Historia Plantarum.* Vol. 1. London.

Ray, J. 1690. *Synopsis Methodica Stirpium Britannicarum.* London.

Reynolds, J.E.F. (ed.). 1996. *Martindale; The Extra Pharmacopoeia.* Ed. 31. Royal Pharmaceutical Society: London.

Richardson, D.H.S. 1975. *The vanishing lichens.* David & Charles: Newton Abbot.

Rimes, C. 1992. Freshwater acidification of SSSIs in Great Britain IV Wales. Countryside Council for Wales, Bangor.

Rodwell, J.S. (ed.). 1991. *British plant communities. Vol. 2. Mires and heaths.* Cambridge University Press, Cambridge.

Rose, C.I. & Hawksworth, D.L. 1981. Lichen recolonisation in London's cleaner air. *Nature (London)* **289**: 289–292.

Rose, F. 1974. The epiphytes of oak. In: **M.G. Morris & F.H. Perring** (eds.) *The British oak, its history and natural history.* pp. 250–73. Classey, Faringdon.

Rose, F. 1976. Lichenological indicators of age and environmental continuity in woodlands. In: **D.H. Brown, D.L. Hawksworth & R.H. Bailey** (eds.) *Lichenology: progress and problems* pp. 279–307. Academic Press, London.

Rose, F. 1992. Temperate forest management: its effects on bryophyte and lichen floras and habitats. In: **J.W. Bates & A.M. Farmer** (eds.) *Bryophytes and lichens in a changing environment.* pp.211–23. Clarendon Press, Oxford.

Rose, F., Hawksworth, D.L. & Coppins, B.J. 1970. A lichenological excursion through the north of England. *Naturalist, Hull* **1970:** 49–59.

Rose, F. & James, P.W. 1974. Regional studies on the British lichen flora. 1: The corticolous and lignicolous species of the New Forest, Hampshire. *Lichenologist* **6:** 1–72.

Sanderson, N.A. 1999. Woodland management and lichens. In: **A. V. Fletcher** (ed.) *Habitat management for lichens.*

Seaward, M.R.D. 1976. Performance of *Lecanora muralis* in a polluted environment. In: **D.H. Brown, D.L. Hawksworth & R.H. Bailey** (eds.) *Lichenology: progress and problems* pp. 323–357. Academic Press, London.

Seaward, M.R.D. 1982. Lichen ecology of changing urban environments. In: **R. Bornkamm, J.A. Lee & M.R.D. Seaward** (eds.) *Urban Ecology: 2nd European Ecological Symposium* pp. 181–189. Blackwell Scientific Publications, Oxford.

Sernander, J.R. 1926. *Stockholms Naturalist,* Almqvist & Wiksells, Upsala.

Seyd, E.L. & Seaward, M.R.D. 1984. The association of oribatid mites with lichens. *Zoological Journal of the Linnean Society* **80:** 369–420.

Sheard, J.W. 1968. The zonation of lichens on three rocky shores of Inishowen, Co. Donegal. *Proceedings of the Royal Irish Acadamy* **66**: 101–112.

Sheard, J.W. & James, P.W. 1976. Typification of the taxa belonging to the *Ramalina siliquosa* species aggregate. *Lichenologist* **8**: 35–46.

Showman, R. E. 1981. Lichen recolonisation following air quality improvement. *The Bryologist* **84:** 492 497.

Siddiqi, M.R. & Hawksworth, D.L. 1882. Nematodes associated with galls on Cladonia glauca, including two new species. *Lichenologist* **14:** 175–184.

Smith, A.L. 1921a. *A handbook of the British lichens.* (Reprinted 1963 by Wheldon & Wesley & J. Cramer) British Museum, London.

Smith, A.L. 1921b. *Lichens.* Cambridge University Press: Cambridge.

Stokoe, R. 1983. *Aquatic macrophytes in the tarns and lakes of Cumbria.* Occasional publication No. 18. Ambleside: Freshwater Biological Association.

Sutcliffe, D.W. & Carrick, T.R. 1986. Effects of acid rain on waterbodies in Cumbria. In: **P. Ineson** (ed.) *Pollution in Cumbria* pp. 16–25. Institute of Terrestrial Ecology Symposium No. 16.

Sutcliffe, D.W. & Carrick, T.R. 1988. Alkalinity and pH of tarns and streams in the English Lake District (Cumbria). *Freshwater Biology* **19:** 179–189.

Tikkanen, E. & Niemela, I. 1995. *Kola Peninsula pollutants and forest ecosystems in Lapland.* Final report of The Lapland Forest Damage Project.

Tønsberg, T. & Jørgensen, P.M. 1997. On the alleged apothecia of *Leproloma membranaceum* (Dicks.) Vain. *Lichenologist* **29:** 597–99.

Topham, P.B. 1977. Colonisation, growth, succession and competition. In: **M.R.D. Seaward** (ed.) *Lichen ecology* pp. 31–68. Academic Press, London.

Tournefort, J. P. 1694. *Élémens de Botanique.* 3 vols. Paris.

Troll, K. 1925. Ozeanische züge im Pflanzenkleid Mitteleuropas. *Treie Wiege vergleichender Erdkunde.* Festgabe an Drygalski, Muchen & Berlin.

Tubbs, R. C. 1968. *The New Forest: an ecological history.* David & Charles, Newton Abbot.

Tubbs, R. C. 1986. *The New Forest.* Collins, London.

Turner, D. & Dillwyn, L. W. 1805. *The botanist's guide through England and Wales.* 2 vols. Phillips & Fardon,

London.

Turner, W. 1568. *Herball, the thirde parte.* Collen.

Venables, U. 1956. *Life in Shetland. A world apart.* Edinburgh.

Vickery, A.R. 1975. The use of lichens in well-dressing. *Lichenologist* **7**: 178–179.

Vickery, A.R. 1995. *Oxford dictionary of plant lore.* Oxford University Press: Oxford.

Ward, S.D. 1970. The phytosociology of *Calluna-Arctostaphylos* heaths in Scotland and Scandinavia. 1. Dinnit Moor, Aberdenshire. *Journal of Ecology* **58**: 847–63.

Watson, W. 1918. Cryptogamic vegetation of the sand-dunes of the west coast of England. *Journal of Ecology* **6**: 126–143.

Watt, A.S. 1955. Bracken versus heather, a study in plant sociology. *Journal of Ecology* **43**: 490–506.

Weddle, H.A. 1875. Excursion lichenologique dans l'ile d'Yeu sur la cote de la Vendee. *Memories Societe des Sciences Naturelles de Cherbourg* **14**: 251–316.

Wells, T.C.E., Sheail, J., Ball, D.F. & Ward, K. 1976. Ecological studies on the Porton Ranges: relationships between vegetation, soils and land-use history. *Journal of Ecology* **64**: 589–626.

Whitaker, J. 1862. *The deer paddocks and parks of England.* Ballentyne, Hanson & Co., London.

White, F.J. & James, P.W. 1985. A new guide to microchemical techniques for the identification of lichen substances. *British Lichen Society Bulletin* **57** (Suppl): 1–41.

Winch, N. J. 1831. *Flora of Northumberland and Durham.* T. & J. Hodgeson, Newcastle.

Winchester, V. 1984. A proposal for a new approach to lichenometry. *British Geomorphological Research Group Technical Bulletin* **33**: 3–20.

Winchester, V. 1988. An assesment of lichenometry as a method for dating recent stone movement in two stone circles in Cumbria and Oxfordshire. *Botanical Journal of the Linnean Society* **96**: 57–68.

Wolseley, P.A., James, P.W., Coppins, B.J. & Purvis, O.W. 1996. Lichens on Skomer Island. *Lichenologist* **28**: 543–570.

Wolseley, P.A. & James, P.W. 1991. *The effects of acidification on lichens 1986–90.* CSD Report 1247, Nature Conservancy Council, Peterborough.

Woods, R.G. 1988. Further neglected habitats: heavy metal outliers. *British Lichen Society Bulletin* **63**: 15–16.

Zopf, W. 1896. Uebersicht der auf flechten schmarotzenden plize. *Hedwigia* **35**: 312–366.

Zukal, H. 1895. Morphologische und biologische untersuchungen über die flechten. *Sitzungsberg Böhmische Gesellschaft der Wissenschaften der Mathematisch-naturwissenschaftlichen Classe Wien* **104**: 1303–1395.

Index

Aberdeen, 246
Abernathy Forest, 86
Absconditella celata, 32, 227
Acaropara fuscata, 96
Acarospora spp., 177
  atrata, 108
  badiofusca, 203, 205
  chlorophana, 163
  fuscata, 59, 98, 101, 107, 152, 245
  glaucocarpa, 144, 205
  heppii, 137, 146, 173, 183, 246
  f. *luteopruinosa*, 173
  impressula, 245, 246
  macrocarpa, 150
  macrospora, 144
  peliscypha, 190
  rhizobola, 201
  rufescens, 246
  sinopica, 33, 175, 179–80
  smaragdula, on acid rock, 108, 160
    on alkali downs, 138
    copper colouration, 158
    on mine tailings, 177, 179
    on shingle, 246
  umbilicata, 153
  veronensis, 245, 246
Acarosporion sinopicae, 108, 179–80
Achnaschellach Forest, 86
acid, bark, 59, 76, 80, 81, 82, 84, 89
  calycin, 41
  defence against grazing, 52
  haemoventosin, 41
  lichen, 28, 40–2
  norstictic, 179
  parietin, 41
  protolichesterinic, 17–18
  rain, 66–7
  rhizocarpic, 41, 207
  rhodocladonic, 41
  rock, 95–117, 138
  salazinic, 20
  squamatic, 124
  usnic, 17, 41
Acrocordia conoidea, 142, 146, 157
  gemmata, 255
  macrospora, 244
  salweyi, 152
Adlestrop (Glos), 184
age of lichens, 38–40
Agonimia astroidestra, 258
  octospora, 71, 79, 256
  tristicula, on acid rock, 107, 112
  on cliff grassland, 252
  near lakes, 226–7
  on limestone, 142, 147, 149
  on machair, 243
  on railway bridges, 184
  on roofs, 154
  on siliceous walls, 152
  vinosa, 258
Aire, River, 217
airfields, as habitat, 185–7
Alantoparmelia alpicola, 40, 196
Alectoria spp., 16, 38, 251
  fuscescens, 98
  nigricans, in Lake District, 210
  on sandstone, 104
  in Scotland, 193, 195, 196, 199
  ochroleuca, 194, 207
  sarmentosa, on acid rock, 103, 104
  at Cuthill Links (Sutherland), 242–3
  in Scotland, 194, 195, 199
alga, green, 29, 48, 62, 166–7, 182
alkali, in bark, 90
  dust, effect on lichen, 65–6
  on rocks, 99–100, 138
Allantoparmelia alpicola, 104, 194, 199, 211
Altar Stones (Leics), 98
Ambersham Common (Sussex), 129
Amelia spp., naming of genus, 196
  andreaeicola, 203
Ameronthrus maculatus, 238
Amygdalaria consentiens, 206
  pelobotryon, 108, 115, 206, 217
Anaptychia spp., 92
  ciliaris, 74, 144, 160, 166, 256, 258
  fuscata, 239
  runcinata, 40, 99, 107, 108, 233, 236, 238
Andreaea spp., 196
Anglesey, 54
Anisomeridium biforme, 256
Aphanocapsa spp., 235
Ardnamurchan (Highland), 106
Ardron, Paul, 229
Arkle (Sutherland), 105
Arlington Park (Devon), 91, 92

Arne Heath (Dorset), 129
Arran, Isle of, 106, 109, 117
Arthonia spp., 30
  anglica, 258
  anomrophila, 258
  arthonioides, 103, 258
  astroidestra, 79
  atlantica, 116
  cinnabarina, 29, 88
  didyma, 88
  elegans, 88
  fuscopurpurea, 29
  impolita, 74, 80, 91, 162, 256
  lapidicola, 163, 164, 175, 245
  radiata, 74, 90
  vinosa, 71
  zwackhii, 258
Arthopyrenia spp., 29
  antecellans, 258
  ranunculospora, 71, 258
Arthothelium ilicinum, 258
  reagens, 82, 88
Arthrorhaphis alpina, 244
  citrinella, 114
Arthur's Seat (Edinburgh), 107
Ascomycetes, 29
Ascophyllum nodosum, 28
ash, colonisation, 65–6, 67, 79
  in deer parks, 91
  ecosystem, 47–9
  pollarded, 75, 89–90
  in Scotland, 82
  in wood pasture, 74
Ashburnham Park (Sussex), 92
Ashdown Forest (Sussex), 129
aspen, 87
Aspicilia caesiocinerea, on acid rock, 100–1, 107, 113
  near lakes, 217–18, 225, 227–8
  calcarea, air pollution, 66
  on flint, 138
  growth, 39, 40
  invertebrate food, 51
  on limestone, 142, 144
  on rocky shores, 239
  on walls, 153
  cinerea, 101, 108
  contorta, 112, 138, 142, 153
  epiglypta, 100, 110, 209
  laevata, 108
  leprosescens, 99, 236
  melanaspis, 225
  recedens, 209, 225
  subcircinata, 144, 159, 163

tuberculosa, 139
Athopyrenia antecellans, 250
aurochs, 72
Avebury (Wilts), 39, 40, 99–100, 159, 254
Avena fatua, 124
Avon Gorge (Bristol), 140
Aysgarth (Yorks), 215, 217, 219

Bacidia arceutina, 112, 146, 251
　arnoldiana, 138, 146, 180, 182, 221
　biatorina, 258
　caligans, 173, 174, 223
　carneoglauca, 221, 227
　chloroticula, 173, 174, 180, 223
　circumspecta, 258
　delicata, 146
　egenula, 146
　epixanthoides, 258
　fuscoviridis, 221, 227
　herbarum, 134, 203
　inundata, 216, 221, 223, 227
　muscorum, 133, 176, 243, 252
　rubella, 161, 163, 166, 256
　sabuletorum, on acid rock, 112
　　on dunes, 241
　　on limestone, 133, 142, 146, 147
　　on machair, 243
　　in mine tailings, 176
　　on railway bridges, 184
　　on roofs, 154
　saxenii, on coke, 174
　　as metallophyte, 164, 173, 175, 176
　　on pine stumps, 87
　　beneath pylons, 180–1
　scopulicola, 236
　subincompta, 258
　viridescens, 175, 176
　viridifarinosa, 110
Bactrospora homalotropa, 88
Badenoch, 85
Baeomyces spp., 114
　carneus, 189–90
　placophyllus, 116, 176, 177, 210
　roseus, 116, 122, 129, 177
　rufus, on acid rock, 97–8, 116
　　on airfields, 186
　　on coastal heath, 250
　　on heathland, 122, 129
　　light requirement, 221
　　on mine tailings, 176
　　submerged tolerance, 217
Baligill (Sutherland), 250
Ballard Down (Dorset), 136–8, 139
Ballochbuie Forest, 86

Barbula convoluta, 171
Bardsey Island, 128, 239, 252
Barkman, Prof. J.J., 72
Barnack Hills (Peterborough), 149
basalt, as substrate, 106–10
Basidia sabuletorum, 152
Basidiomycetes, 29
beard-lichen, 69, 85
beech, 79
Beer (Devon), 249
Belfast, pollution, 54
Bellemerea alpina, 198
　cinereorufescens, 189
Belonia incarnata, 178, 203
　nidarosiensis, 147
　russula, 191, 199, 203, 206, 211
Ben Alder, 191
Ben Avon, 194
Ben Bulben (Co. Sligo), 209
Ben Ghlas, 203
Ben Hope, 198–200
Ben Lawers, 80, 188–9, 198–206
Ben Loyal, 198–9
Ben Macdhui, 193
Ben Nevis, 14, 206
Benfield, Barbara, 163, 249
Berwickshire Naturalists' Club, 104
Biatora carneopallida, 202
　sphaeroides, 71, 82, 90, 258
　tetramera, 191
　vernalis, 203
Biatorella hemisphaerica, 202
Bilsdale (Yorks), 163
biomass, 124
birch, 76, 82, 84, 87, 257
bird, use of lichen, 46–7, 238
Birmingham, 54, 172
Black Mountains (Hereford), 112
Blair Atholl, 163, 189
Blaskets, The, 239
Blatchley, Ishby, 163
Bloxham, Rev. A., 56, 99
Boconnoc Park (Cornwall), 91–2, 257
Bodmin Moor, 110
Bostaurus primigenius, 72
Boston Spa (Yorks), 146
Box Hill (Surrey), 133, 135, 137
Bradfield Woods SSSI (Suffolk), 72
Bradford, lichen mapping, 57
Bradgate Country Park (Leics), 98
Brampton Bryan Park (Hereford), 92
Brandon (Co. Kerry), 208–9

Braunton Burrows (Devon), 240
Breadalbane Mountains (Scotland), 189, 191, 219
Brean Down (Somerset), 143, 239
Breckland (East Anglia), 135, 181
brewing, use in, 23
brick, colonisation, 172–3
Brigantiaea fuscolutea, 191, 203, 211, 252
Brighton, 102
Bristol, 172, 174
British Lichen Society, at Berwyn Mountains, 40
　Churchyard Project, 156
　at Coll, 243
　on Dartmoor, 111
　formation, 13
　Lizard Peninsula, 116
　mining areas, 174
　on Skye, 110
　in Sussex, 165
Broad Down (Hereford), 101
Broad Law (Roxburgh), 108
Brockadale (Yorks), 146
Bryonora curvescens, 203
Bryophagus gloeocapsa, 32, 114, 196
Bryoria spp., 38, 45, 103
　bicolor, 103, 111, 199, 203, 210
　capillaris, 84
　chalybeiformis, 103, 199
　furcellata, 32, 84, 86
　fuscescens, on acid rock, 93, 98, 102–3, 105, 111, 114
　　in Cairngorms, 199
　　in Hawksworth-Rose scale, 255
　　in pinewoods, 84
　　on shingle, 246
　implexa, 85
　lanestris, 199
　nadvornikiana, 104, 105
　nitidula, 189
Buellia spp., 92, 166
　aethelea, on acid rock, 98
　　associations, 180
　　copper colouration, 179
　　in churchyards, 158, 160
　　on flint, 138
　　on railway bridges, 184
　　on shingle, 245
　　submerged, 218
　　in villages, 153, 154
　asterella, 135, 139
　badia, 154
　coniops, 237
　erubescens, 258
　leptocline, 203
　ocellata, 109, 138, 153, 160, 184
　papillata, 208
　pulverea, 65

punctata, 66, 174, 255
saxorum, 100
schaereri, 87
stellulata, 100
subdisciformis, 99
Burbage Edge (Derbs), 103
Burford (Oxon), 153, 160
Burnham Beeches (Bucks), 74
Burren (Co. Clare), 143
Butser Hill (Hants), 136, 139
Buxton (Derbs), 56
Byssoloma leucoblepharum, 250
 subdiscordans, 250
Bywell (Northumb), 163

Caenlochan Glen, 40, 204–6
Cairngorm Mountains, 46, 192–8
Caldbec Hill (Sussex), 15
Caldy Island, 244
Caliciales, 69, 72, 80, 85
Calicium spp., 61
 corynellum, 163
 parvum, 84
 viride, 255
Callunetum, 194–5, 250
caloplaca, snow, 32
Caloplaca spp., on Ben Lawers, 201
 in deer parks, 92
 eaten by molluscs, 51
 growth, 40, 254
 on limestone, 144, 166
 alociza, 144, 239
 ammiospila, 202
 approximata, 202, 205
 arenaria, 107, 110, 186, 225, 228
 arnoldii, 109, 110, 149
 atroflava, 139, 246
 aurantia, on buildings, 153, 157
  on limestone, 137, 142, 147, 148
  on railway bridges, 184
 ceracea, 99
 cerina, 65, 144–5, 202, 251–2, 256
 chalybaea, 51, 149
 cirrochroa, 144, 146, 157
 citrina, on airfields, 185
  on buildings, 152, 155, 158
  on limestone, 136, 146, 147
  on pebbles, 171
 concilians, 203
 crenularia, 101, 110, 153, 184, 185, 228, 236
 dalmatica, 153, 157
 decipiens, 66, 153, 162
 flavescens, air pollution, 59
  on buildings, 153, 158
  on flint, 138
 growth, 40, 161
  on limestone, 142
  on rocky shores, 239
 flavorubescens, 65, 166, 256
 flavovirescens, 96, 109, 112, 153, 185
 granulosa, 144
 haematites, 166
 holocarpa, 66, 139, 147, 152, 172, 218, 251
 isidiigera, 96, 147, 148, 153, 185, 217
 lactea, on buildings, 153, 154, 157
  on chalk, 136
  on limestone, 96, 147, 149
  on shingle, 245, 246
 littorea, 236
 lucifuga, 259
 luteoalba, 74, 137, 166
 marina, 233, 236, 238–9, 247–8
 microthallina, 236
 nivalis, 32, 202
 obliterans, 109
 obscurella, 90
 ruderum, 157
 saxicola, 153, 158, 245
 scopularis, 237
 teicholyta, 32, 158, 184, 244, 249
 thallincola, 236, 239
 variabilis, 52, 142, 149, 153, 184, 239
 verruculifera, 236
 virescens, 166
 zone, on rocky shore, 232
Caloplacetum heppianae, 146–7
Calothrix spp., 29
Candelaria concolor, 61, 162, 166
Candelariella spp., acids, 41
 arctica, 189
 aurella, air pollution, 56
  on airfields, 185
  in lignicolous community, 172
  meaning of name, 32
  on mortar, 152
  on tyres, 173
 coralliza, 98, 100, 107, 111, 186
 medians, 153, 158
 reflexa, 182
 vitellina, on acid rock, 107
  air pollution, 59, 62, 65–6
  on buildings, 152, 155
  in churchyards, 158, 160
  near lakes, 225
  on limestone, 147
  on railways, 183
  in River Coquet, 218
 vitellina f. *flavovirella*, 186
 vitellina f. *vitellina*, 186
 xanthostigma, 90

Canons Ashby Church (Notts), 25
carbon monoxide, effect on lichens, 67
Carbonea assimilis, 107
 vorticosa, 107
cardboard, as substrate, 173–4
Carex arenaria, 127
 humilis, 136
Carlisle, 173
Carreg Wastad (Snowdonia), 115
Castle Eden Dale (Co. Durham), 146
Castlerigg stone circle (Cumbria), 39
Catapyrenium cinereum, 191, 199, 243
 lachneum, 107
 pilosellum, 134
 squamulosum, on cliff grass land, 252
  dunes, 243, 244
  on limestone, 133, 142, 147
  in villages, 159, 161
  near water, 228
 waltheri, 145, 191
Catillaria aphana, 137
 atomarioides, 128
 atropurpurea, 71, 82, 258
 chalybeia, on limestone, 147
  on railway bridges, 184
  on rocky shore, 236
  on shingle, 245
  in villages, 154
  near water, 217–18, 228
 contristans, 193, 198, 203, 209
 gilbertii, 204
 laureri, 79, 258
 lenticularis, 142
 picila, 150
 pulverea, 82
 scotinoides, 203, 206, 212
Catolechia wahlenbergii, 32, 206–7
Cavernularia hultenii, 85
Cawdor Castle (Highland), 92
Cawton-Marsham Heath (Norfolk), 129
Ceratodon purpureus, 171, 183
Certhia familiaris, 47
Cervus elaphus, 72
Cetraria chlorophylla, 114
 commixta, 194, 210
 cucullata, 32, 189, 195
 ericetorum, 207
 hepatizon, 104, 111, 210
 islandica, as animal food, 45, 46
  in Cairngorms, 192–3, 195–7
  as food preservative, 23
  in Lake District, 209

on limestone, 145
in medicine, 17–18
on mine tailings, 176
on moors, 125–6
juniperina, 85
nivalis, 194–5, 208, 212, 251
odentella, 190
pinastri, 74, 128
sepincola, 74
Cetrariella delisei, 193, 196, 207
ericetorum, 193
nivalis, 207
Cetrelia olivetorum, 82, 88, 258
Chaenotheca spp., 30, 61, 258
brachypoda, 164
ferruginea, 57, 93, 255, 258
furfuracea, 88, 164
phaeocephala, 166
stemonea, 93
trichialis, 93
Chaenothecopsis debilis, 29
nigra, 93
parasitaster, 29
pusiola, 29
chalk, as substrate, 131–50
Chambers, Steve, 174, 178
Chatsworth Edge/Park (Derbs), 76, 93–4, 103
Chelsea Physic Garden (London), 61
cherry, bird, 82
Chesil Bank (Dorset), 246–7
Chester, Tom, 151, 154, 163
Cheviot Hills, 110, 219
chimeras, 31
Chobham Common (Surrey), 129
Chromatochlamys muscorum, 244
Chrysothrix candelaris, 61, 93, 255
flavovirens, 84
church, lichen study, 156–64
Churchyard Project, 156–7
Ciste Mhearad snow-bed (Cairngorms), 196–8
Cladina subgenus, 121, 126–7
cladonia, upright mountain, 32
Cladonia spp., acids, 41
in Cairngorms, 192
on cardboard, 174
in churchyards, 162
colonisation by, 86
on Dartmoor, 111
drying, 38
form of, 34
fungal partner, 30
on heathland, 119
near lakes, 227
at Loch Sunart, 82
and metallophytes, 175

used in perfume, 26
in quarries, 181
in sheltered areas, 58
on shingle, 245
alpestris, 189
amaurocraea, 189
arbuscula, in Cairngorms, 197, 207
on dunes, 241
eaten by reindeer, 45
on heathland, 121, 122, 126, 129
bellidiflora, 196–7, 209, 241
botrytes, 85
brevis, 125
caespiticia, 98, 258
callosa, 129
cariosa, 129, 175, 176, 177, 181
carneola, 85, 241
cenotea, 85
cervicornis, 129, 185, 240–1, 246
ssp. *verticillata*, 125, 176
chlorophaea, 120–1, 123, 171, 217, 240–1, 246
ciliata, 241, 250
coccifera, on acid rock, 96
in Cairngorms, 199
on dunes, 240–1
on heathland, 121, 123
as indicator, 168
meaning of name, 32
in urban area, 171
coniocraea, 67, 90, 121, 171, 240–1
convoluta, 143
cornuta, 125, 241
crispa, 121
crispata, 123, 240–1
cyathomorpha, 115, 203, 210
digitata, 93
fimbriata, 120–1, 123, 152, 171, 240
firma, 243, 252
floerkeana, 121, 123, 171, 240–1
foliacea, 102, 149, 240–1, 246
fragilissima, 129, 177–8, 181
furcata, on airfields, 185
on Cairn Gorm, 197
on dumped material, 171
on dunes, 240–1
on heathland, 121, 127
on limestone, 133
near rivers, 217
glauca, 53, 121
gracilis, 121, 123, 124, 127, 240–1, 246
humilis, 169, 171, 174, 183, 241
incrassata, 98
luteoalba, 129
macilenta, 121, 123, 240, 248
macrophylla, 193
maxima, 196, 207

mediterranea, 116
mitis, 241–2, 246
parasitica, 93, 258
phyllophora, 196, 242
pocillum, on dunes, 241, 243–4
on limestone, 133, 142, 147
on mine tailings, 176
parasitised, 37
polydactyla, 121, 241
portentosa, on coastal heath, 250
on dunes, 241
eaten by reindeer, 45
on heathland, 119, 121, 123, 124, 126
as indicator, 168
on limestone, 136
on shingle, 246
pyxidata, 16, 120–1, 217
ramulosa, 120, 121, 123
rangiferina, 45, 126, 207, 210, 241
rangiformis, 17, 133–4, 171, 176, 185–6, 243–4, 246
rei, 129, 183
scabriuscula, 176, 240–1
squamosa, 121, 185, 240
stellaris, 27, 241
strepsilis, 125, 129
stricta, 32, 196
subcervicornis, on acid rock, 98, 111, 114
in Cairngorms, 195
in Co. Kerry, 208
on heathland, 129
near rivers, 217
subulata, 121, 171, 176, 183, 240
sulphurina, 85, 125, 129, 241
symphycarpa, 143, 191, 203
tenax, 244
tenuis, 129
uncialis, in Cairngorms, 196–7
on dunes, 240–1
eaten by reindeer, 45
on heathland, 121, 127
as indicator, 168
on mine tailings, 176
uncialis ssp., *biuncialis*, 241, 243
ssp., *uncialis*, 176–7, 241–3
zopfii, 125, 195, 241–2
classification, 29
Claurouxia chalybeioides, 115, 207, 208, 211
Clauzadea immersa, 96, 136, 141, 142
metzleri, 136, 154, 162
Clauzadeana macula, 110
Cleadale (Eigg), 88
climate, effect on distribution, 74–7

Cliostomum corrugatum, 247–8
Coccomyxia spp., 29
Coccotrema citrinescens, 109, 207, 208
Coelocaulon spp., 127
  aculeatum, in Cairngorms, 192, 195, 199
    on coastal heath, 250
    on dunes, 241
    on heathland, 121
    on limestone, 136
    on mine tailings, 176
    in sheltered conditions, 58
    on Wimbledon Common, 181
  muricatum, 121, 124, 242
coke, as substrate, 174
collecting lichens, 43–4
Collema spp., 144, 174, 220, 221
  auriforme, on buildings, 153, 154
    in churchyards, 161
    on limestone, 133, 141, 148
    on railway bridges, 184
  bachmanianum, 147
  callopismum, 206
    var. *rhyparodes*, 191
  ceraniscum, 191, 202
  conglomeratum, 166
  crispum, 152
  cristatum, 112, 141, 153
  dichotomum, 32, 209, 215, 223, 228
  flaccidum, 216, 227, 228
  fragile, 144
  fragrans, 166, 259
  furfuraceum, 88, 90, 149, 258
  fuscovirens, 153, 154, 184
  glebulentum, 107, 191, 211
  latzelii, 116
  limosum, 171
  multipartitum, 142, 143, 149, 244
  nigrescens, 259
  parvum, 205
  polycarpon, 149
  subflaccidum, 88, 90, 258
  subfurvum, 79
  subnigrescens, 259
  tenax, on cliff grassland, 251
    on limestone, 133, 147, 148
    on machair, 243
    on mine tailings, 176, 178
    beside paths, 162, 171
    water absorption, 38
  tenax var. *ceranoides*, 185
colonisation, 59–68, 80, 170
colour, 40–2, 106
community, 42–3, 59, 60
  climax, 91

competition, between lichens, 107–8
Compton Down (Isle of Wight), 136
concrete, as substrate, 173
Coniocybe furfuracea, 112
Coniston Copper Mines (Cumbria), 178–9
Conizaeoidion, 65–6
conservation, in churchyards, 163–4
Cooke, M.C., 16
copper, 158–9, 178–9
coppicing, effect on diversity, 72
Coppins, A.M., 99
Coppins, Brian, 69, 83, 85, 115, 127, 139, 163, 189, 201, 204–5
Coppins, Sandy, 163
Coriscium viride, 98
Corner, Roderick, 126, 223
Cornicularia normoerica, 45, 103, 111, 194, 196, 210
cotton, as substrate, 174
Coulin Forest (Skye), 86
Cracken Edge (Derbs), 103
Crag Lough (Northumb), 108
Craig Tulloch (Blair Atholl), 150
Cranborne Chase (Dorset), 74
Crawley, Michael, 104
Cresswell Crags (Derbs), 145–6, 147
Cricklade (Wilts), 163
Crombie, Rev. James M., 16, 150, 189, 201
Crombie rarities, 189–90
Cronkley Fell (Upper Teesdale), 144
Cross Fell (Pennines), 104, 149, 211
Crosthwaite Church (Cumbria), 39
crottle dye, 20–1, 27
Crowden Moors (Manchester), 103
Cryptolechia carneolutea, 259
Ctenidium molluscum, 142
cudbear dye, 21–2, 114
Culbin Forest/Sands (Aberdeen), 87, 127, 128, 241–2, 245, 246
cup-lichens, 119–21
Cuthill Links (Sutherland), 128, 242–3, 245, 246
cyanobacteria, 28–30, 51, 67, 235
Cyphelium inquinans, 85, 93, 164–5
  notarisii, 247
  tigillare, 85, 166
Cystocoleus spp., 29
  ebeneus, 98, 114

Dart, River, 216, 217, 219

Dartmoor, 110–11
Darwin, Erasmus, 54
dating lichens, 253–4
Day, Ivan, 67, 90
de Bary, 16
Debenham (Suffolk), 154
Decampia hookeri, 201
Deep Dene (Sussex), 136
deer, 69–70, 72, 78, 80, 90–4
definition of lichen, 28
Degelia spp., 239
  atlantica, 71, 74, 227, 251, 258
  plumbea, 90, 251, 258
Dendriscocaulon spp., 31
Dermatocarpon spp., 225
  cinereum, 251
  intestiniforme, 108–9, 228
  leptophylloides, 216, 228
  luridum, 215, 218–20, 226–8
  meiophyllizum, 216, 227, 228
  miniatum, on acid rock, 107
    in churchyards, 159, 163
    on limestone, 142
    on railway bridges, 184
    near water, 228
    water absorption, 38
Deschampsia flexuosa, 124
Desmococcus spp., 48, 62, 166–7, 182
  viridis, 255
Dictyonema spp., 29
Dimerella diluta, 88
  lutea, on ash, 90
    and climate, 74
    on coastal heath, 250
    in Hawksworth-Rose scale, 256
    in hazel woods, 88
    at Loch Sunart, 82
    in NIEC, 258
    in RIEC, 71
    near water, 227
  pineti, 127, 250
Diploicia canescens, air pollution, 61
  on buildings, 153, 158
  as food, 50
  growth, 39
  in Hawksworth-Rose scale, 255
  increase, 167
  on limestone, 146
Diploschistes gypsaceus, 144
  muscorum, in churchyards, 162
    on dunes, 241, 244
    on limestone, 133
    on mine tailings, 176
    parasitic, 37
    on sandstone walls, 152
  scruposus, 101, 107, 114, 160
Diplotomma alboatrum, 153, 158, 255
  epipolia, 245–6

Dirina spp., 107
　massiliensis, 25
　　f. *sorediata*, 107, 146, 158, 163
Dirinetum stenhammariae, 146
dispersal, 47, 85
distribution, climatic effect on, 74–5
ditches, 164
diversity, factors affecting, 70–7
dog-lichen, common, 17
　ear-lobed, 222
Dolebury Warren (Somerset), 143
Dollis Hill (London), 61
Domelynllyn Park (Gwynedd), 92
Drummond Park (Tayside), 91–2
Dryas octopetala, 145
drying, 37–8
Dryopteris aemula, 98
Duncan, Ursula, 13, 112, 128, 188, 200, 204
dune, habitat, 240–4
Dungeness (Kent), 244, 246
Dunsland Park (Devon), 92
Durdle Door (Dorset), 136, 137
Dutch elm disease, 150
dyeing, 18–22, 27, 114
　crottle, 20–1, 27, 114
　cudbear, 21–2, 114
　orchil, 21–2
Dynevor Park (Dyfed), 92

Earland-Bennett, Peter, 154
Ebernoe Common (Sussex), 74
Echo Crags (Cheviots), 103–4
ecosystems, 47–9
Edinburgh, 107
'Egdon Heath' (in Thomas Hardy), 128
Eigg, Isle of, 87, 88, 106
Eiglera flavida, 205, 216, 246
elm, 82, 87, 91, 150, 166
Ely (Cambs), 158
Endocarpon spp., 163
　adscendens, 216, 226–7
　pusillum, 249
Enterographa spp., 217
　crassa, 71, 80, 168, 211
　elaborata, 79
　hutchinsiae, 221
　sorediata, 72, 79, 258
　zonata, 107
Ephebe lanata, 108, 111, 115, 208, 216, 225, 227
Epigloea soleiformis, 178
Epilichen scabrosus, 177
Ericht, River (Perthshire), 222
Eridge Park (Sussex), 91–2, 97

Euopsis pulvinata, 199, 203, 211
Evelyn, John, 127
Evernia spp., 17, 27, 50
　prunastri, air pollution, 55–7, 60–1, 62
　on coastal heath, 250
　distribution, 74
　on dunes, 241–2
　in Hawksworth-Rose scale, 255
　identification, 69
　in perfume, 25
　spread by birds, 47
Exe, River, 95
Exmoor, 73, 98

Farmer, Andrew, 67
farmland, 164–7
Farnoldia jurana, 141
Feish Dhomhnuill Mine (Highland), 177
Fellhanera myrtillicola, 127
　subtilis, 86, 127–8
fern, hay-scented buckler, 98
　Tunbridge filmy, 98
Ferry-Coul Links (Sutherland), 245, 246
Fiennes, Celia, 101
Fingal's Cave (Staffa), 106
Flannan Isles (Outer Hebrides), 237–8
fluorides, effect on lichen, 64–5
Foinaven (Sutherland), 105–6, 198, 200
folklore, lichens in, 23
food, from lichens, 22–3, 48–53
Forest of Dean, 72–3
Fort William, 64
Fountains Fell (Yorks), 212
Fox, Brian, 109, 191, 192
Freshwater Down (Isle of Wight), 138
Fries Hill (Somerset), 143
fruit body, 36–7
Frullania, spp., 81
Frutidella caesioatra, 193, 196, 198, 211, 213
Fryday, Alan, 110, 178, 194, 196, 201, 204, 206, 208, 213
Fucus pygmaeus, 234
Fulgensia bracteata, 191
　fulgens, 96, 135, 139, 143
Fulgensietum fulgentis, 135, 244
Funaria hygrometrica, 171
fungi, classification, 29
Fungi Imperfecti, 29, 37
Fuscidea spp., 112
　austera, 103, 105
　cyathoides, 100, 105, 113, 138, 199, 239
　gothoburgensis, 105, 113, 193, 206
　intercincta, 105

　kockiana, 96, 105, 111, 113
　lightfootii, 127, 250
　lygaea, 109, 111, 208
　lygaea-tenebrica, 105
　mollis, 116
　praeruptorum, 113
Fuscideetum community, 211
Fyfield Down (Wilts), 100

gabbro, as substrate, 117
Gait Burrows (Morecambe Bay), 143
galls, 53
Gardom's Edge (Derbs), 103
Giant's Causeway (Co. Antrim), 106
Giavarini, Vince J., 111, 128, 163, 206
Gibside Woods (Newcastle-upon-Tyne), 54
Gibson, P.G., 166
glass, as substrate, 174
Glen Affric, 86
Glen Coe, 110, 206
Glen Guisachan, 86
Glen Shee, 128
Glen Strathfarrar, 86
Glenmore, 83–5
Gloeocapsa spp., 29, 141
Glossop, 174
gneiss, as substrate, 112–15
goblin lights, 32
Godlingstone heath (Dorset), 128
golden lichen, 23
Gordon, George, 21
Gower Peninsula, 140, 144, 239, 244
granite, as substrate, 110–11
Graphidion (*Graphidion scriptae*), 43, 75, 79, 88
Graphina anguina, 88
　ruiziana, 82
Graphis spp., 30, 90
　alboscripta, 82, 88
　elegans, 255
　scripta, 61, 88, 168
grazing, by deer, 72, 80, 91
　on heathland, 125–7
　by limpets, 235
　by sheep, 237
　in woodland, 82
Great Calva (Lake District), 126
Greta, River, 219–20
Grimmia spp., 196
groupings, of lichens, 74
growth, forms, 33–4
　rate, 38–40, 161, 253–4
Gyalecta derivata, 259
　flotowii, 256
　foveolaris, 199, 202
　fritzei, 199
　geoica, 203
　jenensis, on Ben Alder, 191
　on buildings, 157, 159

INDEX

on limestone, 142, 146, 149
on machair, 244
meaning of name, 32
on sandstone, 112
ulmi, 112, 150
Gyalectetum jenensis, 146
Gyalidea diaphana, 196
hyalinescens, 208, 209
lecideopsis, 137, 139, 211, 217
roseola, 175, 177
subscutellaris, 175
Gyalideopsis anastomosans, 250
scotica, 199, 202, 208, 211

habitat, suitability, 70–1
Haematomma ochroleucum, 100, 107, 114, 153, 158, 163, 255
ventosum, 32
hair-grass, wavy, 124
hair-lichen, forked, 32
golden, 32, 102
hair moss, 16, 17
woolly, 207, 243
Halecania alpivaga, 191
bryophila, 204
micacea, 204
rhypodiza, 204, 206
Hankley Common (Surrey), 129
hardboard, as substrate, 174
Harland Moor (Dorset), 129
Harlech Castle, 53
Harrold (Beds), 153
Hatfield Forest, 74
Hawksworth, David, 28
Hawksworth-Rose pollution scale, 255
hazel, 82, 87–9
heath, coastal, 249–51
habitat, 118–30
heather, burning, 122–3
colonised, 127
Helianthemum canum, 145
Hemmings, J., 135
Henderson, Albert, 32
Heterodermia isidiophora, 116
leucomelos, 116, 237, 252
obscurata, 116, 252, 258
obscuratum, 237, 252
propagulifera, 250
Heyshott Down (W. Sussex), 137, 139
High Weald, 97–8
High Wycombe (Bucks), 102
Hildenbrandia rivularis, 216
Hitch, Chris, 115, 154
holly, 79, 80
Holy Island (Northumb), 244–5
Homalothecium lutescens, 142
sericeum, 90, 142

Hope Valley (Derbs), 65–6
Hormidium stage of *Prasiola crispa*, 182
horse chestnut, 91
Horton-in-Ribblesdale (Yorks), 153
house, lichen study, 155
Huilia tuberculosa, 40
Hull, 171, 173
Hume Castle (Berwick), 108
Humphrey Head (Cumbria), 239
Hutton Common (Yorks), 149
Hymenelia lacustris, 218–21, 223, 227
prevostii, 137, 141, 159
Hymenophyllum spp., 82
tunbridgense, 98
Hyperphyscia adglutinata, 255
Hypnum cupressiforme, 90
Hypocenomyce anthracophila, 85
caradocensis, 92
friesii, 84
leucococca, 85
scalaris, 61, 84, 157, 164, 247–8, 255
sorophora, 85
xanthococca, 85
Hypogymnia spp., 50
bitteriana, 211
intestiniformis, 194
physodes, on acid rock, 98
air pollution, 60, 61, 62
in Cairngorms, 199
reaction to *Desmococcus*, 167
on dunes, 241–2
eaten by reindeer, 45
Glenmore, 82
grazed by molluscs, 52
in Hawksworth-Rose scale, 255
on heathland, 123
identification, 69
Loch Sunart, 82
on wood, 164
tubulosa, 61, 67
vitata, 189–90

Icmadophila ericetorum, 32, 122, 125, 129, 250
identification, beginning, 69
guides, 42
Immersaria athroocarpa, 108, 199
Imshaugia aleurites, 84, 93, 104
Inchnadamph (Sutherland), 149, 217
indicator species, ancient countryside, 168
pollution, 54–68
woodland, 70–1, 80, 83, 89
Ingleborough (Yorks), 140–2

introduced species, 169
Inverary Park (Strathclyde), 92
Iona Abbey, 25
Ionapsis cyanocarpa, 207
epulotica, 144, 191, 205
heteromorpha, 144, 191, 211
melanocarpa, 149, 217
odora, 206
suaveolens, 105
Ionaspidetum suaveolentis, 219
iron mine tailings, 179–80
Isle of Man, 175
Isle of Wight, 135, 136, 138, 139, 248–9
Isles of Scilly, 237, 250
Ivinghoe Beacon (Bucks), 133

James, Peter, 13, 30, 115, 177, 178, 188–9, 194, 200, 204
jelly lichen, 30, 32
Johnson, Rev. W., 54
juniper, 84

kamenitzas, 143
Keen of Hamer NNR (Shetlands), 116
Keswick, 219–20
Kettering and District NHS, 149
Kielderstone, 104
Kilvert's Diary, 74
King's Lynn, 125
Kirkcudbright, 163
Kirkland (Westmorland), 160
Knaresborough, 146
Knock Fell (Pennines), 211
Knole Park (Kent), 169
Knowles, Matilda, 230
Koerberiella wimmeriana, 191, 203, 211

Ladybower Reservoir (Derbs), 229
Lagopus mutus, 106
Lake District, 209–12, 219, 224–7
Lakenheath (East Anglia), 135–6
Lambley (Northumb), 176
Lambourn Downs (Berks), 99
Lasaea rubra, 234
Lasallia spp., in dye, 21
pustulata, on acid rock, 98, 101, 109, 111
in acid water, 225
in dye, 21
size of, 40
Lauder, Sir T.D., 83
Lavington Common (Sussex), 129
lead, as substrate, 158–9, 182

mines, 175–8
Lecanactidetum
  (*Lecanactidetum premneae*),
  80, 91, 92
Lecanactis spp., 114
  abietina, 61
  amylacea, 80, 91, 258
  hemisphaerica, 164
  lyncea, 71–2, 80, 91, 258
  premnea, 71–2, 80, 91, 258
  subabietina, 258
Lecania spp., near rivers,
  217, 221
  cuprea, 144
  erysibe, 96, 138, 146, 147,
    229, 236
  fuscella, 166
  rabenhorstii, 144
  sylvestris, 144
  turicensis, 164
Lecanora spp., on mine
  tailings, 175, 177
  achariana, 209, 223–5, 228
  actophila, 233, 236, 238
  aitema, 87, 164
  albescens, 51, 152
  andrewii, 107, 244
  atrosulphurea, 199
  cadubriae, 84, 86
  caesiosora, 107, 153
  campestris, air pollution, 66
    on buildings, 153, 158
    growth, 39–40, 254
    on limestone, 146–7
    on shingle, 245
  campestris ssp. *dolomitica*,
    147–8
  carpinea, 127
  chlarotera, 66, 90, 255
  chlorophaeoides, 199
  cinereofusca, 82
  conferta, 164
  confusa, 250
  conizaeoides, on acid
    rock,103
    air pollution, 56, 59–60,
      62, 65
    on buildings, 153
    at Chatsworth, 93
    on conifers, 87
    on dwarf shrubs, 127
    in Hawksworth-Rose
      scale, 255
    as introduced species,
      169
    in medicine, 18
    on roads, 182
    on salt marshes, 247–8
    on street trees, 183
    on trees, 47
    on wood, 164, 172
  crenulata, 149, 158, 239,
    246
  cyrtella, 255
  dispersa, on acid rock, 107
    and air pollution, 56
    on airfields, 185
    on brick, 173
    on buildings, 152, 155
    on calcareous pebbles,
      171
    on dumped material, 173
    on flint, 138
    in lignicolous
      communities, 172
    on limestone, 141–2, 146
    beneath pylons, 180
    on roads, 182
    on salt marshes, 248
    near water, 228
  dispersa f. *zosterae*, 252
  epanora, 98, 103, 108, 175,
    180
  epibryon, 144, 191
  expallens, 247, 255
  frustulosa, 203
  fugiens, 99, 236
  fuscoatra, 184
  gangaleoides, 100–1, 107–8,
    111, 218, 228, 236
  hagenii, 56
  handelii, 175, 180
  helicopis, 233, 236, 238–9,
    245, 247–8
  intrica, 107
  intricata, 114
  jamesii, 258
  leptacina, 196, 198, 203
  muralis, on acid rock, 102,
    107
    air pollution, 55–6, 59, 63
    on airfields, 185
    on car tyres, 173
    growth, 38–40, 254
    near lakes, 225, 227–8
    in lignicolous
      communities, 172
    on limestone, 147
    on railways, 183
    in rivers, 217, 218
    on rocky shores, 237
    on shingle, 245
    on siliceous walls, 152
    on tarmac, 174
  orosthea, 107
  pannonica, 161–4
  piniperda, 93
  poliophaea, 236
  polytropa, on acid rock,
    100, 107, 113, 160
    on airfields, 185–6
    associations, 180
    copper colouration, 158
    on mine tailings, 179
    on railways, 183, 184
    on siliceous walls, 152
  populicola, 166
  pruinosa, 157, 163
  quercicola, 258
  rupicola, on acid rock, 98,
    100, 101, 107
    on airfields, 186
    in churchyards, 160
    on rocky shores, 236
    on shingle, 245
  rupicola f. *sorediata*, 108
  saligna, 164, 172
  soralifera, 98, 102, 180
  stenotropa, 158, 173, 184
  straminea, 237
  subaurea, on acid rock,
    103, 108
    associations, 180
    in churchyards, 163
    metallophyte, 175
  subcarnea, 108
  sulphurea, 102, 153, 184,
    236
  swartzii, 107
  symmicta, 127, 164, 195,
    248
  varia, 164
Lecanoretum dispersae, 173
  epanorae, 180
Lecidea spp., 112, 117, 175,
  211
  auriculata, 110, 246
  berengeriana, 191, 193
  brachyspora, 246
  caesioatra, 197
  commaculans, 103
  confluens, 50
  diducens, 246
  doliformis, 82
  endomelaena, 180
  fuliginosa, 109, 210
  furvella, 107
  fuscoatra, 98, 101, 111,
    153, 245
  griseoatra, 196
  hypnorum, 145, 211, 251–2
  hypopta, 85, 93
  inops, 175, 179
  insularis, 29
  lactea, 105, 113
  lapicida, 180
  lichenicola, 136
  limosa, 193, 198
  lithophila, 29, 108, 113
  orosthea, 107
  paupercula, 203, 206, 209
  pernigra, 103
  phaeops, 115, 210
  plana, 246
  polycarpella, 171
  pycnocarpa var. *sorediata*,
    208
  quadricolor, 123
  sarcogynoides, 110
  silacea, 175, 180
  sublivescens, 258
  sulphurea, 111
  turgidula, 84, 87
  vernalis, 90
  vitellinaria, 29
Lecideion inopsis, 179
Lecidella bullata, 196, 198,
  203
  carpathica, 100, 154
  elaeochroma, 90
  scabra, on acid rock, 101
    on airfields, 186
    on alkali rock, 138, 146,
      147, 148

pollution, 66
  on railways, 183
  on siliceous walls, 152
stigmatea, 218, 228
subincongrua, 236
viridans, 245
wulfenii, 205
Lecidoma demissum, 193, 198, 210
Leeds, lichen mapping, 55–7, 63
Lemmopsis arnoldianum, 143
Lempholemma botryosum, 143
  cladodes, 143, 206, 209
  myriococcum, 142
  polyanthes, 147, 158, 161
  radiata, 203, 206
  radiatum, 191, 199, 212
Lepraria spp., 29, 111, 158, 180
  caesioalba, 103, 113
  incana, 56, 62, 98, 146, 255
  lesdainii, 146, 152
  lobificans, 101
  neglecta, 196–8
Leprocaulon microscopicum, 109, 162, 236
Leproloma spp., 158
  membranacea, 37
  membranaceum, 101, 107
Leproplaca spp., 29
  chrysodeta, 146, 152, 158, 184
  xantholyta, 146, 157, 158
Leproplacetum chrysodetae, 146
Leptogium spp., 82, 144, 174, 220, 221, 227
  biatorinum, 186
  brebissonii, 88
  britannicum, 144
  burgessii, 74, 88, 90, 227, 259
  colcheatum, 259
  corniculatum, 186, 243
  cretaceum, 137, 139
  cyanescens, 88, 258
  gelatinosum, on acid rock, 107
    near lakes, 226
    on limestone, 133, 141
    on mine tailings, 178
    in River Coquet, 218
    on roofs, 154
    on walls, 152
  hibernicum, 82
  intermedium, 134
  lichenoides, 258
  plicatile, 142, 159, 216, 217, 244
  schraderi, 133, 143, 147, 148
  tenuissimum, 134, 203
  teretiusculum, 134, 159, 176, 251, 258
  tremelloides, 251
  turgidum, 147, 171, 184–5, 244
Leptorhaphis spp., 29

lichen desert, 54–7, 63, 164
lichenometry, 39, 253–4
Lichenothelia convexa, 154
Lichina confinis, 232–4, 236, 238–9
  pygmaea, 233–4
Lickey Hills (Worcs), 105
limestone, as substrate, 131–50
  colonised, 51
  pavement, 142–3
Lindsay, Lauder, 13
Linwood Warren (Lincs), 125
Lismore, Island of (Argyll), 150
Lithographa tesserata, 114, 199
litmus, manufacture, 27
Little Dun Fell (Yorks), 126, 144
Liverpool, 60
Lizard Peninsula (Cornwall), 116–17, 128, 250
Lobaria spp., 30, 88, 239
  amplissima, 52, 71–2, 79, 80, 82, 90, 91, 256, 258
  pulmonaria, on acid rock, 98
    air pollution, 67
    link with deer, 69–70
    dispersal, 79
    growth, 38–9
    in Hawksworth-Rose scale, 256
    as invertebrate food, 50, 52
    at Loch Sunart, 81, 82
    in medicine, 16, 17
    in NIEC, 258
    in RIEC, 71
    in Seathwaite Valley, 90
    absent from Wealden, 97
    in woodland, 72, 224
  scrobiculata, 71, 74, 79, 82, 98, 102, 256, 258
  virens, 39, 71, 74, 79, 80, 82, 90, 258
Lobarion (*Lobarion pulmonariae*), 43, 71, 75–6, 79, 88, 89–90, 91
Loch Eribol, 115
Loch Etive (Ardgour), 82
Loch Maree, 86, 115
Loch Sunart (Ardgour), 81–3
Lockeridge Dene, 100
London, 56, 58–60, 180, 182
Longleat Park (Wilts), 92
Loxospora elatina, 258
  elatinum, 71, 82
Lulworth Park (Dorset), 92
Lundy Island, 252
lungwort, tree, 16, 17
Lyngbya spp., 235

machair, habitat, 243–4

Mackinder, Sir Halford, 95
magnesium, tolerance, 148
Maidens Paps, The (Sunderland), 147
Malham (Yorks), 142, 154, 212
Malvern Hills, 101–2
Man, Isle of, 175
Manchester, 56, 103
map lichen, 108
mapping lichens, 57–8
Markfield (Leics), 98
Markland Grips, 146
Marlborough Downs (Wilts), 99
Martin Down NNR (Hants), 133
Martindale, J.A., 115
Massalongia carnosa, 111, 115, 217, 224, 227
Mayfield, Arthur, 164
McKay, Duncan, 88
medicine, lichen used in, 17–18
Megallaria grossa, 259
Megaspora tuberculosa, 258
  verrucosa, 135, 149, 150, 191, 205
  verruculosa, 145
melanism, 50–1
Melbury Park (Dorset), 91–2
Mellanby, Kenneth, 54
Mells Park (Somerset), 92
Melrose, 158
Mendip Hills (Somerset), 143, 175
Mendlesham (Suffolk), 164
Menegazzia terebrata, 88
metal, effect on growth, 158–9
  in mine tailings, 174–81
metallophytes, 175–87
Micarea spp., 111, 112, 164, 217
  adnata, 85
  alabastrites, 258
  bauschiana, 115
  botryoides, 98, 115
  cinerea, 258
  coppinsii, 110
  crassipes, 202
  denigrata, 127, 164, 172, 174, 250
  elachista, 85
  erratica, on dumped material, 173
    on flint, 138
    on iron pyrites, 139
    on pebbles, 128
    on railways, 184
    on shingle, 245
    in villages, 162
  excipulata, 174
  granulosa, 207
  leprosula, 114, 116, 193, 209
  lignaria, on acid rock, 96,

98, 114, 116
  in acid water, 225
  on ash, 90
  on heathland, 125, 127, 128
 lignaria var. *endoleuca*, 208
 lithinella, 128, 171, 173
 lutulata, 115
 myriocarpa, 115
 nitschkeana, 86, 127, 250
 paratropa, 207
 peliocarpa, 127, 128, 250
 prasina, 86, 127, 250
 pseudomarginata, 115
 pycnidiophora, 79, 258
 stipitata, 82
 submoestula, 109
 sylvicola, 115
 tuberculata, 115
 turfosa, 193, 196
 viridiatra, 198
Microcalicium arenarium, 115
 disseminatum, 86
millstone grit, as substrate, 102–4
mine tailings, as substrate, 174–81
Minsmere Lagoon (Suffolk), 247
Minto Craigs (Roxburgh), 108
Minuartia sedoides, 189
Miriquidica spp., 113
 atrofulva, 175, 180
 complanata, 203, 209
 garovaglii, 115
 griseoatra, 203
 leucophaea, 98, 101, 114
 nigroleprosa, 194
mite, lichen association, 52–3
Mniacea nivea, 203
Moccas Park (Hereford), 74
Moeleropsis nebulosa, 144, 236, 251
montane lichens, 188–213
  numerical comparison, 212–13
Monyash (Derbs), 161
moor habitat, 118–30
Morecambe Bay, 143, 161, 239
Morison, Robert, 16
moss, black, 23
  bronze, 23
  chalice, 16
  coral, 16
  grey, 23
  hair, 16, 17
  Iceland, 17, 32, 46, 126
  oak, 25, 26
  reindeer, 26, 27, 32
  tree, 16, 25
  woolly hair, 207
moss of the trees, 16
moth, dependent on lichen, 49–51

Muckanaught (Connemara), 209
Muckle Samuel's Crags (Northumb), 104
Muir of Dinnet (Aberdeen), 124
Mull, Isle of, 87, 103, 106, 109, 117, 243
Multiclavula spp., 29
Mycoblastus affinis, 85, 104
 alpinus, 103
 caesium, 82
 sanguinarius, 74, 84, 103
Mycoglaena myricae, 128
Mycoporum spp., 29
Myxomphalina maura, 171

naming, meaning of names, 31–3
Nardus stricta, 210
Nare Head (Cornwall), 128
National Vegetation Classification, 125, 128
Natural History Museum (London), 60, 189–90
nematode, lichen association, 52–3
Nephroma spp., 239
 arcticum, 208
 laevigatum, 71, 79, 88, 90, 98, 227, 258
 parile, 74, 88, 90, 98, 107, 227–8, 258
 tangeriense, 116, 209, 227
New Forest, 72, 77–80, 125, 128, 129
New Index of Ecological Continuity, *see* NIEC
Newcastle-upon-Tyne, pollution, 54–7
Newquay (Cornwall), 244
NIEC, 71, 257–8
nitrogen, deficiency, 141
  level in rivers, 223
  metabolism, 30
NNR protection, 100
Normandina pulchella, 79, 90, 162, 256
North Berwick Law (East Lothian), 107
North Rona (Hebrides), 234–5
North York Moors, 102
Nostoc spp., 29, 30, 96, 221, 235
number of lichens, 28
Nylander, William, 16, 54

oak, in deer parks, 91, 93
  with *Desmococcus* growth, 167
  in New Forest, 78–9
  in NIEC, 257
  regeneration, 83
  woodland, 74, 76, 82
  as worked wood, 164
oat, wild, 124

Ochrolechia spp., used in dyeing, 21
 androgyna, 21, 81, 88, 98, 184
 frigida, on acid rock, 104
  on Ben Loyal, 199
  in Cairngorms, 192, 195
  on coastal heath, 250
  on dunes, 241–2
  in Lake District, 209
  in Teesdale, 145
 inversa, 74, 258
 microstictoides, 84
 parella, on acid rock, 96, 102, 107, 110, 117
  on bridges, 153
  in churchyards, 160
  early reference to, 15
  on flint, 138
  on rocky shore, 232, 236, 238
  on shingle, 245
  size of thallus, 40
  near water, 228
 tartarea, on acid rock, 98, 103, 114
  in Cairngorms, 193, 199
  in dye, 21
  on trees, 81, 84
 turneri, 93
 upsaliensis, 189
oil-stain parmentaria, 32
Omphalina spp., 29, 114, 122
 chrysophylla, 29
 cupulatoides, 29
 ericetorum, 111
 hudsoniana, 125
 luteovitellina, 29, 125
Opegrapha spp., 90, 111, 114, 144, 158, 217
 corticola, 258
 demutata, 164
 fumosa, 71, 259
 gyrocarpa, 157, 227
 lithyrga, 211
 mougeotii, 144, 146, 159
 multipunctata, 252, 259
 ochrocheila, 112, 164, 252
 prosodea, 72, 91, 258
 saxatilis, 146, 157, 184
 sorediifera, 164
 varia, 255
 vermicillifera, 164
 vulgata, 61, 164, 255
Ophioparma lapponica, 189
 ventosa, 113
 ventosum, 41, 98, 108, 111
orchil dye, 21–2
Orme, Great and Little (N. Wales), 239
Orphniospora atrata, 32
 moriopsis, 194
Orthotrichum lineare, 86
Orton Scar (Yorks), 144
ozone loss, effect on lichens, 67

Pachyphiale carneola, 71, 82, 90, 256, 258
Paddockhurst (High Weald), 98
Padley Gorge (Derbs), 103
Palmer, Ken, 163
Pannaria spp., 82, 239, 256
  conoplea, 71, 88, 90, 258
  hookeri, 191, 199, 203
  leucophaea, 244, 252
  mediterranea, 79, 259
  pezizoides, 144, 176, 199, 227, 244, 250–1
  pityrea, 79
  praetermissa, 201, 205
  rubiginosa, 88, 258
  sampaiana, 79, 256
Parham Park (Sussex), 91, 92
Parmelia sp., 33, 72, 92, 127, 162, 250
  acetabulum, 52, 74, 166, 255
  arnoldii, 258
  caperata, air pollution, 61
    on airfields, 186
    on dunes, 241
    growth, 39
    in Hawksworth-Rose scale, 255–6
    on rocky shores, 233
    in villages, 153
  conspersa, acid rock, 98, 99, 100, 101, 107, 111
    near lakes, 225
    in villages, 154, 163
  crinita, 71, 79, 88, 98, 258
  delisei, 162, 236
  discordans, 37
  disjuncta, 98, 107
  elegantula, 169
  endochlora, 82, 252
  exasperatula, 153, 169, 255–6
  glabra, 189
  glabratula, on acid rock, 107
    air pollution, 61
    on ash, 90
    in churchyards, 160
    growth, 39, 254
    on hazel, 88
    on rocky shores, 239
    on siliceous walls, 152
  glabratula fuliginosa, 40
  horrescens, 259
  incurva, 98, 103, 162
  laciniatula, 169
  laevigata, 74, 82, 88
  loxodes, 96, 100, 101, 107, 111, 186
  minarum, 79, 258
  mougeotii, on acid rock, 100, 101, 104, 110
    on airfields, 186
    in churchyards, 160
    on flint, 138
    on shingle, 246
    in villages, 153
  omphalodes, on acid rock, 99, 100, 102, 103, 114
    as animal food, 45, 53
    in Cairngorms, 199
    in dye, 20
    near lakes, 227
    on rocky shore, 233
  pastillifera, 100
  perlata, 46, 61, 88, 186, 256
  protomatrae, 154
  pulla, 154, 233
  reddenda, 71, 258
  reticulata, 79, 256, 256
  revoluta, 42, 61, 101, 108, 153, 186, 256
  saxatilis, on acid rock, 98, 103
    air pollution, 60, 65, 67
    animal associations, 52–3
    on ash, 90
    in dye, 20
    growth, 39
    in Hawksworth-Rose scale, 255
    on rocky shores, 233, 236, 239
    well-dressing, 23
  sinuosa, 74, 82, 88, 258
  subaurifera, 61, 153
  submontana, 163, 169
  subrudecta, 61, 255
  sulcata, bird association, 46
    on buildings, 153
    colour, 42
    growth, 39
    in Hawksworth-Rose scale, 255
    in lichen deserts, 55
    meaning of name, 32
    on oak, 167
    recolonisation, 60, 61
    on rocky shores, 239
    in salt marshes, 248
  taylorensis, 82, 252, 258
  tiliacea, 153, 186, 255–6
  tinctina, 116, 154
  verruculifera, 108, 184, 186, 217, 228
Parmeliella spp., 239
  jamesii, 258
  testacea, 258
  triptophylla, 71, 90, 227, 251, 258
Parmelietum laevigatae, 76
Parmelion conspersae, 101–2
  laevigatae, 82
Parmeliopsis aleurites, 164
  ambigua, 57, 84, 93, 164, 255
  hyperopta, 84
parmentaria, oil-stain, 32
Parmentaria chilensis, 32, 82
Parys Mountain (Anglesey), 54, 179
Peltigera spp., on acid rock, 116
  on ash tree, 90
  in churchyards, 162
  on coastal heath, 250
  on dunes, 241
  as indicators, 168
  as invertebrate food, 50
  on limestone, 142, 148
  and metallophytes, 175
  in quarries, 181
  on shingle, 245
  aphthosa, 244
  britannica, 108
  canina, 17, 38–9, 243
  collina, 71, 82, 88, 258
  didactyla, on airfields, 185
    on mine tailings, 178
    on railways, 183
    in urban area, 169, 171, 174
  elisabethae, 203, 206, 212
  horizontalis, 71, 81, 90, 258
  lactucifolia, 185
  lepidophora, 222
  leucophlebia, on Ben Bulben, 209
    on Ben Lawers, 203
    in Lake District, 211
    on limestone, 144, 145
    at Loch Sunart, 82
    on mine tailings, 176
  malacea, 242
  neckeri, 175, 176, 185, 186, 243
  ponojensis, 201
  praetextata, 226–7
  rufescens, 176, 243, 252
  venosa, 175, 176, 191, 199, 203, 206
Pelvetia canaliculata, 28, 232, 233
Pennard Burrows (Glam), 244
Penns (High Weald), 98
perfume, lichen used in, 25–7
Pertusaria spp., 81, 90, 117
  albescens, 112, 160, 228, 255
    var. *corallina*, 90
  amara, on acid bark, 90
    air pollution, 67
    climatic effect on, 74
    on farm roofs, 167
    in Hawksworth-Rose scale, 255
    on railways, 184
    on sandstone, 112
  amarascens, 110, 203, 206
  aspergilla, 98, 101, 153, 160
  bryontha, 189, 191
  chiodectonoides, 109, 116, 237
  corallina, 98, 103, 111, 113
  dactylina, 193, 199
  excludens, 110, 111
  flavicans, 109

flavida, 168
gallica, 237
geminipara, 193
glomerata, 191, 203
hemisphaerica, 168, 256
hymenea, 61, 88, 112
lactea, 114
lactescens, 163
monogona, 110, 237
multipuncta, 258
oculata, 191, 196, 199, 207
opthalmiza, 82
pertusa, 52, 90, 112, 160
pseudocorallina, 100, 107, 109, 160, 208, 228
pupillaris, 258
velata, 79, 258
xanthostoma, 195, 250
Peterborough, 149
Petractis clausa, 136, 141, 143, 149, 163
Phaeographis dendritica, 258
inusta, 258
lyellii, 258
Phaeophyscia endococcina, 206, 225
nigricans, 147, 153
orbicularis, 142, 147, 166, 182, 217, 255
sciastra, 99, 186, 218, 225, 228
Phauloppia lucorum, 238
Phlyctis agelaea, 74
argena, 160
Phormidium spp., 141
phosphate, in rivers, 223
Phyllospora rosei, 71, 258
physcia, southern grey, 32
Physcia spp., air pollution, 59, 62, 66
  as food plant, 50
  growth, 40, 254
  near lakes, 226
  on limestone, 148
  on oak, 167
  in parkland, 92
  in Xanthorion, 166
adscendens, 142, 152, 236, 255
aipolia, 61, 256
caesia, 142, 158, 185, 217, 218, 225
clementei, 160, 164
dubia, 108, 154, 160, 228
leptalea, 256
pulverulenta, 160
tenella, 46, 60, 142, 182, 218, 236, 255
tribacea, 162, 255
tribacioides, 32, 166
Physcietum caesiae, 146, 147
Physciopsis fuscovirens, 184
Physconia spp., 92, 160, 166
distorta, 255
grisea, 153, 184, 217, 247, 255
muscigena, 189
perisidiosa, 255

Physodion alliance, 84
Piggle Dene, 100
Pilophorus strumaticus, 109, 111, 207
Pinewood Index, 85–6, 87
pinhead lichens, 36, 69, 72, 84–5, 115
Pinus sylvestris ssp. *scotica*, 83
Pistyll Rhaeadr (Mid Wales), 40
Pitlochry (Perthshire), 154
Placidiopsis cartilaginea, 143, 144, 199, 244
Placopsis spp., 30, 177
gelida, 108, 163, 186, 225, 227
lambii, 177
Placynthiella spp., 86, 122–3
hyporhoda, 175
icmalea, 86, 123, 128, 172
uliginosa, 123, 172
Placynthietum nigri, 147
Placynthium asperellum, 199
flabellosum, 82, 107, 203, 215, 227–8
garovaglii, 144
lismorense, 150
nigrum, on buildings, 150, 153, 159
  growth, 39, 40
  on limestone, 142, 147
  on railway bridges, 184
pannariellum, 199
pluriseptatum, 191, 212
subradiatum, 144
tantaleum, 137
plastic, as substrate, 173
Platismatia glauca, 61, 84, 103, 114, 199, 255
norvegica, 85
Pleasley Vale (Derbs), 146
Pleurocapsa spp., 235
Plymtree (Devon), 167
Poeltinula cerebrina, 144, 217
pollarding, 78, 90
pollution, air, 54–68, 79, 103, 166, 169, 255–6
  indicators, 54–68
  soil, 170
Polyblastia spp., 144
agraria, 178, 244
albida, 137, 149
allobata, 258
cruenta, 216, 219
cupularis, 191
deminuta, 141
dermatodes, 136
gelatinosa, 176, 244, 252
gelatinosum, 133
helvetica, 191
inumbrata, 199
melaspora, 199
sendtneri, 191
terrestris, 191, 204, 212
theleodes, 109, 191
verrucosa, 199
wheldonii, 144, 150, 178, 203, 244

Polychidium dendriscum, 82
muscicola, 108, 115, 211, 224, 227
Polysporina cyclocarpa, 150
cyclospora, 205
lapponica, 107, 110
simplex, 158, 160, 163
Polytrichum piliferum, 127
Porina spp., 111, 114, 217
chlorotica, 184, 216, 220, 223, 227
curnowii, 111
guntheri, 219
interjugens, 227
lectissima, 111, 221, 228
leptalea, 71, 164, 250
linearis, 142, 239
mammillosa, 193, 206, 250
Porocyphus coccodes, 107, 206, 217, 228
kenmorensis, 215
Porpidia spp., on acid rock, 113, 117, 211
  in acid water, 225
  algal partner, 30
  collecting, 43
  on mine tailings, 175, 177
crustulata, 138, 161, 173, 174, 223
glaucophaea, 226–7
hydrophila, 43, 216, 220, 227
platycarpoides, 114
soredizodes, 100, 184, 227
speirea, 96, 98, 108, 109, 228
superba, 203, 211
tuberculosa, on acid rock, 105, 114
  on brick, 173
  on buildings, 152, 153
  in churchyards, 161
  in Co. Kerry, 208
  on dumped material, 173
  on flint, 138
  growth rate, 39
  on railways, 184
  in rivers, 223
  on shingle, 245
Portisham (Dorset), 99
Porton Down (Wilts), 134, 136–40
pottery, as substrate, 174
Praeger, Robert Lloyd, 208
Prasiola crispa, 182
Protoblastenia calva, 141
incrustans, 141
rupestris, on acid rock, 96
  on building, 155
  on chippings, 173
  on limestone, 136, 138, 142, 147, 148
siebenhaariana, 203, 206, 212
Protoderma viride, 216
Protoparmelia ariseda, 115
badia, 101, 103, 163

ochrococca, 84, 85
oleagina, 93
picea, 103
Protothelenella corrosa, 203, 213
  sphinctrinoidella, 198
Prunus padus, 82
Pseudephebe pubescens, 104, 111, 194, 199, 210
Pseudevernia spp., 45
  furfuracea, 25, 26, 84, 103, 114, 199, 255
Pseudevernion furfuraceae, 76
Pseudocyphellaria spp., 74, 82, 239
  aurata, 250
  crocata, 88, 258
  intricata, 88, 209, 227
  norvegica, 82, 88
  thouarsii agg., 259
pseudometallophytes, 175
Pseudotrebouxia spp., 29, 37
Psilolechia leprosa, 158, 163, 164, 175, 179
  lucida, 98, 115, 153, 155, 158
Psora decipiens, on dunes, 244
  on limestone, 134, 143, 148
  in quarries, 181
  in Scotland, 199, 204, 206
  lurida, 112, 142, 148, 161
  rubiformis, 201
Psoroma hypnorum, 199, 244, 250
ptarmigan, 106
Pterygiopsis spp., 216
  coracodiza, 143, 216
  lacustris, 216
Purvis, William, 178, 191
Pycnothelia papillaria, 122, 125, 129, 193, 250
pylons, lichens beneath, 180
Pyrenocollema spp., 233–4
  halodytes, 233–4, 238–9
  monense, 137, 223
  monensis, 173
  orustense, 247
  pelvetiae, 232–3
  saxicola, 137
  sublitorale, 234
Pyrenopsis grumulifera, 107, 203, 211
  impolita, 107, 227
  subareolata, 203, 211
Pyrenula chlorospila, 71, 88
  laevigata, 88
  macrospora, 71, 90
  nitida, 71, 79
  occidentalis, 88

Quantock Hills, 102
quartzite, as substrate, 105–6

Rackham, Oliver, 72, 168
Racodium rupestre, 114
Racomitrium heterostichum, 202
  lanuginosum, 243
radionuclides, effect on lichens, 67
railway, as habitat, 182–5
Ramalina spp., 92, 166, 237–8
  zone, on rocky shore, 232
  calicaris, 256
  canariensis, 112, 157, 160, 256
  capitata, 163
  chondrina, 237
  cuspidata, 236–7
  dilacerata, 189
  farinacea, 61, 69, 127, 241, 255
  fastigiata, 160
  fraxinea, 61, 65, 256
  lacera, 157, 162, 256
  pollinaria, 112, 127, 256
  polymorpha, 107, 108
  portuensis, 237
  siliquosa, on acid rock, 99, 107, 108
    in Cairngorms, 200
    in churchyards, 160, 162
    galls, 53
    on rocky shores, 233, 236–9
  subfarinacea, 108, 160, 200, 236, 256
  thrausta, 189
Ramonia spp., 259
Rannoch Moor, 86, 224
Ratcliffe, Derek, 191
Raven Crag (Cumbria), 115
Ravenglass (Cumbria), 245, 246
Reading, 102
Red Data List, *Alectoria ochroleuca*, 194
  Ben Lawers spp., 204
  Endocarpon pusillum, 249
  spp. near lakes, 226
  montane spp., 188
  New Forest spp., 79
  Parmentaria chilensis, 82
  Peltigera venosa, 176
  pinewood spp., 85
  spp. near rivers, 216, 217
  rocky shore spp., 237
  wayside tree spp., 166
reindeer, fed on lichen, 45–4
reindeer lichen, 26, 27, 120, 241–2
reproduction, 33, 34–7
reservoirs, 228–9
Revised Index of Ecological Continuity, *see* RIEC
Rheidol, River, 176
Rhinns of Kells, 110
Rhizocarpon spp., 113, 177, 211
  alpicola, 39, 194
  amphibium, 219
  cinereovirens, 246
  concentricum, 96, 109, 244
  constrictum, 238
  furfurosum, 175, 180
  geographicum, on acid rock, 96, 100, 103, 105, 107–8, 113
    air pollution, 65
    in Cairngorms, 194
    in churchyards, 161
    colour, 41
    growth, 38–9
    identification difficulties, 254
    near lakes, 227
    on roofs, 154
  germinatum, 113, 199
  jemtlandicum, 196
  lavatum, 102, 216, 221, 223, 227–8
  lecanorinum, 103, 105, 113, 246
  obscuratum, on acid rock, 107
    on brick, 173
    on dumped material, 173
    on flint, 138
    on heathland, 128
    on rocky shores, 239
    in villages, 152–3
  oederi, 128, 179–80
  reductum, 218
  richardii, 233, 236, 245
  simillimum, 211
  subgeminatum, 104
  umbilicatum, 109, 142
  viridiatrum, 101, 109, 113
Rhodes, P.G.M., 244
Rhum, Island of, 116, 117, 238, 243
Ribble, River, 217
Richardson, David, 13
RIEC, 71, 257–8
  scores, 80, 82, 85, 89, 90, 91–2
Rimularia spp., 113
  illita, 113
  intercedens, 107, 113
  mullensis, 113
Rinodina aspersa, 139, 246
  atrocinerea, 99, 107, 111, 236
  bischoffii, 137, 239, 246
  calcarea, 164
  confragosa, 99, 200, 252
  fimbriata, 209, 217, 225
  gennarii, 146, 147, 152
  isidioides, 71, 79, 258
  luridescens, 236
  mniarea, 208
  parasitica, 206
  roboris, 256
  subglaucescens, 252
  teichophila, 184
  turfacea, 189

Risby Warren (Lincs), 129
river, as habitat, 214–29
river jelly lichen, 141, 215
River Jelly Lichen Steering Group, 215
road, as habitat, 181–2
Roccella spp., 21, 27
  fuciformis, 237
  phycopsis, 162, 164, 237
rock tripe, 22
Rogan, Michael, 27
Rollright stone circle (Oxon), 39, 41
Rose, Francis, 69–71, 72, 78, 80, 91
Rotherham, 147
Rothiemurchus Forest, 85–6
Rottingdean (Sussex), 249
Roundton Hill (Montgomery), 109
rowan, 82, 84, 87
rubber, as substrate, 173–4
rush, three-leaved, 193, 196

Sagiolechia protuberans, 144, 149, 191, 206, 211
  rhexoblephara, 199
St David's Cathedral (Pembs), 162, 179
St Kilda, 237, 239, 251–2
salt marsh, habitat, 247
sandstone, as substrate, 112
Sarcogyne clavus, 110
  privigna, 110, 163, 244
  regularis, on airfields, 185
    on chippings, 173
    on limestone, 136, 147
    on pebbles, 171
    on shingle, 246
    in villages, 153
Sarcopyrenia gibba, 137, 163, 164
Sarcosagium campestre, on bonfire sites, 171
  on limestone, 133, 147
  on mine tailings, 176, 178
  beneath pylons, 180
  in urban areas, 169
Sarea difformis, 84
  resinae, 84
sarsen stones, 99–100
Schadonia fecunda, 208
Schaereria cinereofusca, 103
  fuscocinerea, 101
  tenebrosa, 105
Schismatomma cretaceum, 91
  decolorans, 80, 91, 255
  graphidioides, 259
  niveum, 258
  quercicola, 71, 258
schist, as substrate, 112–15
Schistidium apocarpum, 142
Schistostegia pennata, 111
Scilly, Isles of, 237, 250
Scoliciosporum chlorococcum, 56, 127
  umbrinum, on acid rock,

101, 108, 139
  air pollution, 66
  associations, 180
  on buildings, 152, 158
  on car tyres, 173
  near lakes, 225, 227–8
  meaning of name, 32
  near metal, 172, 173
  on railways, 183
  in rivers, 217, 223
Scots pine, 86, 124
Scytonema spp., 29
Seathwaite (Cumbria), 90
Seatoller Woods (Cumbria), 90
Seaward, Mark, 104, 163
sedge, sparse sand, 127
Seil (Oban), 87
serpentine, as substrate, 115–17
Shadonia fecunda, 202
Shakespeare, William, 15
Sheffield, 103, 169, 171, 174
Shetland Islands, 116
Shipmeadow Church (Suffolk), 154
Siphula ceratites, 122, 125
size of lichen, 38–40
Skinner, J.F., 108
Skipwith Common (Yorks), 123, 129, 186
Skye, Isle of, 87, 106, 109–10, 117, 149, 206
slate, as substrate, 112–15
Slieve League (Co. Donegal), 209
slug, association with lichen, 51–2
Smeatharpe (Devon), 186
Smith, Don, 163
snail, association with lichen, 51–2
Snake Pass (Derbs), 86, 103
snow patch communities, 196–8
Snowdonia, 212–13
socket lichen, 142
Solenospora candicans, 144, 147, 149, 153, 184
  holophaea, 236, 251
  liparina, 116
  vulturiensis, 182, 236, 251
Solorina spp., 30
  bispora, 191, 199, 202, 206
  crocea, 196, 199, 209
  saccata, 142, 145, 148, 149, 209, 244
  spongiosa, 144, 145, 244
South Tyne, River, 176–7, 219
Sphaerophorus spp., 16, 103, 233
  globosus, acid rock, 98, 99, 102, 103, 114
    in Cairngorms, 199
    on coastal heath, 250
    Loch Sunart, 82

in NIEC, 259
    as reindeer food, 45
  melanocarpus, 98, 102, 103, 259
Sphaerothallia esculenta, 242
Sporastatia polyspora, 196, 206
spore dispersal, 85
Squamarina cartiliginea, 135, 142, 149, 244
  crassa, 143, 144
  lentigera, 96, 135, 139, 143
SSSI protection, 92, 100
St David's Cathedral (Pembs), 162, 179
St Kilda, 237, 239, 251–2
Staffa, Isle of, 106
Stannor Rocks (Radnor), 109
Staurothele arctica, 207
  areolata, 196
  bacilligera, 142, 149, 205, 217
  caesia, 137, 144, 149
  fissa, 215, 218, 227–8
  frustulenta, 186
  guestphalica, 216
  hymenogonia, 136
  rugulosa, 137
  rupicola, 137
  rupifraga, 141, 142
  succedens, 199, 209
Steep Holm (Bristol Channel), 239
Steinia geophana, 171, 174, 176, 180
Stenocybe spp., 29
  bryophila, 82
  septata, 71, 258
Stereocaulon spp., 30, 34, 113, 114, 177
  condensatum, 129, 175, 177, 242
  dactylophyllum, 99, 175
  delisei, 175, 177
  evolutum, 108, 111, 184, 196
  glareosum, 175, 176, 181
  leucophaeopsis, 175, 177, 179
  nanodes, 152, 163, 175, 180, 182
  pileatum, on acid rock, 108
    in acid water, 225
    on airfields, 186
    on lead/zinc rocks, 175
    meaning of name, 32
    pollution, 64
    on roads, 182
    in villages, 162
  plicatile, 207
  saxatile, 196–7, 242
  spathuliferum, 203
  symphycheilum, 175, 179
  tornense, 207, 212
  vesuvianum, or. acid rock, 115, 116
    in Cairngorms, 196

INDEX

in Co. Kerry, 208
  on lead spoil, 182
  on railways, 183
  vesuvianum var.
    symphycheileoides, 171,
    172, 182, 208
Stichococcus spp., 29, 115
Sticta spp., 74, 82, 88, 239
  canariensis, morphs, 31,
    90, 259
  dufourii, 30, 259
  filix, 31
  fuliginosa, 90, 258
  limbata, 67, 71, 79, 90, 258
  sylvatica, 71, 90, 258
Stiperstones (Shropshire),
  105, 129
Stonehenge (Wilts), 99–100
Stonethwaite (Borrowdale),
  90
Stowe Park, 24, 25
Strangospora ochrophora, 258
Strathspey, 85
Strigula stigmatella var.
  alpestris, 203, 211
structure of lichen, 36–7
Struidh (Eigg), 87–9
Studland Heath (Purbeck),
  128
Suilven (Sutherland), 125
Sulgrave Church
  (Northants), 26
sulphur dioxide, as
  pollutant, 54–63
Sussex Downs, 139
Sutherland, 105
Swale, River, 217
Swansea Docks, 171
Swinscow, Dougal, 177, 188
sycamore, 91
Synalissa symphorea, 143

tarmac, as substrate, 174
tar-spot lichen, 230–2
Tees, River, 95, 217, 219
Teesdale rarities, 144
Teloschistes spp., 129
  flavicans, on acid rock,
    102, 110
  on coastal heath, 250–1
  in Hawksworth-Rose
    scale, 256
  meaning of name, 32
  mollusc food, 52
  in NIEC, 259
  on rocky shores, 237
  on shingle, 246
  on shrubs, 127
Temnostethus pusillus, 238
Tennyson Down (Isle of
  Wight), 135–6, 139
Tephromela aglaea, 108, 209
  armeniaca, 206
  atra, 102, 154, 160, 184,
    225, 232, 238, 245
  grumosa, 102, 107
Test, River, 216
Thamnolia spp., 29

vermicularis, 195, 196, 199,
  207, 210, 242–3
Thelidium spp. 144
  decipiens, 141, 146, 216,
    217, 220
  fontigenum, 216
  fumidum, 203
  incavatum, 136, 137, 141
  minutulum, 137, 171, 173,
    183, 220
  papulare, 142, 154, 162,
    217
  pluvium, 216, 219
  zwackhii, 137, 162, 173,
    216, 220
Thelocarpon spp., 155
  epibolum, 86, 125, 178
  intermediellum, 139, 172,
    173
  laureri, 172, 173, 180
  lichenicola, 171
  magnussonii, 229
  pallida, 137
  pallidum, 139
Thelomma ocellatum, 164,
  166
Thelopsis isiaca, 163
  melathelia, 191, 204, 206
  rubella, 71, 90, 258
  lepadinum, on coastal
    heath, 250
    in Co. Kerry, 208
    as indicator, 168
    in NIEC, 258
    in RIEC, 71
    in woodland, 80, 81, 90
  monospora, 88
  subtile, 88
Thetford Heath/Warren
  (East Anglia), 135, 139
Thixendale (Yorks), 138
Thorpe-by-Chertsy
  (Surrey), 15
Thrombium epigaeum, 178,
  250
Thursley Common
  (Surrey), 129
Tomasellia lactea, 259
Tomintoul (Moray), 180
Toninia aromatica, 158, 184,
  191
  fusispora, 206, 209
  lobulata, 135, 142, 147,
    149, 152, 163, 243
  sedifolia, 96, 134, 142,
    147–9, 161, 171, 243–4
  squalescens, 196, 198, 207
  thiospora, 115, 207-8
  verrucarioides, 153
Topham, Pauline, 188, 191
Tornabea scutellifera, 249
Torrent Walk (N. Wales),
  70
Torridon Hills, 208
Tournefort, designation of
  lichen, 16
Traboe-in-St Keverne
  (Cornwall), 15

Traligill (Sutherland), 217
transect, bluff, 226
  chalk grassland, 134
  Liverpool, 60
  London, 60–1, 62
  river, 218
  rocky shore, 238
  vertical rock, 228
Trapain Law (Lothian), 107
Trapelia spp., on acid rock,
  101, 109, 114, 116
  in acid water, 225
  in churchyards, 161
  on flint, 138
  on mine tailings, 175,
    177
  on railway bridges, 184
  on sandstone, 173
  on siliceous walls, 152
  coarctata, on acid rock, 115
    on brick, 173
    on heaths,128
    near lakes, 227–8
    on pebbles, 171
    beneath pylons, 180
    in rivers, 217, 223
  involuta, 98, 103, 128, 171
  mooreana, 177, 207
  obtegens, on acid rock, 98,
    115
    on brick, 173
    on flint, 128
    on pebbles, 171
  placodioides, 115, 186, 221,
    223
Trapeliopsis spp., 86, 114,
  183
  flexuosa, 172
  gelatinosa, 111, 125
  glaucolepidea, 111, 125
  granulosa, 98, 122–3, 128,
    172, 241
  pseudogranulosa, 111, 122,
    129
  wallrothii, 200, 236, 250
Trebartha Park (Cornwall),
  92
Trebouxia spp., 29–30, 37,
  41, 115
  glomerata, 30
tree, colonised, 47–50, 52,
  59, 60–1
tree lungwort, 16, 17
tree moss, 16
Tremolechia atrata, 106, 108,
  113, 180, 245
Trentepohlia spp., 29–30, 71,
  114, 142, 149
  umbrina, 29
Tromera difformis, 84
tuff, as substrate, 112–15
tulip tree, 91
Turner, William, 15
Twycross (Leics), 56
Tylothallia biformigera, 108
Tyne, River, 176
Tyn-y-Groes (N. Wales), 70

DATE DUE

DUE DATE SUBJECT TO CHANGE IF A RECALL IS REQUESTED